Rocknocker:

A Geologist's Memoir

by
George Devries Klein

CCB Publishing
British Columbia, Canada

Rocknocker: A Geologist's Memoir

Copyright ©2009 by George Devries Klein
ISBN-13 978-1-926585-60-4
First Edition

Library and Archives Canada Cataloguing in Publication

Klein, George Devries, 1933-
Rocknocker : a geologist's memoir / written by George Devries Klein.
ISBN 978-1-926585-60-4
1. Klein, George Devries, 1933-.
2. Geologists--United States--Biography. I. Title.
QE22.K54A3 2009   551.092   C2009-905990-8

Cover photo: George Devries Klein at outcrop of slope turbidites, Atoka Group (Pennsylvanian), Spillway at Paris, AR (Photograph by Kathleen M. Marsaglia, late September, 1980).

Extreme care has been taken to ensure that all information presented in this book is accurate and up to date at the time of publishing. Neither the author nor the publisher can be held responsible for any errors or omissions. Additionally, neither is any liability assumed for damages resulting from the use of the information contained herein.

All rights reserved. No part of this publication may be reproduced, stored in a retrieval system or transmitted in any form or by any means, electronic, mechanical, photocopying, recording or otherwise without the express written permission of the publisher. Printed in the United States of America and the United Kingdom.

Publisher: CCB Publishing
              British Columbia, Canada
              www.ccbpublishing.com

Preface

Since my days as an undergraduate geology student, I read many memoirs and biographies about geologists, both the pioneers of the early days of this science, as well as those I met. The ones I recall in order of their being read are:

Fenton, Carol L, 1952, Giants of geology.

Youngquist, W. L., 1966, Over the Hill and Down the Creek: Caldwell, ID, Caxton Printers, 322 p.

Pettijohn, F.J., 1984, Memoirs of an unrepentant field geologist: Univ. of Chicago Press

Scott, H.W., 1986, The Sugar Creek Saga: Ann Arbor, MI, Cushing Malloy, 308 p.

Rodgers, John, 1999, The company I kept: The autobiography of a geologist: Connecticut Academy of Arts and Sciences.

DuBar, J.R., 2004, Never piss into the wind: Baltimore, MD, Publish America, 343 p.

Friedman, G.M., 2006, Saxa Loquntur (Rocks Speak): The Life and Times of the Geologist: SEPM Foundation Publication.

I also read Dan Merriam's well written and informative biography of Raymond C. Moore, one of my professors at the University of Kansas:

Merriam, D.F., 2007, Raymond Cecil Moore: Legendary Scholar and Scientist: Univ. of Kansas Dept. of Geology and Paleontological Institute Spec. Pub. 5.

Perhaps the most compelling and the best geological memoir I ever read is by a fellow-graduate student at the University of Kansas where I earned my M.A in geology in 1957. It is:

Fisher, W.L., 2008, Leaning Forward: A Memoir: Texas Bureau of Economic Geology.

It details Bill's life from humble origins, to army service, to graduate education, and a distinguished career in scientific research, science administration, and a sub-cabinet appointment. It was, in fact, Bill Fisher who suggested I write my memoirs because as he wrote, "You have a few good tales to tell also."

My approach to this Memoir differs slightly from other memoirs. First, each chapter ends with a section on *Lessons Learned*. One reviewer, Gerald J. Kuecher, recommended including such a section. Second, I added postscripts at the end of some of the chapters. These items are relevant to that chapter and provide additional context.

Because I am writing this book from memory, some details may be sketchy, and some names forgotten or misspelled. Nearly all the dialog and quotes are accurate because some quotes one never forgets. I made an honest effort to be as accurate as I could.

In this memoir, whenever I introduce a person, I add information about their degrees and a brief career description. I do so to provide perspective about them. Again, this information comes from memory so details may be incomplete.

This Memoir is organized into two parts. Part I provides a chronological autobiography. Part II treats several topics that could not be blended successfully into Part I but deserve inclusion.

I thank Robert Isham Auler, William L. Fisher and Gerald J. Kuecher for reading parts or all of an earlier manuscript version of this book and helping improve it. However, all errors and omissions are solely my responsibility. I also thank Peter C. Patton of Wesleyan University for providing critical information about the early history of the Wesleyan University Department of Geology some of which is included herein.

Contents

PREFACE iv

PART I. CHRONOLOGICAL EVENTS AND EXPERIENCES 1

Chapter 1. Early Childhood (1933-1939) 3

Chapter 2. Australia (1939-1947) 8

Chapter 3. High School in America (1947-1950) 16

Chapter 4. Wesleyan University (1950-1954) 22

Chapter 5. Two Summers in Newfoundland and Johns Hopkins University (1954-1955) 34

Chapter 6. University of Kansas and Nova Scotia (1955-1957) 39

Chapter 7. Yale University and Nova Scotia (1957-1960) 55

Chapter 8. The Keuper Marl of the UK (Summer 1960) 73

Chapter 9. Sinclair Research, Tulsa, OK (1960-1961) 77

Chapter 10. University of Pittsburgh (1961-1963) 88

Chapter 11. University of Pennsylvania: The Meyerhoff Years (1963-1966) 105

Chapter 12. University of Pennsylvania: The Faul Years (1966-1969) 126

Chapter 13. Closure: University of Pennsylvania, Interviews, Oxford (1969) 150

Chapter 14. Illinois and Cal-Berkeley (Winter-Spring 1970) 162

Chapter 15. Illinois: The Good Donath Years (1970-Late January 1973) . 175

Chapter 16. The First Donath War (Late January 1973-April 1973) 208

Chapter 17. Illinois: An Uneasy Truce and A Difficult Decision (April 1973-December 4, 1974) 218

Chapter 18. The Second Donath War. Part I
           (December 5, 1974-May 1975) 234
Chapter 19. The Second Donath War. Part II
           (Late May 1975-August 1976) 247
Chapter 20. The Second Donath War. Part III
           (August 1976-May 1977) 266
Chapter 21. Illinois: Transition (1977-1978) 275
Chapter 22. Illinois: The Hower Years, Chicago, Korea (1978-1983) 295
Chapter 23. Illinois: The David Anderson Era (1983-1988) 322
Chapter 24. Illinois: The Kirkpatrick Years (1988-1993) 336
Chapter 25. The New Jersey Marine Science Consortium
           (1993-1996) 363
Chapter 26. Consulting (1996-Present) 393

PART II. ADDITIONAL EVENTS AND EXPERIENCES 407
Chapter 27. First Marriage 409
Chapter 28. Suyon Cheong 417
Chapter 29. Giving Back 424

EPILOGUE 428
END NOTES 431

PART I:

Chronological Events and Experiences

Rocknocker: A Geologist's Memoir

CHAPTER 1
Early Childhood (1933-1939)

I was born on January 21, 1933, and as my parents told me many years later, around 11:00 PM. My arrival startled both the doctors on duty and my parents because the predicted birth date was January 22. I was not expected for another nine or ten hours. Clearly, I wanted to get on with my life right away. This habit of arriving early stayed with me for the rest of my life.

My father, Alfred R. H. Klein, was born on May 31, 1900, in Munich, Germany, where his family lived before moving to Vienna, Austria. On completing high school, he went to a one-year business college, and then immigrated to The Netherlands where he started his career with the Mepel Company. He lived frugally, and saved his money.

Mepel was in the import-export business focusing mostly on waste materials from manufacturing high quality paper. Together with one of his co-workers who he befriended, they bought the business. They sold their product throughout Europe and the USA, including sisal discarded from paper manufacturing. It was used as insulation in automobiles between the carpeting and the floor boards. My dad made several marketing trips to the USA and established a wide network of contacts in New York, Detroit, Chicago and Roanoke, VA. It provided a good income and although the depression caused a downturn, Mepel survived.

My mother, Doris deVries, was born in Krefeld, Germany on April 15, 1906. That location was an accident because of her father's business. He owned a wall-paper manufacturing company in The Netherlands, with a branch factory in Krefeld, Germany. Her family spent time in both the Netherlands and Krefeld while grandfather supervised operations. When my grandmother was due with my mother, they were in Krefeld.

My parents met each other through mutual friends and married in 1928. When they met, she was in university studying psychology but never finished. When both my sister and I left to attend university, she learned ceramics and became a world-recognized potter. She displayed her pieces at art museums in New York, London, and Tokyo. My father built a studio for her in the basement of the last home they owned together.

My birthplace was s'Gravenhage, Netherlands, commonly known as "Den Haag" (The Hague), the seat of the Dutch government. We lived in an upscale suburb, Wassenaar, at 16 Zuidwerfplein. The house was a two-story town row-house built during the early 1920's at the end of the row on a street corner. I also had a sister, Marianne, who was four years older. My grandparents passed away before I was born.

I don't recall much about the house or my life there. Five things I remember. First, I apparently was interested in gravity experiments. We owned a cat (Franci) and I used to pick it up, carry it up the stairs to the second floor, and let the cat fall to the first floor to watch it land on its four feet. Eventually, my parents discovered this and immediately ordered me to stop or ELSE!!

The second thing I recall was at age three, I didn't like vegetables. My parents insisted I eat them. I kept them in my cheeks and after dinner, went to the toilet, spat the veggies out, and flushed them down the drain. Once my parents discovered this habit, it also came to a stop.

I recall an unhappy incident involving my sister. We never enjoyed the most comfortable or filial relationship and one day I provoked her too much while we were playing in a sandbox in the backyard. She hit me over my left eyebrow with a kiddy shovel. My mother witnessed this, rescued me, and provided first aid. No bones were broken and I still have the scar today. But, in time she was forgiven.

Fourth, when I wasn't attending the local Montessori School, I spent my time at a nearby gasoline station. The attraction wasn't the late model automobiles. The owner had two daughters and they were not only attractive, but were much nicer to me than my sister. My parents tolerated my time with the "benzene miesieje's" (gasoline girls) but were less than approving of it.

Then there were the trips to Scheveningen's beach, the Christmas pastries, trips to Alkmaar, a trip to Switzerland, outings to The Feivre, a famous pond in The Hague, and learning ice skating on local canals. In short, I experienced a reasonably pleasant childhood.

Near my fifth birthday, we had visitors. They were relatives, all on my father's side. First to arrive were my aunt Gretel, her husband and two sons. They lived in Vienna and decided to leave. My father arranged immigration visas to Australia and they settled in Sydney, Australia.

4

Next to arrive was my aunt Olga and her husband Willie. They too left Austria and my father, through one of his US business contacts, obtained their immigration visas to the USA. They settled in Yonkers, NY.

I did not know then, but realized later, that these events were to change my life. My pleasant childhood was also about to change. I discovered at an early age that world events do that.

My father, and to some extent, also my mother, had that rare ability to see consequences of their actions and the actions of people and events around them. It was something that I slowly learned much later in life. During one of his business trips to Germany, my father observed the reality of the Nazi consolidation of power.

He read a famous sermon by the Reverend Reinhold Niebuhr in which Niebuhr said about the Nazis "First they'll come for the Jews, then they'll come for the Slavs, and then they'll come for me." This quote made an impression on my father as he told me later. He and my mother decided it was time to leave Europe. He sold his share of Mepel to his partner and explored the family options.

They applied for immigration status to the USA. However, the application was delayed about four years because in those days, immigrants were admitted according to a quota system based on country of birth. Because my parents were born in Germany, the line ahead of them was long. They concluded that if they waited, Germany would invade the Netherlands before their projected departure date. Consequently, we travelled briefly to London in 1938, and my father applied for and eventually was given an immigration visa to Australia.

My parents sold the house in Wassenaar to a family named Vles, packed our belongings, and shipped them to Australia. In 1939, we flew to London, went to Southampton and boarded a Union Castle company liner to immigrate to Australia. Fortunately, our immigration visa application to the USA remained active and we were able to go to the USA much later.

My father returned to the Netherlands on a business trip in 1947 and visited at 16 Zuidwerfplein. He discovered the Vles family sold the house to a new owner. My father went to the front door and explained who he was but the new owner immediately ran him off the property. My sister visited the Netherlands in 1950 and viewed the house from outside. She did not tempt fate by going further. She was the last family member to visit that house until

I visited the Netherlands in 1988.

During 1988, while visiting the Netherlands to do research at the University of Utrecht, I visited a consulting geologist by the name of Steen in The Hague. Late in the afternoon, Mrs. Steen asked if I remembered where I lived in Den Haag, and I told her we lived Wassenaar. I remembered the street name but not the number.

My parents took a photograph of the house and transferred it to a Delft plate which hung in a prominent place in every home they owned. I recalled the house was located on a street corner and its appearance. Mrs. Steen offered to drive me to see if we could find it. On arrival, I could not remember on which street corner the house was located, and nothing jogged my memory.

While standing there, a Volvo drove by, parked in one of the driveways, and the owner stepped out. He asked what we were doing in the neighborhood. Mrs. Steen explained I was from the USA and had grown up in one of the corner houses as a child. Being fluent in English, he told us the history of ownership of every house. When he gave the history of 16 Zuidwerfplein, he mentioned the Vles family who bought the home from my parents and then sold it. That buyer eventually passed away and his son took ownership. The house was now occupied by that son's widow who lived alone. The Volvo owner did not remember my parents' ownership of 16 Zuidwerfplein but suggested we should go and introduce ourselves.

When Mrs. Steen and I entered the property, I noticed a prominent sign from a security company and a system of wires from the sign. We rang the door bell and Mrs. Steen explained to the owner who I was. I gave her my University of Illinois business card, and the owner let us enter.

My first reaction was it was smaller than I remembered. The kitchen still had the same bluish-green Dutch tile on the walls. The staircase seemed less high. But the most surprising thing I noticed was that the house displayed at least 15 original Dutch Master Paintings. That explained the security system. I then realized that many original Dutch Master Paintings were still in private hands.

Although leaving 49 years before, I was the only family member able to gain access to the inside of that home.

I mentioned this visit to Dr. Sierd Cloetingh, then at the University of Utrecht. He explained that when the Nazis occupied The Netherlands, they expropriated people's homes. After the war, people returned to claim their homes, but there were numerous bogus and contentious claims. Cloetingh inferred that when my father visited in 1947, the home owner suspected he was trying to claim the property and therefore ran him off.

LESSONS LEARNED:

1) World or other events with which one may not have any connection will influence one's life and one should pay attention to them. The Nazi rise to power, Pearl Harbor, the atomic bomb, and 9/11, are good examples.
2) It is difficult for a child to hide anything from one's parents.

POSTSCRIPT #1. Dr. Cloetingh also explained to me an aspect of Dutch history I did not know. Before Napoleon's invasion of the Netherlands in the early 19th century, the only people with surnames were aristocrats. The remaining population only used first names. For instance, the famous artist Rembrandt is known only by his first name and signed his paintings accordingly. He was often referred to as 'Rembrandt van Rijn" to indicate he came from somewhere along the Rhine River.

Napoleon was appalled when discovering this practice and required every Dutch citizen to register at the town hall, adopt a last name of their choice, and disclose their address, birth date, religion, parents, and other personal facts. In fact, Napoleon established such registries in every country he invaded and where no registries existed. When the Nazis invaded the Netherlands in 1940 and other countries, they immediately went to the local town halls, confiscated the registries, and rounded up the "untermenschen" for extermination. In short, Napoleon's registration system made possible the mass exterminations by the Nazis in Europe during World War II.

During trips back to the Netherlands, people noticed my middle name and asked if I knew, or was related to their relatives. Clearly, the surname means nothing for genealogy when randomly chosen in a forced registration process. Thus I have no knowledge about mine beyond my grandparents, and what little I know about them came through word-of-mouth.

CHAPTER 2
Australia (1939-1947)

My family boarded one of the ships of the Union Castle Line in Southampton during July, 1939 and steamed into the Atlantic. My parents decided we should take a short vacation in Madeira, and disembarked for ten days. I recall the high relief of this volcanic island with lush tropical vegetation. I also pushed open a glass door at the hotel with my hand, shattering the glass, cutting one of my arteries, and ending up in a hospital emergency room. Once treated, we stayed a few more days and then boarded the *S.S. Arundel Castle* bound for Australia.

En route, we stopped in Cape Town, South Africa. I still remember steaming into the harbor and seeing Table Mountain. We visited onshore, and then proceeded to Durban, South Africa. It was my first introduction to human diversity as I saw my first black people, or Afro-Africans (Native Africans?) to use a politically correct modern term.

We then sailed into the Indian Ocean, stopped at Perth for a shore visit, and continued onto Melbourne. While in the Great Australian Bight on September 1, 1939, we received news that Germany invaded Poland and World War II began. We arrived and disembarked in Sydney a week later.

The city of Sydney was founded by the British colonial government in 1788 to house a penal colony for those committing minor offenses. British losses during the American Revolution required a new location and Sydney was as far away as they could send them. Long after I left Australia, the original settlement was restored as a tourist attraction. During my life in Australia, this historical origin was only mentioned rarely in hushed tones.

My parents found an apartment in Neutral Bay with an outstanding view of Sydney Harbor. During my spare time, I watched all the ships, ferry boats, and tug boats sailing back and forth. As World War II intensified, the *Queen Mary*, the *Queen Elizabeth* and other flagships of allied steam ship companies came to Sydney to pick up Australian troops to fight all over the globe.

Shortly after moving into the apartment, my parents enrolled my sister at the Redlands Ladies College. It also had a co-ed kindergarten. I was enrolled there, but because I was considered a 'big kid' for my age, my parents were told to enroll me in first grade in a state–supported school at the end of the term. Next term, I enrolled in Neutral Bay Elementary School.

Both Sydney and Australia were unquestionably provincial in 1939. Sydney only had one decent restaurant, 'Princes'. Because the war had started, xenophobia was evident. During the seven years I lived there, the most popular song was "Buy British Buy". When people talked about "Going Home", they meant returning to the UK, even though they were third or fourth generation Australians who had never gone back. Being a 'foreigner' meant high visibility and I suffered my share of slurs, epithets, and bullying by older boys in school.

Some of it I brought on myself. My first year at Neutral Bay Elementary School, my class decided to play cricket, and not knowing a thing about the game, I ended up captain of one of the teams. They picked me because I was a 'big kid.' That was a major error. It became obvious when I delivered the first pitch not from the wicket at one end, but from the middle of the pitch. I never played the game again. Cricket was a very boring game and I never understood it.

My father established an import-export business and did moderately well. We bought a car, and drove all over the Sydney area and into the Blue Mountains to the west.

The Dutch ex-pat community in Sydney was very small. My parents became very good friends with the Dutch Consul, Kai Van Der Mandele and his wife, Dora. Kai later became the first Netherlands ambassador to the UN, and in 1960, ambassador to Denmark, where he hosted me in 1963 at their chancery. Their second good friend was a business man named "Appy" Van Roijen who had contacts with every key player in Sydney. He proved to be a valuable resource person for our family.

During the 1940-41 Australian summer, my parents decided it might be a good experience if my mother, sister, and I took a vacation on an Australian version of a "Dude Ranch" located in Porepunka, NSW. We travelled there first on an express train to a certain city (can't remember it) and transferred to the 'Porepunka local' for a 25 mile trip. It consisted of several freight cars with a passenger car at the end. It was hot and there was no air-conditioning. Every time we came to a village, the train stopped, a rail car or two was offloaded or added or both. The train then proceeded to the next town to repeat the process. It involved a lot of shunting back and forth. As I recall, it took almost a half-day to travel the 25 miles to Porepunka, and the same was true on the way back.

At the ranch, I learned horse-back riding, played tennis and quoits, and experienced Australian rural culture. I also witnessed sheep shearing, and one day, they demonstrated how they slaughtered a sheep with a machete. I enjoyed horse-back riding the most.

We received word from the US consulate in Sydney that we were approved to immigrate to America in January, 1942. The Pearl Harbor attack delayed us. Soon after Pearl Harbor, the Netherlands East Indies (now Indonesia) was occupied by the Japanese and the remnants of the Netherlands East Indies Army retreated to Australia. My father sold his business and volunteered to serve with them. He reported for basic training at their Air Force base in Canberra, ACT. Meanwhile, my sister and I were sent to boarding schools in Orange, NSW. I attended Wolaroi College in Orange (now the Kinross-Wolaroi College) for seven months and it was a most unhappy experience. It was run by the Methodist Church and was far too strict for my taste.

During a vacation visiting with my mother in Sydney, we were woken one night by a loud explosion. The entrance to Sydney Harbor was protected by a series of nets which were opened to let shipping in and out. Apparently that night while a ship left, three Japanese kamikaze subs entered the harbor and one torpedoed a ferry boat, waking up the city. The subs were captured and put on display in a local park. We discovered that the submarines couldn't have been more than 12 feet long, and were operated by one person.

After my dad completed basic training, he was given a special assignment in the Dutch East Indies Army headquartered in Melbourne. Although an enlisted man, he was put in charge of their entire financial and business operations, literally telling generals how to spend money. The Netherlands armed forces in Australia completed an analysis of all personnel and because my father was the only person who ran a business, he got picked for this plum assignment.

However, it became obvious to the Dutch East Indies Army officer corps this arrangement had its drawbacks. Whenever meeting with their counterparts from other allied armed forces, it was awkward for a buck private to discuss finances on behalf of officers. Consequently, within a year, he got promoted fast to corporal, a week later to Sergeant and then to

Lieutenant. It improved the business negotiating position of the Netherlands East Indies Army in exile.

Knowing we would be in Melbourne for several years, my parents bought a newly-constructed ranch house on the city's edge. It was a unique experience because during the summer, the paddock across the street was occupied by sheep herds. Some strayed into our front yard and left their 'calling card'. When the sheep left in the fall, the paddock sprouted a large number of mushrooms, fertilized of course by sheep droppings. We harvested a lot of mushroom during those years.

My sister enrolled at the Geelong Grammar Ladies College in Geelong, Victoria, an Anglican (Episcopal) girl's school. She was quite happy to be sent to boarding school and did well. During her final year (equivalent to the US Senior year in High School), she was elected a prefect, a student leader. That made her more stuck up and it stayed with her for her whole life.

My mother took me to visit three 'public' (private) schools, Melbourne Grammar (Anglican), Wesley College (Methodist) and Scotch College (Presbyterian). I did not like Melbourne Grammar and they did not like me. I liked Wesley College the best but they were over-enrolled and turned me down. Scotch College, the largest private prep school in the Southern Hemisphere, accepted me. The School was founded by the Presbyterian Church in 1851 and moved to a new campus during the 1920's. I enrolled in their junior (elementary) School in 1943.

I did above average, learned Australian Rules Football and boxing, excelled in geography and history, did very poorly in penmanship, and made it through English, science, and math. Discipline was strict, reinforced by 'six of the best', namely caning one's rump.

I was promoted to the Senior School in 1946 and completed the standard college prep first-year courses, but not at a distinguished level. The school was sports crazy. In 1946, Scotch College won the "head of the river" a rowing race with five other private schools of comparable size and when the opportunity came to try out for the "crew" I did. I was rejected during my physical because of an undisclosed heart condition.

I watched cricket (boring), Australian Rules football (really interesting), enjoyed the annual rival football game with Melbourne Grammar, and the annual public schools track meet. They also taught students boxing during gym classes. This skill helped me later.

Perhaps the most unusual event I recall occurred when Scotch College played Xavier College (Catholic) on their home turf. In 1946, Xavier fielded a team that literally wiped all the other schools off the floor. During previous years, they usually ended up last place. What changed?

Before the game started, we sat in the visitor's stands while the Scotch College team warmed up. Suddenly church bells pealed from the Xavier Chapel on a hill behind the stadium. The Xavier football team in uniform charged out of the chapel running down the hill and onto what is now known as Chapel Field. They outweighed the Scotch College players by at least 20 lbs, and were clearly about 3 inches taller. The Xavier players were mostly Eastern European and Italian immigrants who survived Nazi concentration camps or in other ways during World War II. They clearly were older. To say they kicked the stuffing out of Scotch College that day is an understatement. They won the Public School Championship that year to no one's surprise.

Their team achievement was just as much a credit to the Xavier coaches. They organized a team of people who were still learning English, came from several cultures, learning the rules of a new game, and develop the necessary skills to communicate and achieve their success.

While at Scotch, I was involved with scouting and participated in scout camp outs, learning some survival skills that came in handy later in life as a geologist. I remember one winter weekend going on a family trip to the Dandinong Ranges where it snowed. I used those skills to start a fire and a few other details.

On weekends, I enjoyed horseback riding. A nearby stable rented a horse for 5 shillings (about $1.00 equivalent) for an afternoon and I rode with a group of riders. The group stopped after about one-and-a-half hours, watered the horses at a creek, let the horses graze, and enjoyed billy tea made by one of the stable's staff. We then rode back. It was a great way to spend an afternoon in the out-of-doors.

While at Scotch College, I enjoyed singing in the school choir. That interest and enjoyment stayed with me throughout my life. I sang in choirs in the USA until 1960. I still enjoy choral music today.

My parents socialized with the officers of the Dutch East Indies Army in exile while living in Melbourne. Some made it out but left their families behind. Their separation placed a heavy burden on these officers. One, Colonel Wim Kniestedt, was a family favorite. He introduced us to the Dutch art of making a fruit preserve soaked in gin. He also told us numerous stories of his experiences growing up in The Netherlands and being posted to the East Indies (now Indonesia). When the war ended, he reunited with his family who joined him in Melbourne.

The Netherlands East Army contingent in Melbourne formed a "Dutch Club". They rented a social club on Saturdays and served "Rijstafel", the national dish of the East Indies. We went every Saturday, met the other officers and their families (who had managed to escape) and generally had a good time. The food was clearly the best in Melbourne and made up for the dreary fare at the Scotch College 'tuck shop.' They also ran a Dutch language school for the children of Dutch service men. I attended, but it did not stay with me.

I made my share of friends, but after leaving Australia, I lost contact. My good friend Gerry Kuecher put it well about growing up in a house of 13 brothers and sisters who moved on into the world: "Time has a way of estranging us from our loves" [1]. The same applies to people we walk with through life at some time, but as we move away and move on, we become estranged from them too.

World War II ended in Europe in May, 1945 and in the Pacific in September, 1945. My father was promoted now to the rank of Captain, and in March, 1946, he was promoted to the rank of Major. He mustered out in July, 1946 and prepared to go to the USA.

My father planned ahead to renew his business activities and printed special announcement cards. He mailed them to all his business contacts in Europe stating that he was moving to the USA to reestablish his business, and wanted to renew his relationship with them. He also asked if there was anything in the way of food and clothing or other items that they needed so we could ship it to them. I helped him by stuffing addressed envelopes with these cards and putting stamps and airmail stickers on them. Together, we mailed them at the post office.

He started receiving replies after five weeks. Half of the envelopes were returned because people moved or died during the war. It was a sober reminder of how transitory life can be. The remaining responses expressed interest in hearing from my dad, doing business with him again, and provided lists of things they needed for their families.

My mother then went to work. She always kept both my sister's and my outgrown clothes including school uniforms. Suddenly, these clothes started disappearing to Europe. She either added some of her and my dad's clothing or bought some. Then she bought canned coffee, canned milk, canned meat, canned vegetables and first aid items. These were packed into boxes, labeled and sent.

When my father started his business in New York, it was very clear these people were only too happy to do business with him again and their gratitude stayed until he passed away.

It was a valuable and unforgettable lesson. It taught me the importance of networking, maintaining a network, and being willing to do kind things for other people. It always pays. In my case, I mentored geologists laid off by large oil companies and completed my share of volunteer professional society committee work. It all started with helping my Dad stuff those envelopes immediately after World War II.

My father left for the USA in August, 1946. He obtained a berth on a former unconverted troop ship, but it was the quickest way to get there. He went to San Francisco and boarded a train to New York. He established a new import/export business, starting over for the third time in his life.

That too was a valuable lesson for me: never be afraid to start over, regardless of one's age. I implemented this lesson several times during my life including opening up my present geological consulting business at age 63.

My mother stayed behind, sold our home, and packed and shipped our belongings to New York. She booked passage for the three of us to travel to the USA in late January, 1947. That departure ended another phase of my life, this time in Australia.

LESSONS LEARNED:

1) When undertaking business networking, remember the business associates one meets are human and under certain circumstances may have critical needs. Be prepared to reach out and help them.

    The incident involving my father's announcement cards to former business associates illustrates this well. Once he was open for business in New York, who do you suppose the people in Europe who received food and clothing from our family wanted to do business with?

2) Never be afraid to start over, whether after a setback, a change in career, or a change in circumstances. But when doing so, stay with what you know.

CHAPTER 3
High School in America (1947-1950)

Travelling from Australia to the USA immediately after World War II took more than fortitude. With Appy Van Roijen's help, we succeeded in getting reservations on the *M.V. Lowlander* owned by the Port Line based in the UK. The itinerary took us from Sydney, Australia, to New York with stops in Tahiti, Panama City, and Curacao. We sailed on January 16, 1947. The trip took 47 days.

The *M.V. Lowlander* was built in Italy as a freighter with cabins for 20 passengers. It operated on the Trans-Atlantic trade. When Pearl Harbor was attacked, it was off-loading cargo in New York and immediately seized as a war prize. The US Government turned it over to the British government which transferred title to the Port Line.

Because the *M.V. Lowlander* was built in Italy, parts were all in metric units. Replacement parts were unavailable. Thus, we would steam ahead and suddenly came to a dead stop. The engineers repaired a damaged part or fashioned one in the machine shop. During these stops, the *M.V. Lowlander* bobbed in the ocean like a cork for ten to 25 hours. Such stops were spaced from three days to one week apart. The last repairs were completed at the end of February, 1947 off North Carolina in the middle of a "Noreaster."

Nevertheless, we made progress. We had an enjoyable shore trip in Tahiti, a scary experience in Panama City, transited the Panama Canal, reconnected with Dutch cuisine in Curacao, and arrived in Hoboken, NJ on March 3, 1947.

Life on board ship was mixed. The passengers included six war brides heading to meet their fiancés in the US. The Ship's officers socialized with some of the ladies and two of them got pregnant during the cruise. When we disembarked in Hoboken, NJ, I noticed both women tearfully explaining their circumstances to angry fiancés. I learned later they were sent back to Australia.

The food was limited, served in small portions and tasted like most English food, namely not very exciting. I never had enough to eat. One day, while on deck, I noticed a crewman throw food overboard. I discovered that the crew, who ate separately, had better food and there was some left over. I negotiated with their cook to let me come by after each meal and get a second meal. They agreed.

Consequently, I came to know the crew, learned what and how such people think, how hard they worked, their complaints about the officers, and their slant on life. That experience served me well in later years during two cruises of the Deep Sea Drilling Project and the Ocean Drilling Program, one as a co-chief scientist (Chapters 17 and 21).

The ship docked in Hoboken, NJ, where we were met by my Dad who had purchased a new 1947 Chevy Coupe. We drove to Manhattan. My father found housing in Greenwich Village in New York City.

We reached the apartment and unpacked. Our living quarters were in an old house chopped up into apartments and rooms. We had a living room, three bedrooms and a bathroom on the second floor, but the kitchen and dining room were in the basement. Housing was scarce in post-war New York City so we were fortunate to have something. Other tenants mostly were artists and musicians who practiced at all hours of the day and night.

One musician came by one evening and offered four free tickets to attend the world premiere of an opera written by a good friend. He explained the friend needed a large audience (I suppose to get a good review). I was not eager to go, and I didn't find the experience that exciting. The event was held in a medium-sized off-Broadway theater. The composer was Gian Carlo Menotti, and it was indeed the world premiere of "Amal and the Night Visitors" which was well received. The significance of the experience didn't register until I saw it on a TV Christmas special years later.

Within two weeks after arriving, my mother and sister went on a frantic search to find a college for her. They visited Radcliffe, Wellesley, Smith, Mt. Holyoke, Vassar and Bryn Mawr. Because we arrived in the USA in early March, application deadlines had passed. Finally, Bryn Mawr College accepted her and she enrolled in the fall of 1947, graduating with a degree in English in 1951.

She received a very good education, book-wise, but her attitude did not change from her prefect days at Geelong Grammar Ladies College. During her senior year, she was nominated to be president of her dorm but lost the election. She told us she was disgusted with the back-biting that occurred during the short election campaign. It left her with a disdainful attitude about most of her classmates.

My father enrolled me at Haaren High School in Manhattan as a freshman and I took college prep courses in English, Latin, history, algebra, and science. I coasted, yet earned high grades because most of the material duplicated what I had learned at Scotch College in 1946. However, the school served a critical purpose. It socialized me to the American way, American civics, and American history. That was always the mission of the American public school, namely to socialize immigrants, including me and for that I am grateful.

The student population was diverse, and some students came from economically poor homes. I recall one incident during a science class. The class was shown a movie and two black kids sat behind me. They kept pestering me for money. I told them to shut up and watched the movie. They persisted. So, not necessarily knowing the norms of an American High School, I stood up, turned around put up my fists and gave one an uppercut to the jaw with my right fist. He went sprawling over the floor.

The teacher stopped the movie and asked what happened. I explained the problem and because she viewed me as her 'star student', she accepted my explanation and sent the other two kids to the Principles office. They were expelled.

My parents started looking for a home in the suburbs. They bought one at 409 Weaver Street, in Larchmont, NY, and we moved during June. They sent me to a summer camp in New Hampshire, a co-ed Quaker camp devoted to socializing immigrant children to America. Fellow "campers" originated from France, the UK, Germany, Netherlands, Belgium, Norway, Italy, Yugoslavia and Poland. Generally we got along.

The high point of the summer was to climb Mt. Washington. We drove to its base and spent the day climbing to the summit. We took the cog railway back to our cars. An elderly man approached us while at the observation deck and asked who we were. Because we were immigrants, he invited us to his house for a coke and cookies and pointed to its location. He owned the Cog Railway and clearly was a wealthy man. What impressed me was that he drove in a Chevrolet no different from my father's. Clearly, the lesson learned was that despite his wealth, it was best to spend money wisely. Wheels are wheels; it's the chrome, price tag and nameplate that are different.

I enrolled that fall as a sophomore at Mamaroneck Senior High School, a mile walk from our home. The student body came mostly from affluent and upper middle class homes. Moreover, because I skipped most of a year of school, I was the youngest member of my graduating class. I took the usual college preparatory courses, sang in the high school choir, made it on the swimming and track teams, and joined the current affairs discussion club. I made several friends too.

My activities continued through my Junior and Senior year. During my senior year, I was selected to host a student from Malaysia sponsored by the *New York Herald Tribune* International Student Exchange program. He stayed for five weeks, attended my high school, and participated in a variety of events. His stay concluded with a ceremony and forum at the General Assembly Hall at the United Nations and featured all the exchange students from overseas and their American student hosts.

I also renewed my interest in scouting and was active in the local Boy Scouts of Larchmont. During my junior year, I was appointed "Police Chief for the Day," had my picture in the local paper, and learned much about the Larchmont Police. On the day I served as "chief," a marked patrol car came to my home and the chief of detectives took me on patrol. I then went to the Police Station to sign in, met the real police chief, was shown the jail, their crime lab, and allowed to operate their radio dispatch to other patrol cars.

The chief of detectives drove me home and asked if he could inspect our home. I checked with my mother and she approved. First he went to the attic where we had noticed a rusty stove to which were connected wires. He then went to the basement where we noticed similar wires and a buzzer button connected to them.

The chief of detectives explained that his first assignment in 1925 was to watch the house. They noticed trucks backing into the driveway at all hours of the night, loading cases and driving off. The Larchmont Police Department staged a raid and discovered a bootleg operation there and arrested everyone on site. The attic stove was part of a still. The wires went to a basement lookout who signaled if trucks or suspicious people were coming. He was surprised that some of the original equipment was still there. It taught me that community police departments have very long memories.

A college fair was held every year at Mamaroneck High School and I met admission officers from colleges and universities that my parents suggested I consider, and to which I had seen better students matriculate. By the time my junior year ended, I decided I preferred attending a smaller college. I focused on Amherst, Williams, Wesleyan, Swarthmore and Oberlin. At the Fall College Fair during my senior year, I met their admissions directors and dropped Williams from the list. I visited Amherst, Swarthmore and Wesleyan (CT). My parents were not too enthusiastic about Oberlin because of distance. I applied to all four. Only Wesleyan and Oberlin accepted me. I decided to enroll at Wesleyan University. Graduation was on June 25, 1950, the same day as the outbreak of the Korean War.

Four of my classmates joined the National Guard during their junior year. Within a month after graduation, they were called to active duty and shipped to Korea. Two came back in body bags, one returned badly wounded, and one returned intact.

All my friends went off to college except one, Charlie Albert, the star halfback on the football team. Charlie was always full of life with a positive outlook. He was also African-American. His girlfriend, Charlotte Latten, was a cheerleader and extremely attractive. During the summer of 1949, he got her pregnant and they married. Charlie finished high school and then took a job driving a delivery truck. I saw him during a vacation from college. He became a beaten-down man and requested I never contact him again. Regrettably, I honored that request. During my life I met many people with potential who were derailed from their goals because of financial status, unfortunate events, or serious mistakes.

During the summer of 1948, I attend Camp Pocono, a Quaker boy's camp on Lake Wallenpaupak in the Pocono Mountains. It was a beautiful setting. We lived in tents. I learned canoeing, did a lot of swimming, and a lot of woodcraft.

The following summer, I returned as a junior councilor. I instructed canoeing and swimming. I returned after High School graduation as a regular councilor. Things had changed and I left during the middle of the summer. The enjoyment of the place disappeared. On my return home, I prepared to go to college.

LESSONS LEARNED:

1) How to function and make it in the USA.
2) How to avoid mistakes and which mistakes are likely to cause the most long-term problems (e.g. the Australian fiancées who got pregnant on

ship; Charlie Albert).

3) During long voyages on board ship, relationships can become strained over time, particularly because space is limited and it is difficult to hide from others.

4) When choosing a college, pick one where one is likely to be sufficiently comfortable and can achieve success.

5) Try to avoid fights with minorities in today's culture. Find a way to back away. The incident in the science class would turn out very different in today's politically correct climate.

CHAPTER 4
Wesleyan University (1950-1954)

I arrived in Middletown, CT, to start my college career at Wesleyan University, an all-male institution, in Mid-September, 1950.

Wesleyan University was founded in 1831 by the Methodist Church. The university focused on liberal arts, including science, and broke its ties with the Methodist Church in 1937. Founded as a male institution, it was coeducational between 1872 and 1912. Alumni pressure terminated the arrangement and it reverted to a male institution in 1912. The ousted female students established The Connecticut College for Women in New London, CT to continue their higher education. Wesleyan became coeducational again in 1968 and likely will stay that way in perpetuity.

My dorm room was in Clark Hall and consisted of a three room suite. At the front was a larger room with four desks and in the back were two bedrooms with bunk beds. Arriving early, I took the lower bunk bed in one of the rooms.

Soon the three other roommates arrived. Andy Maggatt was a party boy who flunked out by mid November. John Stacey was a quiet individual. He left Wesleyan after a year to attend business school. The person who shared my room and took the upper bunk was Mowbray (Mo) Dietzer. He came from Syracuse, NY, and graduated from a private school, Pebble Hill in upstate New York. Mo and I graduated and Mo made his career with *The Wall Street Journal*.

The freshman orientation included registration, opening up a local bank account, hearing pep talks from Alumni Trustees about Wesleyan's virtues, a physical education screen, a cursory medical exam at the university health center, and participating in Fraternity Rush. This was an experience for which I had no preparation as I moved from fraternity to fraternity trying to scope it out.

My three roommates were pledged to various houses. Maggatt was pledged to Delta Kappa Epsilon, Stacey to Sigma Chi, and Dietzer to Alpha Delta Phi. I was not invited to pledge any fraternity but received invitations to join the eating clubs of Delta Tau Delta (DTD) and Delta Upsilon (DU). Although initially disappointed, it gave me more time and options to assess the fraternity system.

Wesleyan fraternities were limited by the number of freshman pledges they could accept so those who were considered "alternates" were invited to join their eating club (one dined there and participated in all their social activities). If all went well, one was pledged the following year. I chose DTD for the Fall Semester, and then switched to DU during the Spring Semester because I liked it better. DU told me at the end of my freshman year that I was on their sophomore pledge list and to return in the fall.

I signed up for the usual distribution requirements expected of freshman: English, Spanish, Humanities, Biology (with lab), and a general social science course. The general social science course overlapped a lot with what I had learned at Scotch College. The Humanities course was of marginal interest mostly because we were trying to divine what the professor in charge wanted us to think was important.

In reality, I was untypical of the students at Wesleyan. I was one of two immigrants in the class of 1954. The other, Sigmund Franczak, spent most of World War II in a Nazi concentration camp in Poland and later in Germany. He was separated from his family and did not know if they survived. An American Army Captain arranged to adopt him and he went to high school in St. Charles, MO. He was accepted to Wesleyan on a scholarship studying pre-med.

Sig spent a lot of time trying to locate his family. During his sophomore year, he finally heard they were all alive. He and his adoptive father arranged for them to move to Middletown where his father got a job.

Most of the student body at Wesleyan during the early 1950's came from small New England towns. Sig and I were the first immigrants they ever met. During Freshman Orientation, the university president, Victor Lloyd Butterfield, told the class that the aim of a Wesleyan education was to acquire a liberal arts background so we could return to these small towns and become community leaders. My goal was different. I originally planned to become a marine lawyer which required Law School and practice it in a major port city like New York or Los Angeles.

The student body at Wesleyan was also very turn-of-the-20th-century, 'old-style' Republican. During the 1952 Presidential Primary season, a mock primary was held. 79% of the students voted for Robert Taft, a staunch Ohio Republican conservative Senator. Except for one black classmate who later became a federal Judge and three black students who arrived in 1955, the campus was anything but diverse. In fact, diversity was a concept that was not on the minds of the Wesleyan Faculty, administration, or student body. It was a New England, white, Protestant campus.

I auditioned for and was accepted to the chapel choir, a plum assignment that paid $1 per hour, and simultaneously enabled me to meet the college chapel requirement (10 per semester). I also tried my hand at debate, and found the faculty member in charge a phony and dropped it.

My English professor, Dr. Cowie, told me after a month of classes that I had absolutely no writing ability. Thus, he was transferring me to a special English class to improve my writing. Colloquially, it was called "Bonehead English". I discovered much later that it was a game-changer for me and one of the best things that could have happened. I took the course both semesters. It was taught by a Dr. Cochran who tutored us individually as well as in class and did a great job.

By the middle of the fall semester, my academic performance was not good and I received a warning letter of impending probation. Although I applied myself more rigorously, I could not turn it around and ended the semester on probation. That meant I had to drop my extra-curricular activities. I filed an appeal to continue with the choir because I earned part of my expenses that way and it was approved.

The choir experience was a good one. To balance it, ladies from Middletown participated. We went on local tours with them. However, the men's section also did an annual joint home-and-home engagement with a choir from one of New England's many women's colleges. Some of my choir-mates met their wives that way.

One cloud on the horizon was the Korean War. I observed my fellow students of all classes facing draft notices. They chose instead to enlist in Officer's Candidate School (OCS). Many left Wesleyan during the middle of the semester and returned to complete their studies after I graduated. I estimate 15% of the student body departed that year to Navy or Air Force OCS schools.

As an immigrant, I was required to register for the draft. If called up, I had the choice of being drafted and receiving citizenship quickly. If I declined, I would be ineligible for US citizenship. A system of deferments for science and engineering students and professionals was established during the summer of 1951. Such deferment was at the option of the local draft board. My draft board deferred me until age 35 when my eligibility for military service expired.

When the second semester started in 1951, I resolved to improve my grades and compete for the prize of "The most improved freshman." I worked hard studying in the library to avoid distractions. The only memorable thing that semester was that in the biology class, we dissected a baby piglet. Biology labs for the freshman course were all on Wednesdays, and that evening all campus fraternity houses served roast pork for dinner. It was no accident.

During some party weekends, I periodically worked as a bar tender at fraternity parties. The job had a downside. Some of the Wesleyan men drank too much and one or more passed out while their dates watched frantically and hysterically. One of the tasks of the bartenders was to calm them down. Moreover, whenever dates came to Wesleyan on weekends, housing was usually found for them in local private homes.

At closing time, the dates of these passed out boyfriends were escorted to their temporary homes by the student bartenders because Middletown was not known as being a university-friendly community. It was mostly a Catholic working class community (Middletown had three Catholic Churches, one for those Italian descent, one for those of Polish descent, and one for those of Irish descent). Town-gown relations were abysmal. Often, stranded dates didn't know their way back to the private home where they stayed. I started a system of logging the local addresses of everyone's date early in the evening while their Wesleyan hosts were still sober.

There were recriminations and finger pointing when the guys sobered up. Most of us who bartended were accused of spiking drinks to "bird-dog," go to bed, or 'make out' with their dates. The recriminations arose because next morning some of these ladies delighted in teasing these guys with exaggerated stories. I often told my compatriots, there's nothing desirable, romantic, or amusing about a partly stupefied, half-drunk woman.

The semester ended and my GPA went up from a D plus to a B plus. However, another classmate earned the "Improved Freshman Prize."

Because Wesleyan was small, it lacked certain course offerings. I was interested in taking anthropology and sociology courses and attended summer school at Northwestern University during the summer of 1951 to do so. It was a great summer and I enjoyed the coursework. I met one graduate student in the dormitory, Frank Hoodmaker. He was earning a MS degree in geology and I looked at his books and found them of interest.

Over the summer, I concluded that if I went to summer school the following summer and took a lab course each of my remaining semesters at Wesleyan, I could earn enough credits to graduate in two more years.

During the fall of 1951, I took a third semester of 'bonehead English' and chose geology for my lab science. I don't recall what else I took, but the geology course changed my life. I did very well and chose to major in it. That meant a complete reorganization of my course work. It ended my plan to graduate in 1953 because I had cognate science courses to take, and complete a major from a late start.

The 'bonehead English' course was taught by a visiting professor, Charles A. Muscatine. Muscatine was an English Professor at the University of California, Berkeley (UCB), earned all his degrees from Yale, and was an internationally recognized Medievalist. He refused to sign a loyalty oath and under California laws of the time, he was fired. Wesleyan offered a visiting appointment and he stayed until the California law was repealed three years later (See Chapter 14).

Muscatine was particularly irritating. He kept lording over us that he was a Yale graduate and implied a Wesleyan education was inferior. He acted as if he didn't want to be at Wesleyan, although to earn his livelihood, he was fortunate to have a job there. I crossed swords with him a few times in class but I earned a B plus. I was happy that I did not have to take another course with him. Little did I know then that I would visit him at his office at UCB in 1970 (Chapter 14).

Overall, my sophomore year was a great year and I scored high grades in all my courses. I decided not to pledge at DU and instead joined the independent counterpart to fraternities, the John Wesley Club.

During my sophomore year, I competed for a campus wide public speaking prize, the Parker Prize. I intended it to be a trial run to compete more effectively my junior year. However, I won and was also told that because I did so, I could not compete for it again.

To make up for my change in major, I spent the summer of 1952 at

Harvard Summer School to complete a year's freshman chemistry course in eight weeks. I passed, but not well. I also was disappointed in Harvard because it was clearly unfriendly to undergraduates. Nor did I care for Boston. The attitude of the people there reminded me of the 'going home" attitude I observed in Australia.

When I chose to major in geology during 1952, the Wesleyan Department of Geology was small, consisting of two professors, a technician and a secretary. The first scientist on the Wesleyan faculty was John Johnston, a Bowdoin chemistry graduate with an interest in mineralogy. He taught Natural history, including geology between 1837 and 1873. The department was founded in 1867 when its most prominent early geology graduate, William North Rice, was appointed. Rice was the first person to earn a PhD in geology from a US university (Yale) around 1870. When Rice retired in 1918, he was replaced by William Foye as professor of geology. Foye was replaced in 1935 by Joe Webb Peoples (BS, Vanderbilt, MS. Northwestern, PhD Princeton, economic geology; Mahoning Coal Company, Lehigh, Wesleyan) who stayed until he retired during the mid-1970s. He was department chairman during my student days.

Rice, Foye, and Peoples were assisted by non-tenure track junior faculty. They included Gilbert Cady from 1909 to 1910 who went to the USGS to become a distinguished coal geologist, and Ralph Digman (1944-47) who founded the geology department at Binghampton University. Norman Herz taught at Wesleyan from 1950 to 1951 and went to the University of Georgia where eventually he became department chairman. Rueben J. Ross (BA, Princeton, PhD Yale, paleontology; Wesleyan, USGS, Colorado School of Mines) was hired on the faculty in 1948 and taught me Historical Geology. Rube left in 1952. He was replaced by Elroy P. Lehmann (BS, MS, PhD, Wisconsin, stratigraphy; Wesleyan, Mobil Oil eventually VP of Exploration) who left in 1955. John Rosenfeld served on the Wesleyan faculty from 1955 to 1957 before moving on to UCLA. In 1958, Joe Weitz (BA, Wesleyan, PhD, Yale; structural geology) was added and left in 1960 to go to Colorado State University. In 1959, Gordon P. Eaton (BA Wesleyan, PhD, Cal Tech, metamorphic petrology; Wesleyan, USGS, Texas A&M Dean and Provost, Iowa State University President, Lamont-Doherty Geological Observatory Director; U.S. Geological Survey – Director), Joe People's star undergraduate major, returned and left in 1962 for the USGS.

In 1972, the department's name was changed to 'Department of Earth and Environmental Sciences' and began to expand. It now has a faculty of eight tenure-track professors and two research professors.

During the fall of my junior year, I enrolled in Mineralogy, Stratigraphy and Sedimentation, Physics, Calculus, and Philosophy of Religion. It was a tough grind and Physics proved to be a particularly difficult course. I liked the Stratigraphy and Sedimentation course, and decided that if the opportunity came, I should pursue a career in sedimentation, or as it is now called, sedimentology. I earned B's in all my courses except Physics where I scraped by with a D minus.

I still sang in the Wesleyan Choir. During my junior year, our exchange arrangement was with Pembroke College in Providence RI (they later merged into Brown University). They came to Wesleyan first and on the Saturday afternoon when they arrived, we completed a rehearsal. Because by this time I was an officer of the choir, I had certain duties to complete after the rehearsal. I witnessed a conversation between the choir director, Dick Winslow, and the chapel organist, Bill Prentice. Winslow said with a panicked look on his face, "They can't sing!"

The concert took place that evening and a party was held afterwards. Some of the choir members really liked those Pembroke ladies. One of my colleagues put it well when he mentioned that those girls couldn't sing but they sure could make out.

The following Tuesday, the student newspaper appeared and one of my buddies at the John Wesley Club was its music critic. He reviewed the concert and severely criticized the Pembroke ladies for their inability to sing. Because the Pembroke choir had a high opinion of themselves, they arranged for the paper to mail them the issue with that review. On arrival at the Pembroke campus the following Saturday, we received a frosty welcome and a few choice words were exchanged. Suddenly the warmth and friendliness of the ladies disappeared.

Several of my choir colleagues were furious and wrote harsh letters to the student paper about the critical review and how it undermined their socializing. All were published, and so was a reply by the reviewer. He provided a primer on reviews and concluded by writing, "Always remember a review is the opinion of the reviewer only." I used that comment during later years when reviewing scientific manuscripts for journals or book publishers, proposals for funding agencies, or writing book reviews for scientific journals.

During the Spring Semester, 1953, I took petrology, structural geology, second semester physics, second semester calculus, and a seminar in Marine Ecology. Again, Physics was a struggle although I did well when we reviewed optics and acoustics. Still, I earned a D. I did better in the other courses, earning a mix of B's and "A's".

The Marine Ecology seminar was basically a tutorial and I did well. The instructor, Dr. Haffner, gave me some interesting advice. He advised that when I started publishing papers, I should stop, reassess, and write a summary review paper and cite my own papers extensively. I discovered when doing so that more people read such review papers than the original detailed one.

During spring vacation, we took a one-week field trip to southern New York State and the coal mining regions of eastern Pennsylvania, including a trip down a mine-shaft to examine underground operations. Prior to the spring field trip, we went on afternoon field trips in the Connecticut Valley. During every class before the field trip, Joe Peoples reminded us to bring hand lenses on the spring trip, which we did.

The first stop on the spring trip was in the Peekskill Norite. After using our rock hammers to get samples, we peered at them with our hand lenses. However, "Doc' Peoples was wandering around with a sheepish look on his face. He called us together and said, "Gentlemen, I have a confession to make. After haranguing you all week about bringing hand lenses, I left mine at home. Can I borrow one of yours please?"

We traveled in two cars. I rode with Doc Peoples. On the way back from Pennsylvania, we drove past New York City to drop people off so they could take trains home to visit their parents. I rode to Middletown because my parents were in Europe. I used the time to inquire about how to develop a career in geology, what my options were, what would be required to get there, and above all, what it would take to be successful. He answered all my questions and gave me lots of good advice. It was a good mentoring session. Joe Peoples taught me not only valuable things I would not have learned any other way, but also how to develop mentoring skills in the future.

After my parents returned from Europe they called. My sister, Marianne, was getting married in July to someone I had yet to meet. He was H. George Mandel, a Yale PhD in chemistry, who was an assistant professor of pharmacology at George Washington University Medical School. George was born in Germany. His father had been a director of the Deutsche Bank, but left in the late 1930's. His parents lived in Scarsdale, NY, and were both pretty haughty and not the most pleasant people to know. Fortunately, Yale had rounded and humanized George Mandel and he was, in fact, a reasonable gent.

I attended their wedding where Georges Mandel's mother was trying to match me to some of the daughters of her friends. They were very high maintenance and didn't interest me. George's mother berated me on a regular basis but her continued efforts to pair me up always ended in failure, for her.

That summer, I took a required geology field camp to learn geological mapping and field methods. I enrolled at Northwestern's field camp because their course started in latest July and ended just before Labor Day. That enabled me to first accept a summer job as an engineering helper with the Connecticut Highway Department. The state was building a highway bridge in Middletown. My job was to stand next to a steam-driven pile driver, count the strokes and when the number of strokes reached six per inch, tell them to stop and go to the next one.

Sig Franczak had a job with the contractor at the same work site so we ate lunch together and visited on weekends. One day, he walked on a crosswalk over an empty sewage holding tank. He fell, landed on his head and ended up in the local hospital. I visited after work while a Catholic priest was administering last rights. I explained to the nurse that Sig was Methodist; she notified the priest who left. Sig died that night. It was a great shock to all of us.

I left in late July to join Northwestern's field camp in Duluth, MN, where I rendezvoused with the two professors teaching the course and the other students. We examined the Keweenawan basalts along Lake Superior, went to the Hibbing open pit Iron Mine and reached Ely, MN after three days.

There, we boarded chartered canoes and paddled and portaged our way to our first camp site. We mapped geology from canoes and bushwhacked inland to complete our maps. Because I had been a canoeing counselor, I helped instruct the other students on canoeing.

The other participants came from a variety of universities: DePauw University in Greencastle, IN, University of Cincinnati, LSU, Franklin and Marshall, and Northwestern. The best prepared student was Bill Rush, who attended LSU. He spent the previous summer on an oil company field mapping crew. I learned a lot from him as well as the instructors, and was shocked to discover later Bill never submitted his final report.

After completing the course, I returned to Evanston and met the department chairman, Art Howland. Northwestern had, at that time, a nationally-recognized team of three professors, William C. Krumbein (PhD, Chicago, sedimentology, geostatistics), Laurence L. Sloss (PhD, Chicago, stratigraphy) and Edward C. Dapples (PhD, Wisconsin, sedimentary petrology). I met all three because I planned to apply there for graduate work. Frank Hoodmaker warned me that Northwestern's program was tough and he left and completed a Master's degree at the University of Wyoming.

Krumbein was busy so we talked briefly, Sloss talked with me but at the same time was busy drafting, and Dapples spent time rambling about his work. Perhaps I caught them at a bad time or perhaps they met many applicants so interviewing one more became a routine and boring exercise for them.

I returned to Wesleyan and had two months, in addition to regular course work, to write my field camp report, draft the diagrams and submit it. I did with a week to spare and earned an "A". Years later, at the 1958 annual meeting of the Geological Society of America (GSA), Ed Sullivan who graduated from Northwestern and took the course also, told me that I was the only one to earn an "A."

Wesleyan's financial fortunes also suddenly improved. An alumnus, Davidson, bequeathed his fortune of $6 million. Davidson's wife predeceased him and there were no children. The endowment suddenly doubled and Wesleyan's expectations changed.

During my senior year, I took a year's course in statistics, a semester of Paleontology, a year's senior geology seminar, a philosophy course and a year's worth of credit to complete a senior thesis. I graduated with three other geology majors. They were:

- Dana Schrader who became a Navy pilot and then flew for Eastern Airlines.
- Tom Rogers, who earned an MS from the University of California, Berkeley, took a job with Gulf, was laid off, and then worked for the California Department of Mines.

and

- Lou Wilcox, who took a job with the Army Map Service, and later earned a PhD in geodesy from St. Louis University. Lou died in 1972.

I lost contact with Dana and Tom.

A week before graduation, we received our yearbooks which included a profile of the editors' recollections of each class mate. Mine stated "Well-Assimilated Immigrant."

Graduation was held on a hot day. My family joined me. Joe People and his family hosted a reception for the four geology majors and their families. We then went to graduation where Earl Warren, Chief Justice of the Supreme Court, delivered the commencement address.

At the close of Commencement, the college President, Vic Butterfield, delivered his "Charge to the Class of 1954." Starting with a quote from Alfred North Whitehead that "great ages are also dangerous ages", Vic discussed how

- the atmosphere in the USA changed with McCarthyism,
- the American role in the world became more dangerous with the Cold War, and
- so many of the world's people increased their demand for freedom and well-being. He discussed also the raging battle of ideas of the time and closed his remarks saying,

"These are the times for these forces *(of moderation, justice, freedom)* to speak out, if quietly yet very firmly. You are about to join them. With judgment and courage you can become effective spokesmen for the cause of civilization and freedom. *You may get hurt if you speak.* But you won't be worth your Alma Mater's faith and investment in you if you falter from fear. This is no time for timid men. *There is no place in the hearts of Wesleyan men for this kind of fear.*

I charge you therefore to join the battle with courage, working with all the judgment and skill you possess for the harmony and the mutual confidence of men based on a passion for freedom and tolerance, forever strong to insist, firmly, rationally, respectfully, on all that makes for a free and civilized society, on all that can encourage the fulfillment of men in terms of their nature, their conscience, and the will of their God."

A framed autographed copy of that charge hangs in my office. Looking back, it still strikes me that during my time at Wesleyan something (perhaps the Davidson Bequest) changed Butterfield's view about the role of Wesleyan graduates. It changed from just returning to New England small towns and becoming pillars of the community to something where I was perhaps more main-stream with respect to Wesleyan's aspirations for its graduates than when I arrived.

LESSONS LEARNED:

1) Be willing to mentor those who follow you and help them achieve their success.

2) Be aware of how the values and goals of an institution change. The Davidson bequest changed Wesleyan forever.

3) When attending university as an undergraduate, explore what the place offers and be prepared to follow your interests, even changing majors to do so. In my case, switching to geology changed my life and I never regretted it.

4) If told to remediate deficiencies, look at it as an opportunity to improve oneself. My bonehead English experience provided me with a critical skill. Professor Cowie did me a favor advising me to take it.

5) Never be afraid to ask questions or get advice from people who are more experienced. At the same time, be willing to do the same for those who are younger.

6) If you need to earn expenses through part-time employment, find out what's available in the department of your major. I did, curating mineral and rock specimens, mounting maps, and storing laboratory materials. The experience reinforced my desire to continue in geology, particularly during moments when I had momentary doubts.

CHAPTER 5
Two Summers in Newfoundland and Johns Hopkins University (1954-1955)

During my senior year at Wesleyan, I explored my career options and graduate schools. I decided I wanted to do research and that meant earning a PhD. I focused on sedimentology and sedimentary petrology and narrowed my possibilities to Princeton, Johns Hopkins, Northwestern and Wisconsin.

During winter break, I visited Princeton and met Franklin B. Van Houten (PhD, Princeton, sedimentary petrology; Princeton) who was their sedimentologist/sedimentary petrologist, and Harry H. Hess (BA, Yale, PhD Princeton, Petrology, marine topography, plate tectonics; Princeton) the chairman of the department and a distinguished scientist. Hess was a Princeton fellow graduate student of Joe Peoples. When my visit ended, Hess suggested I look at their bulletin board to see if any summer jobs were of interest. If I found any, Hess suggested I mention he suggested I apply. I appreciated his interest. It was a kindness I never forgot and subsequently learned to extend to others.

Next I visited Johns Hopkins. I met Francis J. Pettijohn (BS, MS, PhD, Minnesota, sedimentary petrology; Chicago, Johns Hopkins) whose textbook I had read. I found it difficult to communicate with him. I also met the department chairman, Ernst Cloos (PhD, Breslau, structural geology; Johns Hopkins University), a famous structural geologist, German immigrant, and incoming president of the Geological Society of America.

I chose not to visit Wisconsin.

I was accepted by Johns Hopkins with no financial aid. Ditto for Northwestern. Princeton and Wisconsin turned me down. I found out later that Wisconsin's sedimentologist, William H. Twenhofel (PhD Yale, sedimentology; Kansas, Wisconsin) was retiring in two years and was not accepting new students.

After visiting Princeton and Johns Hopkins, I wrote letters about the summer jobs I jotted down during my Princeton visit. One was a senior field assistantship with the Newfoundland Geological Survey. Within a week, I received a telegram offering me a summer job and I accepted.

Closer to the departure date, I received a letter with instructions to meet my field party in St. George, Newfoundland, enclosing an airplane ticket. I flew first to Montreal from New York, connected to Halifax and then connected on another flight to Corner Brook, in western Newfoundland. After an overnight stay, I took a combination freight train with a passenger car to St. George.

The train reminded me of the Porepunka local. A 60 mile train ride took five hours with shunting and off loading of freight cars. There were no amenities so I arrived in St. George extremely hungry. I was met by my party chief, Bill Fritz, a PhD candidate at Michigan, and the cook. The two junior assistants, who were undergraduates at Memorial University of Newfoundland in St. John's, joined us two days later.

The province of Newfoundland was Canada's youngest province. Originally a British colony, it was transferred to Canada in 1948. It was a neglected region subsisting on fishing, lumber and mining. The people were hardy but on a level comparable to the working poor of the USA.

Western Newfoundland was originally settled by French prisoners who were off-loaded and left to fend for themselves. They intermarried with the native ladies and were considered a wild bunch. The French lost their claim on Newfoundland during the French-Indian war.

Our assignment was to make detailed maps of Mississippian-age gypsum deposits with a plane table and alidade. Because I learned plane table mapping at Wesleyan, that became my job. The assistants did sampling, took turns holding the stadia rod for me, and other scut work. Bill completed regional maps to go with the detailed mapping I was doing. We established camp on the edge of St. George's and went to work. During the summer we moved camp twice to map two other gypsum deposits.

The Director of the Newfoundland Geological Survey, Don Baird, visited. He was a knowledgeable geologist but too bureaucratic and jocular. He arranged rapid payment of our monthly salary checks which had fallen behind, so I couldn't complain. Baird later left Newfoundland to head the geology department at the University of Ottawa. Ten years later, he took a job in the Canadian government.

Field work was completed in early September, just as it got cold and ice appeared on the water buckets in the cook tent. I returned home and enrolled at Johns Hopkins a week later.

As a first year student, I was required to take Cloos's Maryland geology field course consisting of Saturday field trips around Baltimore and other parts of Maryland. He used it to teach his detailed mapping methods, including orientations of deformed oolites, cleavage-bedding relations, fold axes, faults and so forth. The emphasis was on metamorphic and igneous geology. We examined two outcrops of Triassic red beds, one of Ordovician carbonates, and one of Carboniferous sandstones.

In addition, I enrolled in year-long courses in stratigraphic paleontology (regional stratigraphy and index fossils) taught by Tom Amsden (PhD. Yale, paleontology; JHU, Oklahoma Geological Survey) during the fall term, and Harold Vokes (PhD, University of California, Berkeley (UCB), Cenozoic molluscan paleontology; JHU, Tulane) during the spring term, geophysics with Byerly (PhD, UCB, Seismology; Johns Hopkins, USGS), and a semester course in crystal morphology and crystal chemistry taught by Jacques Donais (PhD Louvain, Belgium, crystallography; JHU).

Donais' course was a killer and I barely passed. On one exam, Donais wrote next to one of my answers, "Will you admit to yourself this is pure bluff and you do not know the answer?"

During the second term, I took a hand-specimen petrology course taught by a fellow graduate student, Ollie Gates. Ollie was a World War II vet, and had worked at the U.S. Geological Survey. His course was well-taught.

I wanted to take Dr. Aaron Waters' (PhD, Yale, petrology; Stanford, Johns Hopkins, UC-Santa Cruz) course on petrogenesis. Waters required every student signing up for it to take a rock quiz first. When I went to his office, he handed me two dark olive gray rocks. I looked at them and identified one as basalt and the other as an altered greenstone. Turned out the "basalt" was greywacke, and the "greenstone" was a basalt. Moreover, they came from the Keweenawan Peninsula in Michigan on Lake Superior, near my field camp area. Waters told me to take Gates' course before taking petrogenesis.

During the late winter, I applied for a summer job with the Geological Survey of Canada (GSC) indicating a preference for an assignment in the Canadian Rockies. I was hired as a Senior Geological Field Assistant, but sent back to Newfoundland.

I took a quick trip home to go to the White Plains NY Federal Court

House in late April, 1955, to receive my citizenship papers and was sworn in as a U.S. Citizen.

Hopkins turned out to be a very unhappy experience. In May, about two weeks before the semester ended, Cloos announced that all first year students were required to take a written prelim exam, although earlier he told me I should take it my second year. I appealed and was turned down. I took it and did badly. Cloos asked me to find another graduate school.

The following week, I visited Lehigh and Penn State. Because it was late in the year, they turned me down. I then applied to the University of Kansas, the University of California at Berkeley, the University of Illinois, and the University of Minnesota. Only the University of Kansas accepted me and in time, I appreciated their acceptance more and more.

Coincidentally, Amsden, Byerly, and Vokes also left that year. Amsden went to the Oklahoma Geological Survey, Byerly went to the U.S.G.S, and Vokes became chairman of the department of geology at Tulane University.

The semester ended, I passed all course exams, and went to Newfoundland. I flew to Gander and took the train to Terra Nova on the eastern side of the Island. I was the last to arrive and was met by the party chief, Stuart Jenness (PhD, Yale, field geology; GSC), Reg Bates (a local Boatman), Colin Bull (local cook) and Frank Nolan (Junior Assistant). Frank, an undergraduate at St. Francis Xavier University in Antigonish, Nova Scotia, was Nova Scotia's junior tennis champion.

GSC provided a jeep and I was taken to the base camp. There was also about 3 inches of snow on the ground and it was still snowing. We pitched our tents despite the conditions.

Frank and I were tent mates, sharing a two-man tent. Reg and Colin shared a tent. Stu had his own tent which also doubled up as a field office. We also had a rifle in Stu's tent which could only be used on his approval, literally "in the name of the Crown" in case provisions ran low.

We mapped a mixed Precambrian and Lower Paleozoic metamorphic terrain. Most of our mapping was completed by running traverses through the Newfoundland bush, muskeg and rough terrain. Stu assigned Reg to work with me because he felt Frank Nolan needed a lot more supervision. We plotted our outcrop data on our base map and transferred them every night to Stu's master map in camp. Camp meals were acceptable, but Colin had the habit of overcooking roast beef. Moreover, Stu arranged to have grapefruit shipped in for breakfast and Colin kept cutting them vertically instead of across the mid-section. We tried to explain it to him and it never registered.

On weekends, Reg went home, and Stu gave me the jeep to do roadside mapping. One week, Frank returned to Nova Scotia for family reasons, so Stu and I worked together. It was a good learning experience for me. He also told me a lot about the graduate program at Yale and how it was structured with a series of hurdles.

During August, Reg and Colin decided to brew homemade beer. Frank and I helped get it started and after the beer was bottled, we left it in a nearby creek to mellow and cool. Frank didn't want to wait and got roaring drunk. In the middle of the night he got sick and threw up all over the tent. I helped him up clean up the mess. Frank was, of course, very embarrassed and profusely apologized to all of us for days.

We finished mapping by Labor Day and said our good-byes. I went home to visit my parents and made preparations to go to Lawrence, KS.

LESSONS LEARNED:

1) An academic environment is extremely unpredictable and guidelines or verbal understandings and instructions, particularly with students, can be reversed almost capriciously.

2) A graduate student must prepare for all contingencies.

3) When faculty advise taking preparatory courses (such as the hand-specimen petrology course), accept that advice.

CHAPTER 6
University of Kansas and Nova Scotia (1955-1957)

The University of Kansas (KU) was founded by New England settlers to the Lawrence, KS, region. Instruction began in 1866 to 29 male and 26 female students. It was the first university established on the Great Plains. Joe Naismith started a basketball tradition at KU. The inventor of Vitamins A and D is a KU graduate. The first extraction of Helium as a gas was completed in a KU chemistry lab.

The history of geology at KU is partially summarized from two books by Merriam[2]. From inception, natural science was taught by a professor of mathematics and natural sciences, Francis H. Snow, an entomologist who also held interests in geology and paleontology[2]. In 1890, he became Chancellor of KU. That year, Samuel W. Williston (MD, PhD Yale, paleontology) was hired to teach geology, paleontology and biology to replace Snow. He left later for the University of Chicago.

In 1892, a new department of Physical Geology and Mineralogy was established and Erasmus Haworth (BS, MS, Kansas, PhD, Johns Hopkins, economic geology; Kansas, Private Consultant) was hired as its first department head. He also became Director of the Kansas Geological Survey in 1895. In 1910, W. H. Twenhofel was hired as a faculty member and became State Geologist in 1915. He left a year later[2]. Raymond C. Moore (BA Denison, PhD, Chicago; paleontology and stratigraphy; KU) was appointed to the geology faculty in 1916. He was promoted to a full professorship in 1919, serving also as Department Head until 1940[2].

The Department expanded, especially after World War II. By virtue of Moore's working style and its location, KU became a globally recognized center of paleontology and stratigraphy during his service. Moore died in 1975.

KU began offering a Master's degree in geology in 1875, and a PhD program was approved in 1894[2]. The first PhD in geology (paleontology) was awarded in 1899 to Joshua W. Beede.

After arriving in Lawrence in early September, 1955, I rented a room and settled in. That evening, I decided to find Lindley Hall, the geology building built during the early 1940's, and met a micropaleontology graduate student, Quinn Lockel. Quinn showed me around and told me some things about his first year there.

Next morning, I returned to get office space and select courses with the help of the department chairman, M. L. (Luke) Thompson (BS, MS, Mississippi A&M (now Mississippi State University), PhD, Iowa, micropaleontology; Kansas, Wisconsin, Kansas, Illinois Geological Survey). Luke was a world-class micropalentologist specializing in *Fusulinidae*. I enrolled in aerial photograph and geomorphology (both taught by Dr. H. T. U. Smith; PhD, Harvard, geomorphology; Kansas, Univ. of Massachusetts), economic geology (taught by Bill Hambleton, B.A. F&M, MS., Northwestern, PhD, Kansas; also Associate Director of the Kansas Geological Survey), and a groundwater course with Frank Foley (BA, Toronto, PhD, Princeton; hydrogeology, Director of the Kansas Geological Survey). I then registered and completed a routine medical exam.

After the medical exam, I was directed to a door which opened to a large room with ten desks, each with an attractive, well-groomed co-ed. I was motioned to sit with one of them and she asked me where I was from, why I came to KU, and where I spent the summer. I thought this odd so finally asked, "Why are you asking me these questions?" She laughed and explained that KU admitted a large number of students from rural areas and all new students had to complete an elocution screen. That screen was to identify students who needed an elocution course and make their spoken English more main-stream American. She explained she majored in elocution and I passed, and could leave.

Because I had no financial aid, I inquired about part-time work and the department secretary networked me to the Kansas Geological Survey which needed a student draftsman. I spent the year drafting measured columnar sections that were on file and thus quickly learned Kansas geology. My supervisor was Bill Ives, who also worked part-time on his PhD. I also met other Kansas Geological Survey geologists, learned how such an organization functions, and its mission. I valued the experience, although the pay was not great.

The semester started and I discovered that both the undergraduate structural geology and the mineralogy courses ran field trips to the Arbuckle

Mountains in Oklahoma, Magnet Cove, AR, and the Eagle Picher Mine area of southwest Missouri. I joined them, saw great geology and collected some nice mineral specimens. It was well worth the time.

In Oklahoma, we stayed at an old hotel in Sulphur, OK, and Louis Dellwig (BS, MS, Lehigh, PhD Michigan; structural geology; KU; served as an Army Captain during World War II and had most of his left shoulder blown away during the Battle of the Bulge) invited the teaching assistants and me to his hotel room to share a bottle of scotch he brought with him. Oklahoma was a dry state and KU did not permit alcohol at any campus functions. Louis told us that after the sacrifices he made during World War II for his country, no bluenoses were going to tell him where, when and what to drink. Louis was a lovable, yet crusty individual, and he became my Master's thesis supervisor.

The geomorphology course also had a one-week field trip to the Rio Grande Valley of New Mexico. We camped the first night within a V-shaped highway intersection near Guymon, Oklahoma, and were kept awake most of the night by passing trucks. I remember having dinner at a Mexican restaurant in Santa Fe with Ralph Lamb (BS. MS, Kansas, Chevron - later Exploration manager in Latin America) and Billy D. Holland (BS, Texas, MS, Kansas, Humble, chief geologist for Pogo Petroleum, president of his own oil company). They helped make decent choices for the first Mexican meal I tasted.

We camped on the eastern rift shoulder of the Rio Grande Valley and while completing a field mapping exercise, H.T.U. Smith pulled out a revolver from his car and started firing away across the valley. No one was around and it gave a wild western touch to the trip. I also realized that the departmental faculty included some colorful people.

Back on campus, Bill Hambleton helped my confidence by letting me teach his classes whenever he was away on state business. It also earned the respect of the students in the class.

At the end of the semester, I earned two A's, one B, and a C in aerial photography. That C grade resulted from a team-exercise to prepare a geological map from air photos and I was paired with a student who did not fulfill his end of the bargain. I completed a hurry-up job at the last minute.

As good as KU was, it was also extremely uneven in quality and standards. One had to pick and chose selectively to maximize the experience.

When discussing this with H.T.U. Smith before he left in 1956 to move to the University of Massachusetts, he said, "It's very hard to set standards higher than those around you." His statement stayed with me since then. I saw firsthand what he meant in every university where I taught as a permanent faculty member.

I returned to complete my first semester, and worked at the Kansas Survey during the between-semester break. I also experienced the first of many ice storms which made walking to campus difficult. I moved to a room closer to campus, and because there was an unrented room in the building, I arranged for a new PhD student, Stuart ("Duffy") Grossman to rent it. Duffy completed his undergraduate and Master's at the University of Illinois and just returned from Military duty.

Another new student whom I befriended, John C. Mann, also arrived after serving as an Army Captain in Korea. I spent time with these older students and they taught me much. Lloyd Foster, a graduate of VPI (Now Virginia Tech), was a third new arrival. He was just discharged as an Army Captain having been in combat in Korea. Lloyd was likeable and I enjoyed his company, but he hadn't quite discarded his army fatigues, talked loudly, ordered people around, and did his share of stepping on peoples' toes. Usually a good sense of humor brought him around.

Despite the year at Hopkins, I was only the second youngest person in my entry class.

I also befriended three other graduate students. Charles Dodge was a second-year Master's student working with H.T.U. Smith. Charles earned a BA from Princeton, and on completing his Master's moved with Smith to the University of Massachusetts as an Instructor while working on a PhD. Jim Sorauf was awarded the Pan American Fellowship. He earned a BS and MS from Wisconsin and was undertaking a PhD with Raymond C. Moore in paleontology. He made his career as a professor at the University of Binghampton.

The third graduate student I befriended was Alistair (Al) McCrone. A Canadian, he earned a BS from the University of Saskatchewan and a MS from the University of Nebraska. He spoke with a Scottish brogue and completed a PhD in stratigraphy, also with Moore. He taught at NYU, serving as its Dean of their Graduate School, and moved on to become President of the University of the Pacific, and, later, President of Humbolt State University.

I also met Daniel F. Merriam (BS, MS, PhD., Univ. of Kansas; stratigraphy and mathematic geology, Kansas Geological Survey (KGS), Syracuse University, Wichita State University, KGS; William Smith Medalist of the Geological Society of London) who worked at the Kansas Geological Survey and was completing a PhD with Raymond C. Moore. Dan worked for many years at the KGS, then headed the department of geology at Syracuse University, then did the same at Wichita State University, and returned to the KGS. Dan was responsible for pioneering the field of mathematical and computational geology. In his later years, he also wrote books about Raymond C. Moore[2] and the history of the first hundred years of the KU geology department[2]. As both our careers advanced, I came to appreciate Dan's contribution more and more in time.

One of the differences between KU and both Wesleyan and Hopkins was that the majority of graduate students were married; some also had children. A large contingent of students came from Texas and all were married. I knew them around Lindley Hall and in class but because they went home at the end of the day, I seldom saw them outside of working hours. I discovered later they tend to marry much younger in The South, and also in rural areas. I also became good friends with Ed Gutentag who graduated from Brooklyn College and served in the Military. He contracted TB while on active duty, was cured, but as he explained later, he received disability payments for life which covered his education expenses along with the GI Bill.

Mann, Foster, and Grossman also received GI Bill benefits but they were also awarded teaching assistantships. In Mann and Foster's case, they had earlier offers of TA's from KU but went into service first. KU was obligated to give them their TA's on return.

Second semester I enrolled in Advanced Structural Geology (Dellwig), Economic Geology (Hambleton), Petroleum Geology (Walter Youngquist, BA, Gustav Adolphus, MS, PhD. Iowa; paleontology; International Petroleum; KU, Univ. of Oregon; Consultant), and field stratigraphy taught by the eminent Raymond C. Moore[2]. Moore earned a PhD from the University of Chicago in 1916 and joined KU afterwards. He was chairman of the department twice, Director of the state geological survey once, and spent most of his time while I was there in his office editing "The Treatise of Paleontology", a multi-volume compendium of all fossil groups.

Moore was an austere individual who could be both gruff and kind at the same time. Stories about him were legend and he put the fear of God into the hearts of many generations of Kansas students. However, if a student worked diligently and did their homework, he took kindly towards you. He later became president of the Geological Society of America and was definitely one of the major figures of stratigraphy and paleontology in the world. He taught me much and I admit that I modeled some of my career after my perceptions of him.

Moore was an awe-inspiring leader, responsive to critical needs, and if one demonstrated that you were focused on your goals, he respected a student. He was very thoughtful in his replies to me, and extremely helpful.

Moore also imparted a sense of professional commitment and how important commitment was in developing a career.

After a month, I dropped Petroleum Geology. Youngquist had a mean streak and graded idiosyncratically. In discussion one day he was unable to consider alternate points of view advanced by students, particularly those whose fathers were in the oil business, and unjustly penalized them. I was not learning much.

It also became obvious that the faculty was badly split. Thompson and Youngquist basically took over the department and rammed things down the throats of the other faculty. Both were paleontologists and wanted to make KU a paleontological powerhouse. When Smith left for the University of Massachusetts that summer, he was replaced by Charlie Pitrat, a paleontologist teaching geomorphology. Thompson also hired Ed Zeller to start a program in thermoluminescence geochemistry but Zeller earned a PhD from Wisconsin in paleontology, as had Doris, his wife.

This split impacted graduate students. Those who weren't part of the paleontology group were not as well treated as the paleontology graduate students. After earning my Masters in 1957, Thompson left to become principal geologist at the Illinois Geological Survey in Champaign, IL, Farquar (petrology) left for the University of Massachusetts, and Youngquist left for the University of Oregon.

Frank Foley (KGS Director) became both KGS Director and Department chairman in the fall of 1957, and Hambleton basically ran the KGS. Louis Dellwig became Associate Chairman and ran the department. Pitrat taught paleontology, and Wakefield Dort was hired from Penn State to teach geomorphology. Under Foley's leadership, the department settled down.

The format for the field stratigraphy was definitely atraditional. Early in

March, 1956, we received a notice that Moore wanted to meet the class one evening to give instructions. The class would measure stratigraphic sections during spring vacation in newly blasted road cuts along the Kansas Turnpike. We were to meet on the first Saturday of spring vacation, start work and return to Lawrence that night. We did the same on Sunday. Then we were to bring changes of clothing to complete a road trip, spending the night in local hotels. On the first three days, we brought lunches, and from Monday night until we returned, we ate in restaurants.

On that Saturday, I arrived slightly later than the rest of the class and the two carryalls were already full. I therefore rode in the car that Ray Moore drove. That was a blessing in disguise. The others in the car were John Mann, and Wayne Bates. Bates, Mann and I were all completing Master's thesis work with Dellwig.

It was perfect spring weather: Clear skies, cool, crisp air, and the spring wheat had just sprouted giving the landscape a coat of green.

In some ways, the week was a defining one for me. I was able to have the same mentoring type of conversation with Moore like the one I had with Joe Peoples after a field trip in 1952. Moreover, we were measuring sections on Pennsylvanian cyclothems. Moore taught us how to keep track of each stratigraphic unit by observing the topography and identify them on successive ridges as we headed west. Not once during the trip did Moore lecture so we never discussed as a class the origin of cyclothems. I learned his interpretation by asking questions in the car.

Ray Moore clearly taught me much including:
- the art of scholarship,
- the importance of preparation and the meaning of terms,
- how to become a committed scientist,
- the importance of long, hard work,
- the importance of setting and achieving goals,
- how to staying focused,
- the importance of brevity,
- the importance of establishing and maintaining high standards professionally and ethically,

and

- the importance of never giving up on my goals, my dreams, or myself, no matter how tough sometimes things can be.

Be assured, I owe Ray Moore a lot because his example guided me during much of my career even though he was neither my thesis advisor nor on my Master's committee.

I recall during one lunch stop at a family-style restaurant in Larned, KS, Moore gave me some unexpected advice. I sat across from him on the inside seat of a booth, with two other students. We were served an entree on our plate, Dutch-style fried potatoes in an urn, and vegetables in a separate urn. Dutch-style friend potatoes were a big favorite, and I helped myself with seconds and thirds. Moore looked at me and said nothing.

We each received individual checks for the meal and the two students on the outside of the booth filed out to pay their bill. I followed standing behind them. Suddenly, I felt someone grab my arm and turned around. There was a red-faced Dr. Moore and he said, "Klein, I think you'll make a great geologist, but if you eat like that you'll never make it."

I replied, "Thank you sir" and remembered being told he survived two heart attacks. I never ate Dutch fried potatoes again.

There was a second incident that occurred at an outcrop at a road cut off the Kansas Turnpike where Moore knew there would be good fossils. I left the car a little late and everyone was crowded around a thin bed. Clearly, there was no room. So I walked down the road and examined cross-bedded sandstone exposed on a ledge. While looking at it, Moore and the rest of the class trooped by and Ray said,

"Klein, you're wasting your time there. The fossils are down below."

It was a memorable week. On our return, we had five weeks until the end of the semester to complete our reports and turn them in. Moore gave me a B plus and on the copy he returned he wrote "Very good. You are getting close to an improved understanding of stratigraphy."

Before the semester ended, a seminal event occurred that changed my life. For me one of the biggest turning point in my career was a colloquium offered by Ed McKee of the U.S.G.S. in April 1956, and a second one by Harold N. Fisk, Vice President of Exploration Research, Humble Oil Research Laboratory (also in 1956).

McKee showed how different assemblages of sedimentary structures characterized some modern depositional environments and it caught my attention. I had not seen any papers published using this approach. I concluded there had to be more to it in terms of depositional process than what McKee presented.

Later, Harold Fisk came to talk on his work on the Mississippi Delta and demonstrated sedimentology was predictive.

Those two colloquia were game-changers for me (See Chapter 29).

The semester ended and I was ready to undertake field work for a Master's thesis. While working in Newfoundland the previous summer, I observed Precambrian red beds and asked Stu Jenness if the Geological Survey of Canada could support me the next summer while I did a Master's thesis on them. He told me that might be difficult.

However, Frank Nolan told me there were spectacular outcrops of Triassic red beds along the coast of the Bay of Fundy and suggested I work there. I completed a literature search and discovered also that M.I.T. ran a geology field camp in Antigonish, Nova Scotia. The route for an American to do field work in Nova Scotia was through them. I wrote their field camp director, Dr. Walter Whitehead who arranged funding from the Nova Scotia Research Foundation (NSRF) and Nova Scotia Dept. of Mines (NSDM). I broached the subject with Lou Dellwig. He approved the topic and agreed to supervise me.

I arrived in Wolfville, Nova Scotia (NS) in the middle of June, 1956, and lived the entire summer in the only hotel in town. I arranged a favorable long-term rate for room and board, including packed lunches. I discovered the place was occupied mostly by retired widows and a few widowers.

Nova Scotia was far more prosperous than Newfoundland. It was well-networked with Canadian and subsidiary US businesses, phone service was adequate, roads were paved, and people were better off. They also seemed better educated.

The part of Nova Scotia where I completed my Master's thesis was known as "Evangeline County." The original French settlers were driven out, forced on British ships after the French-Indian War and shipped to Louisiana. Their descendents are the original "Cajuns" of Louisiana. Longfellow's poem "Evangeline" was inspired by these events.

Every summer, this part of Nova Scotia was visited by people from Louisiana. The Cajuns make a pilgrimage at least once during their lifetime. They either drove individually, or as part of 'Airstream' caravans organized in those bays by Wally Bynum

The Bay of Fundy has the highest tides in the world (highest tide of 52' at Burntoat Head, Hants County, NS) so my work was scheduled around tide tables. On days of mid-day high tides, I generally worked on inland outcrops. On days of early morning and evening high tides, I worked along the coast which had spectacular outcrops.

I enjoyed good weather most of the summer and completed my mapping, establishing a type section for the Triassic. Using sedimentary structures, I identified possible depositional settings.

Dr. Whitehead visited me together with Dr. Robert R. Shrock, head of the Geology and Geophysics Department at M.I.T. late in July, 1956. Shrock earned his PhD at Indiana University completing a definitive, widely cited published thesis on Silurian reefs in the upper Middle West. After earning his PhD, he was appointed to the faculty at the University of Wisconsin. Several years later, one of his colleagues, L. T. Meade, was appointed head of the geoscience program at M.I.T and invited Shrock to join him. Shrock agreed and eventually succeeded Meade as department head. He later became president of the Society of Economic Paleontologists and Mineralogists (SEPM, now the Society of Sedimentary Research), and served on a variety of significant national committees.

Both stayed in my hotel and I gave them a field trip to show the progress of my work. During lunch, Shrock told me they liked my work and asked about my future plans. I told him I wanted to earn a PhD, do research, preferably in a research university, and pursue my interests in sedimentology. I mentioned the colloquia by McKee and Fisk with which Shrock was familiar. He told me that given my background and interests, M.I.T. would not be a good place for me, but he would help me get into places wherever I applied. Whitehead offered to do the same. Shrock suggested I apply to Yale and work with John E. Sanders who had similar interests. Shrock also made it clear he would write a letter of recommendation only if I got one from Raymond C. Moore.

On completing my field work, I returned to KU for my second year with the goal of completing my Master's and pursuing a PhD elsewhere. I enrolled in two courses that fall, with the third course being thesis credits. I enrolled in Dr. Andrew Ireland's course in sedimentary petrology. Ireland earned a PhD with Pettijohn at the University of Chicago, worked in the U.S. Department of Agriculture and Standard Oil Company of Texas before coming to KU. He was much loved by most graduate students. However, his course was a total waste of time. Instead of learning sedimentary petrology, it was a lab techniques course and most I knew. I earned an "A", but got little out of it.

The second course I took was "Geological Development of the World" taught by Raymond C. Moore. The course focused on the type stratigraphic sections of Western Europe. The course met Saturday morning at 8:00 am, with a break around 10:00 am, and finished at noon. Each student was assigned a topic to research in the library and then present a 45 minute lecture to the class. We then discussed that paper before going to the next student presentation.

I knew in advance that Moore had a reputation of being very tough in his questioning and if one didn't know the answer, he had various ways to make a student know that he was unimpressed. The first class meeting we were assigned topics and dismissed. During the second meeting, four presentations were made ahead of me so I wasn't called on.

Given his reputation, I watched how Moore handled himself with students when he asked questions and a pattern emerged. He wanted to know if student knew the meaning of terms, and where localities occurred on a map. With that in mind, I reviewed my presentation for the next class checking definitions of critical terms, and map locations. During my presentation, I mentioned a term and he interrupted and asked what it meant. I answered and he was satisfied. He did it again after a few minutes, and he was satisfied.

During the rest of the semester, he NEVER asked me to define a term again. Clearly, he was trying to instill in each student the importance of doing one's homework and knowing what one was talking about. If you did, he left you alone.

I live now in the greater Houston area and one of the city's most prominent citizens is James Baker, former US Secretary of State. Baker, according to an article in "The Houston Chronicle," came from a family of lawyers. The family firm had a paradigm known as 'the Five P's – Prior Preparation Prevents Poor Performance.' Moore was trying to instill a similar paradigm with KU geology students.

Later in the semester, during one of my presentations, he interrupted and asked me to locate something on a map. I did, but was off five miles. He got up, walked to the map, pointed to the correct spot, sat down and said nothing. I used that technique during my PhD dissertation defense several years later (Chapter 7).

Every now and then, Moore would talk off the cuff and outline critical principles and insights. I took notes on his comments.

Moore also discussed professional and career matters. His best advice: "Always go into an uncrowded field" to get early notice and develop a research reputation.

A week before the last class meeting, Moore explained that we would take a final exam during the last class. It consisted of everyone making a 15-minute presentation on one of the geological systems of the world. He assigned me the Permian System, so to prepare the 15 minute talk I reviewed all my course notes, and went to the library.

After a day of library research, I said to myself *'There is no way I can review the Permian geology of the world in 15 minutes. Gee, George, you're sunk! What are you going to do?'*

I went through my notes again and compiled all the off the cuff comments Moore made about stratigraphic principles. Some were repeated many times.

My compilation of those principles was less than two pages, including his repeated comments.

Suddenly I realized if I talked about the Permian in terms of those principles I could review it in the time allotted. I planned accordingly.

On the last day of class, my fellow students successively got up to talk about their assigned geological system. They discussed the type sections and proceeded to describe the world's geology of their system. When the 15 minutes were up, none finished. Moore cut them off. We started with the Precambrian and it was that way with every talk about the Precambrian and Paleozoic. My turn came and I quickly explained the type area of the Permian system. I then talked about the boundary problems of that system, explaining it in terms of all the stratigraphic principles Moore had discussed. I glanced at Moore and he sat there with his usual expressionless, poker face but was focusing on every word I said.

I finished in 13 minutes. Moore said "Well Klein, you have more time. Don't you want to use it?"

I replied "No sir." and sat down.

Jim Sorauf, who also took the class, told me afterwards, that I was the only one who talked about all the principles Moore reviewed during the semester. Given Jim's experience, it left me with a good feeling.

Earlier during the semester, I made an appointment to see Moore and told him about my conversation with Shrock. I asked if he would write a letter of reference to another university.

Moore immediately asked, "What! Leave Kansas?"

I explained that I would complete all courses of interest to me when I would complete my Master's degree in 1957. I also explained I found the language requirements at KU, as then promulgated, far more onerous than other places.

Moore replied, "Now you know why so many of my good students like Ellis Yochelson and Norman Newell went elsewhere. I'll be sorry to see you go, but I'll be happy to write those letters."

After the final exam presentations were completed, I stopped by Moore's office with all the reference forms for a National Science Foundation (NSF) pre-doctoral fellowship, and the four graduate schools to which I applied. He assured me he would back me. He gave me an "A" for my work in "Geological Development of the World."

During the 1956-57 academic year, I earned my expenses as a Teaching Assistant in 'Western Civilization.' I applied for an assistantship in the geology department but Thompson and Youngquist turned me down. I saw an advertisement in the student newspaper that the Western Civilization program needed instructors. I talked to the director, Paul Heller, and applied. I received an appointment, including a tuition and fee waiver.

It turned out I was not the only one so treated by Thompson. I found out recently that one of my fellow students who arrived that fall, Bill Fisher, was similarly treated[3]. While an undergraduate at Southern Illinois University, Bill applied to Kansas and was offered a TA. However, he was drafted and went on active duty explaining in a letter to KU he would return in two years. When he arrived at KU, Thompson did not give him the TA promised before. Two weeks later, Moore gave him a research assistantship[3]

Bill Fisher and I also got acquainted. He was married so I only saw him during the day. He came from Southern Illinois University and then served for two years in the US Army. He earned a PhD with Ray Moore and went to the Texas Bureau of Economic Geology. There he pioneered the concept of Depositional Systems, Systems tracts, and contributed to the development of sequence stratigraphy. He also became Bureau Director as well as a part-time professor of geology at the University of Texas at Austin. He served for two years in the Ford Administration as Assistant Secretary of Energy. Later, he went full-time at the University of Texas, was chairman of the geology department for a brief period, and also served as Director of its Geology Foundation. He was directly instrumental in the negotiations that led to the $253 Million bequest from John A. Jackson which led to establishing the Jackson School of Geosciences there. Bill was its 'inaugural' dean.[3]

That fall, I befriended some of the newly admitted graduate students. One was Mahlon Ball who returned from Military Service. He earned a BS and MS from Kansas and returned to work with Ray Moore on a PhD. Mahlon worked for Shell Research, and then made a career as a marine geologist with the U.S. Geological Survey.

Over Christmas break, I applied to Yale, Princeton, Northwestern and the University of Wisconsin.

On my way back from Nova Scotia during early September, 1956, I met Richard F. Flint (PhD. Chicago, Quaternary Geology; Yale), director of graduate studies in the geology department at Yale. He spelled out the ground rules if I was accepted. If I did well, financial aid was assured for three years only because I would come in with a Masters. In other words, I had three years to get the PhD done. I raised the possibility of continuing my research in Nova Scotia and explained the grants I had received from the Nova Scotia government. He assured me that the department would likely approve the thesis topic because few Yale geology graduate students came with financial support for their thesis field work. We had a good meeting.

I also revisited Franklin Van Houten at Princeton and explained my research goals and discussed the McKee and Fisk colloquia. He was supportive of my continuing my Nova Scotia research if accepted.

Northwestern and Wisconsin were basically back-up applications. I heard that Wisconsin hired someone to replace Twenhofel, but was given no details.

Second term I took Hambleton's economic geology course (he offered three semesters) and two credits of Master's thesis.

I interviewed oil companies in the fall and Humble offered me a job in Midland, TX. I declined after receiving a letter from Yale offering me graduate admission, a teaching assistantship, a tuition-and-fee waiver, and an additional stipend of $1,000.00. I immediately accepted Yale's offer and also wrote Princeton, Northwestern and Wisconsin to withdraw my application. Northwestern and Wisconsin wrote back wishing me well and thanked me for letting them know promptly.

I spent most of my time during the second semester writing my thesis and drafting illustrations. I finally completed it near the end of March and gave it to Louis Dellwig. Although it took a month, he returned it after a long working session during which he spent considerable time showing how to rewrite the text. The grammar was fine. The organization and tortuous logic needed refining. It was a valuable lesson that stayed with me my entire career. Late in May, 1957, I completed a general Master's oral exam.

With my master's degree completed, I packed my belongings into my car and drove home. I did not attend commencement and the degree was mailed later.

After briefly visiting my parents, I returned to Nova Scotia to expand the scope of my work on the Triassic of Nova Scotia. Support was continued by the NSRF and the NSDM, but was told I may not get such support in the future. A spring election resulted in a change in government. A different political party took power and because I was funded by the previous regime, support was not guaranteed. I expanded my mapping and research into the Annapolis Valley and into Hants County.

At the end of the summer, I returned home and prepared to pursue my PhD at Yale.

LESSONS LEARNED:

1) Raymond C. Moore clearly taught me the art of scholarship, the importance of commitment as a scientist, the importance of long, hard work, the importance of setting and achieving goals, the importance of doing one's homework and knowing what one was talking about, how to stay focused, the importance of brevity, the importance of establishing and working towards and maintaining high standards professionally and

ethically, and the importance of never giving up on my goals, my dreams, or myself, no matter how tough things sometimes can be.

2) After meeting with Bob Shrock and Walter Whitehead, I realized that I was given a second chance after my experience at Hopkins. I learned then and later, second chances often come throughout one's life and one should rise to meet them.

3) When a student encounters a split faculty as I did at Kansas, it's best to be polite to everyone and not allow oneself to be caught up in any warfare that might occur. It was better to duck, hide, and keep working and stay with one's objectives and hold true to one's aspirations. I saw several students hurt by their involvement in faculty fights.

4) In a department with a faculty and course offerings that were variable, it was critical to pick and chose in terms of one's goals.

5) Attend colloquia. They are the gateway to broader opportunities.

6) Moore told his class "Always go into an uncrowded field" to get noticed quickly for the research one completed. I did. Sedimentology as practiced today was in its infancy in the 1950's through the middle 1960's. I capitalized by starting and publishing my research during those early years.

POSTCRIPT #1. Within a month of arrival at the Illinois Geological Survey in 1957, M.L. Thompson suffered a stroke which disabled him permanently. During 1970, I was walking in the Lincoln Square mall in Urbana, IL, and saw Thompson and his wife. I approached them and asked if they remembered me. Thompson stared at me and had no recollection. Mrs. Thompson remembered me. I told her I was now on the faculty at the University of Illinois. I never saw them again, but read Luke's obituary in 1985.

CHAPTER 7
Yale University and Nova Scotia (1957-1960)

Yale University was originally founded in 1701 as the "Collegiate School." The name was changed to Yale University when a Welsh merchant, Elihu Yale, donated proceeds from the sale of 17 bales of goods, and a portrait of King George I to endow the institution. He did so on the condition that they change their name to reflect his bountiful generosity.

Yale is the third oldest institution of higher learning in the USA and over time grew in stature and reputation. The Yale geology department awarded the first American PhD in the field to William North Rice, a Wesleyan University graduate who returned there to teach (Chapter 4). The Yale geology department had many renowned scholars on its faculty including Othneil Marsh, a vertebrate paleontologist, Charles Schuchert, an eighth-grade self-educated paleontologist who earlier assisted the legendary James Hall, Benjamin Silliman, a mineralogist, and James Dwight Dana, a mineralogist who also published in the field of tectonics.

Prior to 1950, famous geologists teaching there included Chester R. Longwell in structural geology, Allan Bateman, who discovered the Kennecott Mine, in economic geology, Adolph Knopf in petrology, Carl Dunbar in paleontology and stratigraphy, and Richard Foster Flint, a quaternary geologist. Dunbar, who retired in 1959, and Flint were, in effect, my only contact with Yale's illustrious past, although I met Bateman and Longwell. Flint was much younger than the other three and retired in 1971.

Yale geology graduates were amongst many leaders in the field. They included Stuart Weller, a professor at the University of Chicago, Edward Sellars who directed the Texas Bureau of Economic Geology, William H. Twenhofel a professor at both the University of Kansas and the University of Wisconsin, Samuel Williston, a professor at both the University of Kansas and later the University of Chicago, Paul D. Krynine, a pioneering sedimentary petrologist who taught at Pennsylvania State University, and Tom Nolan, Director of the USGS, amongst numerous others. I realized when admitted to the PhD geology program at Yale, I was walking on hallowed American geological ground and it required a strong commitment on my part to attempt to live up to the reputation of my forebears.

Prior to moving to Yale, I read materials they sent me. It included a letter of welcome, a statement about wisdom and maturity, and instructions. As a new student, I was expected to pass before registration a series of rock, mineral and fossil identification tests and a test on editing based on the U.S. Geological Survey publication "Suggestion to Authors." I read that publication over the summer. To pass the other three tests, I arrived a week earlier and examined the Yale teaching collections.

I also obtained information about the faculty. My advisor was John E. Sanders. He earned a BA in geology from Ohio Wesleyan University where he was captain and quarterback of the football team and student body president. He earned his PhD at Yale and completed a post-doctoral fellowship with Ph. H. Kuenen at the University of Groningen, a leading sedimentologist who developed the turbidite concept. Sanders returned to Yale in 1955.

Richard F. Flint was now department chairman. He earned all his degrees from the University of Chicago. His expertise was Quaternary and glacial geology and had written the definitive textbook (at that time) on it. He was widely known for his research and an internationally-recognized scholar in the field.

John Rodgers was considered a rising star in structural geology. He earned his undergraduate degree at Cornell and PhD from Yale. Before returning to Yale, he worked with the U.S.G.S.

At the Peabody Museum, three faculty members comprised the paleontological side of the department. Carl Dunbar was director of the museum and an expert in *Fusulinidae.* Dunbar earned his BS in geology from Kansas. His father was the first farmer in Douglas County, KS, to use a mechanized tractor in 1911. Carl earned his PhD at Yale with the legendary Charles Schuchert and after two years at the University of Minnesota returned as Schuchert's successor. Karl Waage was a Mesozoic paleontologist and stratigrapher. He earned his PhD from Princeton and worked at the U.S.G.S. before coming to New Haven. Joseph Gregory, now graduate advisor, was a Vertebrate Paleontologist focusing on dinosaurs and earned all his degrees from the University of California at Berkeley (UCB).

Other faculty in the department included Matt Walton, a petrologist with a Columbia PhD, Leroy Jensen, an economic geologist with a PhD from

M.I.T., Karl Turekian, a geochemist who earned a BS at Wheaton College and a PhD from Columbia, and Horace Winchell, a mineralogist with a Harvard PhD.

From the outside, it looked like a powerhouse faculty.

Graduate students were expected to complete a core program of a year of Geomorphology and Pleistocene Geology (Flint), a year of Structural Geology (Rodgers), a year of Stratigraphy (Dunbar and Sanders), a semester of Mineralogy (Winchell) and a semester of Igneous and Metamorphic Petrology (Walton). Other courses could be taken as electives.

To earn a PhD, one was also expected to pass a Comprehensive General written exam administered during spring vacation of the first year, a Qualifying combined written and oral exam during the second year if one had entered with a Master's degree, or the third year if one entered with a BS degree, and a thesis defense. PhD candidates were required to pass two language exams administered by the department. Sanders administered the German exam and Rodgers the French exam. To advance from the first year to the second, one was required to earn a grade of "Honors" (equivalent to an "A") in at least one's year's worth of course work.

All this was spelled out in no uncertain terms in the materials they sent me at the beginning of the summer. They were marching orders for success to complete their graduate program and to ignore them invited disaster.

In short, Yale had a structured program. Graduate students were on their own. It was not exactly a student-friendly environment and mentoring was unheard of. One was expected to be mentored before arrival or figure out where to get it. Fortunately in my case, the two years at Kansas prepared me to hit the ground running, and in the end, it paid off. I have often said that without the two years at Kansas, I would never have completed the PhD program at Yale.

I made the operational decision that the next three years were to be focused on course work, thesis research and passing critical exams. To do so, I carefully scheduled my hours in an appointment book on a weekly basis, and stuck to them. There would be little room for social life. It also meant be nice to everyone to the extent you could, avoid arguments, and stay focused. One of the graduate students who I got to know well, Dick Heimlich (BS Rutgers, petrology; Kent State University), once commented that few do it that way at Yale. I replied I would adjust as I saw how things developed. One could always scale down, but it would hard to rescale up again. In short, I chose to work for three years in a focused manner.

Because the written Comprehensive and the Qualifying written and oral exams were administered at the end of spring vacation, I knew I had to begin reviewing for both on January 2, well before each exam.

I arrived at Yale immediately after Labor Day, rented a room, and went to the department in Kirtland Hall, an old brownstone building. I selected a vacant office carrel in a room with 16 such carrels (it was called the "Rats Nest"), and started looking at the collections. In the process, I met Waage, Flint, Winchell, Gregory and Walton. Two weeks later, I was able to get room and board at the Hall of Graduate Studies, a better arrangement, and moved immediately.

Slowly, other first year students arrived. One earned a geology degree and then went to theology school, didn't like it and tried geology again. He flunked out at the end of the semester. A second student from India was gone by the end of the year.

A third, David Doan (BS, MS Penn State), was on leave from the Military Geology Branch of the U.S.G.S. where he acquired extensive experience in the Pacific mapping many of the islands that comprised the 'island hopping' strategy of World War II. I learned a lot from him, but in November, he returned to the U.S.G.S. While in Guam, he married a lady who was previously married successively to two Air Force pilots. Both died in plane crashes. When Dave arrived, she made a play for him hoping to get back stateside. She also had two children, and was clearly "high maintenance," having lived on military bases where living costs were cheap and she could afford a maid. The adjustment to life as a graduate student wife on a fellowship was difficult for her.

Dave invested in the stock market but lost money. He tried commodity investments which did not work out. So he left. It was a great loss because he shared his extensive geological experience with us.

The other five members of our "entry class" eventually earned PhD's. First to arrive after me was Peter Robinson (BS, Yale, MS, University of New Mexico) who completed his PhD in Vertebrate Paleontology with Gregory and took a job at the University of Colorado Museum. Then Steve Porter (BS, Yale) arrived. He returned from service on an Aircraft Carrier in the Pacific fleet. He showed me his slides from Japan, Korea, the Philippines and Australia and it made me want to visit (and revisit) these places. Steve lived in the Hall of Graduate Studies and we became close friends. During his second year he married one of Turekian's lab techs so I saw less of him. He went to the University of Washington and made a life-long career there, becoming director of their Quaternary Institute.

Tom Williams (BS, Dickinson, MS, SMU) studied with Dunbar in paleontology and returned to teach at SMU. Last to arrive was Roger Ames (BA, Williams College, 2 years U.S. Army service in Korea) who earned his PhD with Leroy Jensen in economic geology. He joined the Amoco Research Lab in Tulsa where he worked in organic geochemistry.

We initially had the building to ourselves because everyone was still away doing field work. The only returning student was Mike Carr (BS, Imperial College, University of London). Mike was a geochemistry grad student and made his career at the U.S. Geological Survey in Menlo Park.

I visited Joe Gregory to let him know I was there. During our meeting he asked if I was married and I told him I preferred to wait until I was earning a living and could afford to support a wife. He replied, "Mr. Klein, a wife will be a handy thing to have around when you need your thesis typed." At Yale everything was formal and faculty did not address students by their first name until they passed the first year comprehensive general written exam and were approved to return for their second year.

After a week of preparation, the eight of us who were new took the rock, mineral, and fossil tests and edited a piece of geological prose. I was never told outcomes but assumed I passed because those who didn't were told to retake them a month later.

The next afternoon I registered for courses, but this was handled by the department. I met with the entire faculty, was introduced by Joe Gregory, and selected the following courses: Geomorphology and Pleistocene Geology (Flint), Stratigraphy (Dunbar and Sanders), Mineralogy (Winchell) in the fall and Petrology (Walton) in the spring. I was assigned as a TA to teach "Science II" a general geology course for non-scientists taught by Flint.

By this time, the other graduate students were back. I made friends with Chuck Ross (BS, MS, Colorado; Paleontology, Dunbar), Don Eicher (BS, MS, Nebraska; Paleontology, Waage), Lee McAlester (BS, SMU; paleontology, Dunbar), Art Bloom (BS, Miami of Ohio, MS, Otago; Geomorphology, Flint), George Moore (BS, MS, Stanford; Sedimentology,

Sanders), Chuck Ellis (BS. University of Texas, Austin; sedimentology, Sanders), and Dick Heimlich (BS, Rutgers, petrology, Walton). I maintained contact with many of them throughout my career.

Ross worked at the Illinois Geological Survey, Western Washington University and Chevron. He became an internationally-renowned fusulinid micropaleontologist and was the first to propose Pennsylvanian and Permian sea level curves. Eicher went on to teach at Colorado.

Art Bloom taught at Yale for one year and then accepted a faculty appointment at Cornell. He was one of the first to evaluate uplift rates, choosing an area in Papua-New Guinea where uplifted coral terraces were exposed and datable. George Moore was on leave from the U.S.G.S and returned there. Chuck Ellis went to work with Conoco Research, moved to Sinclair Research, and after Sinclair's merger with ARCO, worked for a small company in Colorado. He died in a plane crash in 1969. Heimlich taught at Kent State for his entire career.

Of these, George Moore clearly had extensive (five years) research experience with the U.S.G.S. George played the role of mentor very well, always emphasizing that "an idea was the most important thing." As scientists, we should focus on new hypotheses, new ideas, and challenge existing paradigms. Because he also lived in the Hall of Graduate Studies, meals with him turned into seminars discussing many current paradigms of geology and what was weak about them. It was as much a learning experience as we obtained in the classroom. Those discussions taught me the importance of tying one's research to the major themes and paradigms of geology, contribute to their understanding, and utilizing new observations and analysis to challenge them.

He also advised something else. If one can't complete a project in five years, it's not worth pursuing. In general this is true for work involving projects completely on land. In areas of coastal restriction, such as those with high tides, lack of available field time prolongs the investigation. When undertaking marine geology, it takes longer because of limited ship availability, or availability of particular data sources such as provided by the Ocean Drilling Program which may not drill in certain areas for four or five years. One compensates in such situations by undertaking different projects simultaneously.

Art Bloom and I worked together during the first semester because he was the head TA for Science II. He spent a year as a Fulbright Scholar in

George Devries Klein

New Zealand and I drew on that experience.

I met two recent PhD's, William B.N. Berry (Paleontology, Dunbar) and V. Rama Murthy (BS, Madras, petrology, Walton) who dropped by to visit. Bill Berry eventually joined the paleontology department at the University of California, Berkeley (UCB). Rama, after a bad year in India where he saw no future, returned to the USA, completed a post doc at the Scripps Institute of Oceanography, and went to the University of Minnesota. He eventually became department head, dean of the college, and Vice President of Academic affairs there.

Because Rama was from India, I asked how he adjusted to the US. He (and one of his friends) told me the following:

He arrived at Yale in 1954 and soon was recognized by the faculty as a promising student. His faculty advisor, Walton, suggested he map an area in Vermont for his PhD thesis because the Vermont Geological Survey only had a state geologist on staff. The Vermont Geological Survey devised a way to geologically map the state by hiring geology graduate students to map in the summer for their thesis work, and provided a paltry salary, all expenses, and field equipment. Each student provided their own car.

Rama neither had a car, nor knew how to drive. Buying a car was easy. Yale's geology department had a fund (donated by a wealthy alum) to buy used field cars for graduate students with the understanding that when field work was completed, the fund was reimbursed for whatever price the car sold. Rama was voted funds to buy a car and his friends helped him.

The driving lessons were a little more complicated. Alan Bateman a retired faculty member who was independently wealthy paid for Rama's driving lessons. Rama passed his driver's license test.

Rama wrote his family in India (from the old "Brahman" caste) and told them about his good fortune since coming to America. Not only did he buy a car six months after arriving but also one of the professors paid for his driving lessons.

About six weeks later, his mother wrote back, "Well Rama, that's very nice that Yale University paid for your car. But tell me, who pays for the chauffeur?"

The next day, I saw a notice written on the Bulletin Board in French about the French language exam. It was scheduled for the following

Saturday. I took it and Rodgers told me that I completed a perfect translation but it was too short. I would have to take it again a month later and increase the amount translated. I did, and this time I passed but I had made mistakes because he wanted more volume. Three days later, a notice appeared in German and when I met with Sanders afterwards, it was a repeat of the experience with the French exam, and I passed on my second try. Within a month, I had crossed off all the first set of hurdles (rock, mineral, fossil tests, editing test, both language exams).

Because Flint's and the Dunbar-Sanders courses required term papers and Winchell required an independent project, I knew I had to spread them out. If I didn't everything would catch up with me at the end of the year, something I wanted to avoid. One picked a topic according to the lecture schedule and as I looked at the course outline, I noticed that in mid-October, the stratigraphy course scheduled a lecture on cyclothems. I picked that topic. That meant I not only had to complete the final paper, but also present a lecture to the class at the scheduled time. Dunbar was extremely pleased with both the written paper and the class presentation as was John Sanders. What was unknown to me was that the faculty shared student progress during a weekly luncheon. Consequently, I had a head start in terms of positive perceptions.

Geology graduate students would take coffee breaks every day at a nearby restaurant, George and Harry's, at 10 am, 3 PM and 9 PM. Karl Turekian (being single) usually joined us and often converted them into geochemistry seminars.

The only grade recorded at the end of the first semester was Mineralogy where I earned a "High Pass" (equivalent to a B plus). The following semester I took Walton's petrology course with a required laboratory where one had to describe 40 thin sections with a petrographic microscope. In the past, students resisted and only completed about half. I set to work and did them all which student colleagues did not appreciate, but Walton also gave me a "High Pass" which he seldom did for those who were not his students.

Towards the end of the first semester, I noticed a group of graduate students who were most unhappy with Yale's geology graduate program. One was Lucian Platt (BS Yale, Structural Geology with Rodgers) who had more money than most people. He was self-supporting, arrived every morning at 9:00 am in his Mercedes, and started the day reading the *Wall Street Journal*. He married a Swiss lady from a prominent family whom he met while on active duty in the U.S. Army in Germany, and managed her finances too. The others were Cy Field (BS Dartmouth, economic geology), John Cotton (BS Dartmouth, petrology), Dick Berry (BS Williams, petrology), Jim Allen (BS, Toronto, economic geology) and Ernie Dechaw (BS Witwatersrand, South Africa, economic geology). I ignored them and whenever they complained to me, I responded, "Look, I just got here. I'm a mere wheat kernel blown in from Kansas during the last tornado." John Sanders told me that summer that the faculty was aware of their carping, and enjoyed my response.

I'll always remember Jim Allen's last day at Yale. When he gave his PhD thesis defense presentation and passed, he immediately left the building. His wife met Jim at the front door. She had sat in their car during his defense with all their belongings and they left right then and there.

At the end of spring vacation, I took the written comprehensive exam and passed. When the semester ended, I earned "Honors" grades in both Flint's course and the stratigraphy course. I was awarded the Sterling Fellowship by the graduate school for the next year. It included a tuition and fee waiver, and paid enough to live on, or as one of my friends put it, "starve graciously".

During the academic year, we had a weekly colloquium. One of the speakers in February, 1958, was Dr. L. M. J. U. Van Straaten from the University of Groningen, Netherlands. We read his papers on the tidal flats of the Dutch Wadden Sea in the stratigraphy course. Sanders arranged for me to meet him to show some slides from the Bay of Fundy, but he wasn't interested. Sanders and the stratigraphy class took him on a field trip to the Connecticut coast and he did not mix with the graduate students. We concluded that graduate students were of no interest to him. That attitude cost him dearly later when the Dutch government closed the department of geology at the University of Groningen during the early 1970's.

When classes ended, I took a field trip led by Carl Dunbar to all the major North American Paleozoic type sections in New York State. It was a memorable experience with Dunbar recounting many stories about the pioneer paleontologists who worked there and also saw some excellent geology.

The day after the term ended, everyone undertaking thesis field work was expected to leave for our field areas. I left on returning from Dunbar's field trip.

I returned to Nova Scotia, settling in Digby to complete field work at the south end of the Annapolis Valley. Once that was completed, I returned to Hants County and worked east towards Truro in Cobequid Bay. NSRF continued my support, but the NSDM said they would send me monthly payments. After receiving the first check, they terminated funds for all US geologists supported under the arrangement with M.I.T. When John Sanders visited to provide field thesis supervision, I explained the funding cut off. John advised I write Flint for a supplemental grant. Flint sent a check a week later from the Donnell Foster Hewitt fund.

Sanders and I visited all critical locations. He observed other things that were outside my thesis topic and suggested I do research on them. I realized if I did, I would never finish. He did suggest that I expand my reconnaissance of Bay of Fundy intertidal zone sediments which I did and in time this paid off. If I learned anything from John it was that one should take advantage of all the research opportunities an area offered.

At the end of the summer, I returned to Yale. Fifteen new graduate students were enrolled, including one third without financial aid. Flint mentioned that enrollments increased because of the recession. Half of these new students were gone by the end of the year.

I developed a friendship with B. Clark Burchfiel (BS, MS, Stanford, Structural Geology) who later became a professor at Rice University and moved in 1975 to M.I.T. Later, Clark was elected a member of the National Academy of Sciences and served on the Board of Directors of Maxus Oil Company. He is the recipient of the 2009 Penrose Medal, the highest award of the Geological Society of America. I also met Gil Benson (BS, MS, Stanford, Structural Geology) who had an educational leave from Texaco. He taught me a lot about the operational side of an oil company. He taught at Portland State University.

Also in that class was Brock Powers (BS, Southern California). He was on a fully-paid leave from Aramco to earn a PhD on Aramco data because he was to be their next chief of geological research. His predecessor was ill and Brock was picked to replace him. Aramco required their research directors to have earned PhD's. Brock made a tour of the US geology PhD programs to find which would accept a thesis on Saudi Aramco's data on Saudi Arabian petroleum geology. Yale was the only university to agree with this stipulation. Brock worked under Sanders' supervision on carbonate petrology of the "Arab D" reservoir at the famous Ghawar field. Once a month, a team came from the Aramco office in New York to meet with him.

I reconnected with another member of the entry class, Ed. Belt (BA Williams, MS, Harvard). Ed was a TA at the M.I.T field camp in 1956 where we met. After service in the army, he decided to go to Yale to work with John Sanders. Ed and I spent a lot of time together reviewing common interests. Ed taught first at Villanova, and then at Amherst College.

Also among that group was Larry Ashmead (BS, MS, Rochester, petrology) who was terminated after two years and became a senior editor with Random House. Ed Hansen (BA, Princeton, structural geology) was a late arrival after the rock and mineral exams were completed. Ed was a bright guy (probably the brightest of all the graduate students) but lacked discipline. He finished his PhD, earned a Post-doc at the Geophysical Laboratory, and then left geology to become an artist in Greenwich Village. He died young.

Ashmead and Hansen palled around with Mike Carr, and ingratiated themselves with John Rodgers. Neither Hansen nor Ashmead were comfortable with my self-disciplined approach to graduate work partly modeled after my perceptions of Ray Moore's work style. They thought it unscholarly. The three spent considerable time talking about contemporary literature at the Hall of Graduate Studies dining hall. Whatever their concerns, clearly my determined working style learned from Ray Moore put them off.

I enrolled in Structural Geology with Rodgers, invertebrate Paleontology with Dunbar, and thesis credits. The structure course was difficult because we had to translate a book by Jean Goguel written in French. The paleontology course was problematic because it required a lot more memory than I could to handle. I passed both courses.

At the end of spring break, I took the written part of my qualifying exam and my orals a week later. Sanders asked questions to start the exam and focused on topics that he knew with which I would do well. Next was Gregory who asked questions in an off-putting monotone. After adjusting, I answered them. Rodgers was next and I answered his questions. Walton was the last member of the committee and I had difficulty understanding his questions. I passed, but felt I could have performed better. Walton said as he left the room after the committee vote, "George you did very well." Cy Field told me later that Jensen told him I had scored the highest on the written qualifying exam because I had also done well on the section on economic geology in addition to fields I knew. Jensen was unaware I completed three semesters of economic geology with Hambleton.

Because of funding uncertainties in Nova Scotia, I applied for a Penrose Grant from the Geological Society of America (GSA) for $2,800.00. Bob Shrock at M.I.T chaired the committee that year and GSA awarded me $2,000.00. When the grant list was published, I received the largest grant nationwide and more than Ernst Cloos at Johns Hopkins, and the famous stratigrapher, G. Marshall Kay, a distinguished professor at Columbia.

Near the end of the year, we received notices about our financial aid package. Mine was cut because of my less than stellar performance in paleontology and structural geology, and I wondered how I could make it. It consisted of a tuition and fee waiver and a quarter time assistantship teaching field methods with Gil Benson as instructor. Tom Williams' funding was also cut, and because he had a wife and two children to support and the GI Bill didn't cover all of it, he accepted a faculty appointment at SMU. However, it delayed his PhD by three years, which had a negative cascading effect on his subsequent career.

I read a posted notice about a faculty opening at Willimantic State College in Willimantic, CT, and discussed applying for it with Joe Gregory. He told me every grad student took some cuts because endowment income dropped in a recession climate. He added that my situation was unlike Tom Williams who had a family to support. If I needed to borrow money from Yale, Gregory offered to countersign the loan. His comment was a morale and confidence builder. I took his advice.

I returned to Nova Scotia and finished my field work. Around Labor Day, I returned to Yale knowing all I had to do was complete lab work and write my thesis. My goal was in sight.

I still needed additional funds. Art Bloom was now on the faculty and taught Science II. He hired me to grade exams. I discovered that the Law School dining room offered waiterships. If one waited on tables for one meal, one received three free meals a day. I applied and took the breakfast shift until December. Flint arranged a private office for me in an adjacent building where the department had empty office space.

Because I worked exclusively on my thesis, I had fewer interactions with the entering class than during previous years. I met Pierre Biscaye (BS, Wheaton College) who worked with Karl Turekian and became a senior scientist at Lamont-Doherty Geological Observatory of Columbia University,

Terry Offield (BS, Virginia Tech, MS. Univ. of Illinois) who worked with John Rodgers and returned to the USGS from where he took an educational leave, and Brad Hall (BS, Univ. of Maine, MS. Brown), who worked with John Rodgers and John Sanders and then taught at the University of Maine.

Yale drew geology students from all over the USA and many parts of the world. Yale was a national university in the geosciences. It was a unique place with a great faculty, good equipment, outstanding library resources, and outstanding collections. What it lacked then was a decent geology building. Kirtland Hall was a late 19th century period piece that needed major renovation or replacement if Yale remained competitive during the rest of the 20th century and beyond. The department moved into a new building in 1964.

The annual meeting of the Geological Society of America in 1959 was held in Pittsburgh, PA. I presented my first scientific paper to a professional society dealing with the sedimentary structures in the Blomidon Formation. I characterized it as a lake deposit and explained why.

The speaker before me was Louis M. Cline of the University of Wisconsin who tried to present a 10 minute paper on Carboniferous turbidites of the Ouachita fold belt with 40 slides shown in parallel on two screens using two projectors. In those days, slides were manually placed into and extracted from metal holders and pushed in and out. The union projectionist was having a difficult time. I handed him my slides for a single screen.

While Cline talked, I introduced myself to the session chair and told him I might need more than ten minutes. By this time, Cline had already passed 16 minutes and the session chairman assured me that I could take all the time I needed.

As I walked to the podium, I saw Ray Moore walk into the room where the session was held. That gave me a real morale boost because I knew how busy he was, yet he was willing to come hear the first paper presented by a former classroom student who left Kansas.

I presented the paper and Bruce Heezen (BS, Iowa, PhD Columbia, marine geology; Lamont-Doherty Geological Observatory), the eminent marine geologist at Columbia's Lamont Doherty Geological Observatory, commented that this was an important paper because lakes acted dynamically like ocean basins and I proved it. I thanked him for his remarks.

I ran into Moore later that evening and with his usual expressionless face, he said, "Klein that was an excellent paper. Keep up the good work." John Sanders was with me when he made that comment. Both Moore and Heezen literally made my day.

Steadily, I worked on my thesis. I wrote two chapters at night while in the field and Sanders' critiqued the drafts. By January, I finished lab work and only had to write the remainder of the thesis. The faculty changed thesis submittal ground rules and sent a memo stating that candidates had to turn in final drafts by March 10 for the committee to review and make changes in order to graduate in June, 1960. Once cleared, we revised, got the thesis bound and defended.

Basically, I wrote my thesis in a month. I wrote a chapter and drafted illustrations from Monday through Sunday working steadily from 8:00 am until midnight. Sanders always was away from his office, so I checked his teaching schedule. He taught an 8:00 am class and at 8:45 am on Monday, I was at his classroom door. When class was over, I handed him a chapter. To his credit, he reviewed it immediately and returned the draft with suggested changes by 2 PM that afternoon. It was hard work, but I finished in time and turned the final draft of my thesis in to John who then circulated it to the committee.

John Rodgers was on sabbatical in Paris in 1960, and his place was taken by a visiting professor, S. Warren Carey (PhD, Sydney; structural geology and tectonics) from the University of Tasmania. Carey was controversial in science and behavior and within a month of arrival, antagonized the entire faculty and most of the graduate students.

Carey replaced Rodgers on my committee and after reading my thesis draft, he asked me to meet him on a Saturday morning in mid-March to review it. His office was on the third floor of Kirtland Hall. Carey spoke with a booming voice that was heard all over the building. We discussed the thesis. We reached a point when he said to me, "Look young man, if you want to be impertinent, you can take your thesis and go." I told him no impertinence was intended, suggested that because the issue was in petrology, why not defer the matter to Walton who was also on the committee, and would read the thesis next. He agreed.

We finished our discussion and I went down the stairs not knowing I attracted an audience. The "Rats Nest" was on the second floor and Burchfiel, Benson, Ashmead, Carr, Hansen, Field, Platt, and Ames were at the door having heard Carey's remarks about impertinence. As I approached they burst into applause.

I met with Walton and he sustained my position on the issue Carey had raised.

During the year, I looked for a job in a research university. I was interviewed at Brown University and the University of South Carolina. Alonzo Quinn, the chairman at Brown, wrote they hired someone else, but stated that "your talk was presented in a masterful way." South Carolina verbally offered me the job but retracted it a week later. Jobs were few during the 1959-60 recession.

With no academic job offers, I decided to look for a job at an oil company research lab. I interviewed with Conoco's Research lab but they hired Charlie Ellis. The American Association of Petroleum Geologists (AAPG) held its annual meeting in Atlantic City in Mid-April, so I attended. I was offered a job with Lion Oil, and interviewed with Pan American Research. The Pan-Am representatives assured me an offer would be made but I found out later that the lab management fired A. F. Frederickson, Vice President of geological research, and hiring was frozen.

My conversation with the Pan-Am people had an eaves-dropper, Bernie Rolfe of Sinclair Research Tulsa, OK. He wrote a letter inviting me to apply. I did. Nearly two weeks after my thesis defense I was flown to Tulsa, interviewed, and left with a letter of offer in hand. I accepted starting on September 15, 1960.

When returning to New Haven, I received startling news. Ashmead failed his qualifying written and oral exam and was told to leave. Rogers wasn't around to save him, and Walton didn't back him. A year earlier, Hansen failed the comprehensive written exam but was given a reprieve and passed on his second attempt. It made him a little more focused.

My thesis defense was in the middle of May, 1960. The format at Yale was one presented a 45 minute talk followed by a 15 minute question period. Then the faculty voted in committee. Two defenses were held each day. Gregory, as graduate advisor, introduced each candidate.

On the day of my defense, Lucian Platt defended first. He gave his talk about structural geology in the Taconic region in New York. It was acceptable research. However, he and Carey had reached a major disagreement, and Carey started going after him hard. After five minutes, Flint stood up and turned to Carey (who was to Flint's left and behind him a few seats over) and said, "Mr. Carey, perhaps you may not know that in an American University we allow a student the latitude to develop any hypothesis and interpretation they chose. As a faculty we must see if that student can properly defend it with facts and a reasonable analysis and careful thought. Mr. Platt has done so." Flint sat down and there were no more questions.

I was then introduced by Gregory and walked to the podium to give my thesis defense lecture. It was a bitter-sweet moment. I presented talks in that lecture hall for Journal Club, introduced a colloquium speaker, attended colloquia and previous PhD thesis defenses, and was facing my collective professors and graduate student colleagues as a group for the last time. I proceeded to give my talk. It went smoothly having presented it at two interviews and having practiced it.

I also spent time anticipating questions beforehand and organized some slides which were coded with a batch number so they could be shown if an anticipated question came up. Ed Belt projected my slides and I showed him what was needed if relevant questions arose.

The question period started and I anticipated the first one. I said, "I can answer that with a few slides. Ed, Batch 2." Everyone laughed, and I answered the question. Then Carey and asked me an arcane question about my fault mapping. He was on my right, my map was hung on a wall to my left in the far corner of the room and I knew he probably couldn't see the details. I walked over there and in a fashion reminiscent of how Ray Moore pointed at a map, I pointed to a spot. I asked Carey if that was where he saw a problem. He said, "Yes." I then said, "Well, Sir, if you could come over here I can show that . . ." and proceeded to explain it. Flint looked at me and he had an unusual nonverbal way of showing approval and also smiled (because Carey was behind him) giving that nonverbal signal as I answered. My answer satisfied Carey. Turekian, who had cross words with Carey earlier that year, asked about red-gray color changes, and I told Ed Belt to go to another batch and showed more slides. Karl was satisfied. There were no more questions and we adjourned.

Flint told me I had given the best defense he had ever seen by a Yale PhD during the 27 years he had been there. That too made my day.

I waited outside the lecture hall while the faculty met in a conference room across the hall. After three minutes, Sanders came out and congratulated me. He then returned to the conference room and eventually they passed Platt as well.

Several days later, I drove to Dartmouth College for an interview. On arrival, they told me that they hired a clay mineralogist, so I returned to New Haven, checked my watch, and went to George and Harry's. That evening, only Karl Turekian came and I told him about my wasted trip to Dartmouth. He said, "George when interviewing, you must understand they don't know what they want. We just interviewed three geophysicists, all good people, and we can't decide what we want. What you need to do is not go as a candidate. Go as a consultant trying to find out what the department needs and then show them how you fill that need." I used that advice many times later in life.

Karl then asked if I was attending commencement and I told him that Yale could mail me the degree. He explained that graduation is for the parents and I owed it to them, particularly with my immigrant background. He explained he missed his PhD graduation ceremony because a research cruise left three days before. His mother raised him and his sister as a single parent. She worked scrubbing floors at Altman's department store, had looked forward to seeing her son march in the graduation procession, and be awarded his PhD from Columbia. She never forgave him.

I called my parents and they said they would come. I got their tickets, arranged to rent a cap and gown, and on June 13, 1960, received my PhD. However, my father went to Europe on a critical business trip three days earlier. I caught up with him ten days later in Paris.

I was now ready to make a career in research geology.

LESSONS LEARNED:

1) When things go well, don't rest on your laurels. I took a slightly more relaxed approach my second year and it created some financial difficulties my third year.

2) Even though my third year was financially tight, I never gave up on my goal to earn a PhD in geology from Yale.

3) When things were financially tight my third year, I looked for alternative ways to finance my education by knitting together other work for which I was paid in cash or in kind.

4) Always seek alternative ways to stay on course. My GSA grant my last

field season was a better arrangement than what the Nova Scotia government awarded me in previous years.

5) Always tie one's research to the major themes and paradigms of geology, contribute to their understanding, and utilize new observations and analysis to challenge and improve them.

6) Take advantage of every research opportunity an area offers.

7) During job interviews, always assume the potential employer is looking at you for guidance in their decision and by doing so, they are more likely to offer a position.

8) When giving oral presentations at meetings, the perception of delivering a good or great paper is often enhanced by a poor presentation or paper by the previous speaker.

POSTSCRIPT #1. When Carl Dunbar was offered a faculty appointment at Minnesota in 1917, he went home to discuss it with his father, as he told us during the spring, 1958 field trip. The Dunbar family owned the largest farm in Douglas County, KS. After listening, his father thought a while and said, "Carl, that's fine. Just make sure they pay you at least $300.00 a month."

POSTSCRIPT #2. Richard F. Flint was considered by many to be extremely vain and many stories circulated around the department. My favorite was about his return to Yale after World War II ended. He served as a Captain in the U. S. Army supervising a weather station in Greenland. On returning, he wore his uniform, including to his Science 2 class. One day he stepped into the Science 2 classroom and noticed the students all wore their military uniforms. He discovered that ten students in the front row out-ranked him through battlefield commissions. He never wore the uniform again.

CHAPTER 8
The Keuper Marl of the UK (Summer 1960)

Walter Whitehead retired from M.I.T. in 1957. His job as M.I.T's director of field camp in Nova Scotia and Professorship was accepted by Arthur J. Boucot, a paleontologist at the U.S.G.S. Art earned a BS and PhD in geology from Harvard. He fought in World War II as a tail gunner in allied bombing missions over Germany and as a result, developed a crusty, fearless, and direct style.

During the summer of 1959, Art invited me to the M.I.T camp in Antigonish to give an evening colloquium on my research. When I was finished, a guest, Dr. W. Stuart McKerrow (BSc, Glasgow, PhD Oxford, paleontology; Oxford) asked questions. Stuart was an Oxford lecturer who kept asking how similar my findings in the Nova Scotia Triassic were to the Keuper Marl of the UK. I answered some of his questions and then said, "Sir, let me ask you a question. Has anyone done any sedimentology on the Keuper Marl?" Stuart responded, "No, why don't you raise some grant money and come over and do it yourself."

Art, Stuart and I then adjourned to Art's cabin for drinks with Art's wife, Bobbie (Barbara), and we discussed Stuart's suggestion. I reviewed it later with John Sanders and applied to Sigma Xi for field expense money. They awarded me $750 and my parents, as a graduation gift, offered to pay for my plane ticket and give me another $500.

I bought a cheap charter flight through Yale University on Trans International Airlines (TIA). After leaving New York in June, 1960, and a refueling stop in Iceland, we detoured to Shannon, Ireland, for engine repairs.

We arrived in Shannon before sundown and the landing strip had about 12 fire engines parked and ready to roll. TIA provided meal coupons for food no better than the tuck shops at Scotch College. We waited. I sent Stuart McKerrow a telegram to let him know I was delayed.

We left at 7:00 am next morning and arrived three hours later at Gatwick Airport. From there, I took connecting trains to London and in Oxford I took a cab to the University Museum (Oxford's Geology Building). On the way, the driver asked (my phonetics), "Oh Sai, what univoisity in America did you goe to, Harvard or Yale?" I told him Yale and realized if it had been elsewhere he would have been lost.

Oxford University, as the reader knows, is one of the oldest universities in the western world. Its origins were in the classics and liberal arts. Oxford is a confederation of colleges where the power of the institution resides. Admission as an undergraduate is to one of the colleges where nearly all their instruction is given. Thus, much duplication of effort exists. Instruction is through a combination of lectures and weekly individual tutorials with faculty, even in the sciences. Exams are given only at the end of each term, and an overall general comprehensive exam is given to candidates to earn a BA- or BS-equivalent degree.

Science Departments appeared at Oxford during the 19th century. Their acceptance was slow because most science faculty lacked a connection to an established college. Much political effort was required to appoint science faculty to various colleges so degrees could be conferred through the colleges even though academic work was done in these new departments crossing college lines. Graduate programs as known in US universities were established later and also required connections into an existing college to confer degrees. Eventually, Oxford established special graduate colleges during the 1960's.

I met McKerrow and stayed at his house. He had a nice wife and three sons. I spent a few days reading literature and planning places to work. I bought a second-hand moped and learned to drive it so I could travel in the field.

I was now ready to start field work. That meant taking trains to major communities, and riding my moped to outcrops, staying overnight in small hotels. I loaded the moped on the baggage car, entered a passenger compartment and on arrival, offloaded the moped. It was the same drill when connect with several different trains.

My first stop was the coastal cliffs at Sidmouth, a beach resort community on the Dorset Coast. The outcrops were spectacular and I got off to a good start. Similarities to the Blomidon were obvious.

I returned to Oxford because my father invited me to meet him in Paris for the weekend. I went there on Friday. We went to his favorite museums and restaurants, and at night, his favorite night clubs in Montmartre. I admit I was not all that impressed. There were still signs, 16 years later, of damage from World War II and the recovery was hardly complete. The shows at Montmartre were not much better than the strip joints in St. Louis and Kansas City which I visited occasionally with friends during graduate school. I said little not wanting to offend my father. I ate well while in Paris.

Returning to the UK, I proceeded to look at Keuper Marl quarries and outcrops in the Midlands. I also took a side trip to the Welsh Coast to examine Ordovician turbidites.

Later, I attended a weekend field meeting of the Yorkshire Geological Society focusing on Permian Rocks. Sam Carey attended also and told Sir Kingsley Dunham, head of the geology department at Durham University, and later Director of the British Geological Survey, that I was the best geology graduate student he met at Yale. I was flattered, but wondered if the Brits also had the same view of Carey as the Yale geology faculty.

My last field area was on the Island of Arran. Afterwards I went to Edinburgh and drove my moped to Sicar Point to see the "Great Unconformity" described centuries ago by Hutton.

That year, the International Geological Congress was held in Copenhagen. I wrote to register and was put on a wait list because I registered late. Eventually the organizing committee agreed to let me come but I had to prove that I arranged housing on my own. The Danish tourist office found me a private bed-and- breakfast. It was 45 minutes by trolley from the convention site.

I attended technical sessions and met many leaders of international geology. John Sanders and John Rodgers were there. Rodgers acted surprised that I found a job. One of Sanders's graduate school classmates, Heike Ignatius (BS University of Helsinki; PhD Yale, Quaternary geology) also attended. He worked for the Geological Survey of Finland and eventually became director during the 1980's.

As a government official, Heike stayed at the Finnish Embassy and decided the Finnish Embassy should host a Yale alumni cocktail party. Six of us attended. They offered canapés and aquavit. After 15 minutes, the canapés disappeared and they only served acquavit. To say I was feeling warm and no pain was an understatement. I made it my bed-and-breakfast place with no difficulty, however.

I recall a paper by Dr. Kingma, a New Zealand geologist who challenged the turbidite paradigm. Ph. H. Kuenen challenged the speaker during the question period. The speaker interrupted Kuenen at which point Kuenen responded "I did not interrupt you when you gave your silly talk, so don't interrupt me when I'm talking." There was a hushed silence. When Kuenen finished, the session chair, Dr. Rudolph Trumpy of Switzerland said words to the effect that there are still differences of interpretation on this topic and we needed to pay attention and debate them. Apparently, not all European professors were dogmatic or believed they held a monopoly on truth.

Later Sanders introduced me to Kuenen, but Kuenen lived up to his reputation as being totally uninterested in students and young PhD's and barely talked with me.

I saw John Sanders and Heike Ignatious again at lunch. They were talking about one evening in 1951 in New Haven. Both attended a movie and were stopped by some hoodlums demanding money. Heike stepped up to one, lifted him off the ground by the scruff of his neck, and threw him at the others saying, "Go back to your mother's milk." The hoodlums ran away.

John then explained that Heike fought on the Russian front during the Russian-Finish War of 1941 and sustained a serious head injury. A major part of his skull was a metal plate.

I returned to Oxford, packed, went back to Gatwick, and returned to the USA.

LESSONS LEARNED:

1) When working overseas, time schedules are different and things move more slowly. My mother told me later, "What you save in money, you lose in time."

2) When budgeting for overseas work trips or other travel, expect to spend at least 20 to 30% more and plan accordingly.

3) When attending International conferences, expect everything to be more formal. In particular, expect academicians to be somewhat more dogmatic. Allow time to develop collegial working relations, particularly with continental Europeans. It is easier to establish collegial relations with the British.

CHAPTER 9
Sinclair Research, Tulsa, OK (1960-1961)

The Sinclair Oil Company was founded in 1916 by Harry F. Sinclair and grew to be the seventh largest oil company in the USA at one time. It was primarily an exploration company. Harry Sinclair was a colorful gent and spent time in jail for his involvement in the Teapot Dome Scandal. The company was extremely conservative and risk adverse. Their exploration paradigm was the "hind tit" approach, namely, let someone make the initial discovery and then lease offset acreage if available. The company was acquired by ARCO in 1969.

Sinclair Research was formed during the 1950's and individuals from the operating company were transferred to staff it. Their only PhD on staff when I interviewed was Bernie Rolfe who hired me. In 1960, when Glenn Visher and I were hired, it was as much of an accident as to fill a need. The operating company, applying for a loan from a major New York bank, was asked, "What is Sinclair doing in the way of expanding its research to find more oil and gas?" Sinclair management responded that they just hired two PhD's, one from Yale and one from Northwestern. The loan was approved.

After returning to the USA, I called Bernie Rolfe to make arrangements to start work. We mutually agreed I should rent a car and drive to Tulsa. I did so and drove west. I arrived around 1:00 PM on September 15, 1960, reporting to Bernie. After filling out forms to get on the payroll, health insurance, and few other formalities, I was asked to attend a staff meeting that afternoon at 2:00 PM.

Bernie then took me to another building and introduced me to my manager, Robbie Robinson. Robbie and I quickly established rapport particularly because he earned both his BS and MS at the University of Nebraska. We discussed the KU and Nebraska athletic rivalries. I also asked Robbie about his experience in the petroleum industry and he told me he had spent time on a Sinclair geophysical crew in Louisiana before the lab was established. He was transferred three years before I arrived and completed work on classifying and identifying visual porosity in carbonate rocks.

Robbie then took me to a large room which I was to use as an office and lab. I shared it with Glenn S. Visher (BS, MS Cincinnati, PhD Northwestern, stratigraphy; Shell Oil, Sinclair Research, Univ. of Tulsa, BNJ Exploration). Glenn worked for Shell Oil and left because they did not transfer him to their research lab, despite assurances they would do so.

I then attended the staff meeting and was introduced to everyone, including a computer systems person, a geochemist (Bill "Jake" Jacobsen), a water chemist (Nat Sage), three palynologists (Ned Potter, Bill Meyers, and a tech), a regional geologist (John Rogers), Jim Westphal (geophysicist) and a petroleum engineer. Also present was Jimmy Johnson, the Vice President of Exploration Research. Jimmy was a geophysicist. Bernie Rolfe reported to him as manager of geological research. The meeting was chaired by George Fanshaw, president of the lab, and included another Vice President who was an engineer.

Fanshaw opened the meeting to announce that the lab would handle all in-house training. We had a month to get ready for the first course. Assignments were given and because I had just arrived, Fanshaw asked if I could prepare a presentation. I offered to give a two-hour talk on the application of sedimentary structures to determining environments of deposition. Fanshaw commented that this would be new and encouraged me to put it together

Afterwards, Bernie and Robbie and I went to Bernie's office. They assigned me to work on the reservoir sedimentology of the Minnelusa Formation (Permian) of eastern Wyoming and western North Dakota. At Sinclair Research, most project work was contract work for the operations division. The Casper, WY, office requested this project and funded it. I was the lead geologist and would coordinate with Chuck Tenney in Casper. During the previous summer, Chuck, Robbie, and John Rogers collected samples and thin sections were cut. I was first to complete a petrographic study, point counting all four hundred thin sections (all 400 samples were quartz arenites as per Bob Folk's classification). Bernie and I were to fly to Casper in late October to meet Tenney and do reconnaissance field work, because I also was scheduled to give a paper at the GSA in Denver beforehand. I was to return next summer to continue detailed field work.

Bernie (BS CUNY; PhD Penn State; clay mineralogy; Cities Services Research, Sinclair Research, Consultant) then brought up something else. He explained Robbie did not have a PhD and asked if it would be a problem for me to have Robbie as my supervisor. I explained that Robbie had more experience in the oil industry than me. Robbie and I agreed he would train me about the oil business and I would be his resource person in areas where he wanted more expertise.

Bernie was a tough individual. He grew up in the streets of New York City and was drafted into the US Army. He saw action in the Battle of Bulge and survived without injury. The GI Bill provided him with the opportunity to receive a university education and achieve a better life. However, the hard street side occasionally flared and one had to be careful with him. He also liked calling me "kid" because I was the youngest of the professional research staff.

Sinclair Research's facilities were located on the northwest side of Tulsa in the old Pan Am Research Lab across the street from the Carter Oil Research Lab. When Pan Am built a new lab in South Tulsa, they tried to sell the old lab but no buyer was forthcoming. Pan Am finally donated it to the University of Tulsa who rented the space to Sinclair. We were located in four buildings, an administrative building with a small library, a core storage building, a maintenance building for trucks and equipment, and a lab building. My office was in the lab building.

My office served another function. It was where the only coffee pot in the lab building was located. There was no refrigerator. Only black coffee was available. When tasting my first cup, I discovered its unique taste. It was Luzianne Chicory coffee. When Jake, Robbie, Jimmy Johnson, and John Rogers were working in Louisiana on geophysical crews they acquired a taste for it. They assured me that I would too.

I spent two nights at a nearby motel and searched for a furnished apartment. After finding one and moving in, I returned the car to the rental agency and had no car. Jacobsen lived nearby and picked me up every morning and took me home at night. Grocery stores were in walking distance so I could manage until I bought a car.

After receiving my first paycheck and opening a local bank account, I bought a 1954 Ford station wagon. My bank provided a loan. I planned to camp out on weekends if I wanted to see something different and used it to do some extracurricular field work in the Ouachita Mountains of Arkansas and Oklahoma that fall.

Slowly, I began work on the Minnelusa and prepared my two-hour lecture for their October short course. We also offered the course in

November and December. The October course was for regional managers and Sinclair housed them at a motel with its own private club and a conference room. In Oklahoma, the only way to buy liquor was from a state store and drink at home or be a member of a private club. The motel provided club membership for its guests and during the training sessions, for Sinclair instructors.

Bernie suggested I stay each evening and become better acquainted with the regional managers, even joining them for dinner. That enlarged my network into the company. I also obtained a close-up view of what happened to career geologists at Sinclair, how it influenced their lives, their attitudes, their politics, and their morale. It was a mixed picture.

I presented my lecture the first afternoon of the course. During the middle of it, an older gentleman appeared at the door and Bernie went outside the room to talk with him. I continued and they returned. During the discussion period, that visitor asked me questions, most of which I answered without difficulty. When discussion was over, Bernie came to the podium and said,

"Thank you George. By now you know we have a surprise visitor, Fred Busch, Sinclair's Chief Geologist. Fred, would you care to say a few words?"

The older gentleman who arrived in the middle of my lecture was Fred Busch. He made the usual remarks about the importance of the new training programs at Sinclair and even said that ". . given what I heard so far from George Klein, this program is off to a strong start."

Busch stayed the rest of the afternoon and invited me for a drink so we could get better acquainted along with some of the other participants. Bernie took me aside and told me that normally, he would have interrupted so Busch could give a pep talk but Busch insisted on hearing what I had to say first. In the corporate world, breaks like this seldom come in a lifetime, let alone during the first month on the job.

However, I observed a sobering side-show. While having a drink with Busch, Rolfe and some of the others, I noticed three attractively dressed women in their late twenties at the bar. They were having drinks with different patrons and now and then, snuck off to the patron's motel room. One didn't have to think hard about what was going on. Underneath America's bible-belt family town, there was, indeed, a dark side.

I also joined the Tulsa Geological Society and attended most dinner and luncheon meetings. Consequently, I met a lot of well-known geologists in Tulsa and reconnected with Gerry Friedman (BSc Imperial College, University of London, PhD, Columbia; sedimentary petrology; Univ. of Cincinnati, Uranium Consultant, Pan Am Research, RPI, CUNY, Northeast Science Foundation) and Harry Werner (PhD, Syracuse, petroleum geology; Pan Am Research, University of Pittsburgh) at Pan Am Research. They apologized for not arranging a plant trip and explained that when A. F. Frederickson was fired, all hiring stopped. Frederickson found a new job as head of the department of geology at the University of Pittsburgh. I developed a lifelong friendship with Gerry Friedman and his family.

I attended GSA in late October and presented a paper on Triassic sandstone petrology challenging the paradigms of tectonic associations and sandstone petrography. The paper was scheduled in a general session and I was the second speaker. The speaker before me was Larry Sloss who presented a paper on "Cratonic Sequences." The room was packed. When I reached the podium to make my presentation, the room was half full. Instantaneously, I resolved to give the best talk I was capable of presenting and make sure I never spoke to a half-full room again. Later, I met Sloss in the halls who apologized for leaving and said he heard I gave a great paper.

Bernie and I met at the Denver Airport and flew to Casper, WY via Cheyenne, WY, the first stop and then Laramie, WY. We flew over the "gangplank" as the ground rose to meet us. We arrived in Casper and were met by Chuck Tenney. It was cold, snow was falling and we went to the Sinclair office where Bernie met everyone and introduced me.

After meeting in Chuck's office reviewing maps, cross-sections and Chuck's perspectives of the project goals, we drove to Beulah, WY to stay at "Ranch A" a dude ranch. We arrived on October 31 for dinner. The place had many hunters from Minnesota, Nevada, and elsewhere. Some hunters also brought along their girlfriends or mistresses (not sure which). Bernie commented over drinks that "they'll shoot poorly with their rifles during the day, and if they don't drink too much, they might do better with their pop guns in bed." Hunting season started next morning.

After breakfast, we drove into a box canyon as the sun rose. It snowed overnight and it was beautiful. I suggested we stop and examine an outcrop on the canyon wall. On opening our car doors, shot guns fired away and the noise echoed around the box canyon. Bernie said, "Sorry George, let's get out of here. This is worse than the Battle of the Bulge."

We spent two days in the field examining outcrops. We heard gunfire all day. At night, the hunters were getting tanked up at Ranch A but none bagged a deer.

On Sunday, we drove back to Casper. During the entire trip, Bernie and Chuck talked about other company people, their strengths, their foibles, their weaknesses, their intelligence, their ethics, and so on. I listened. Finally, Bernie said to me, "George, you must think we do nothing but gossip about everyone in the company." After a non-committal response, Bernie said, "Look kid, you're still wet behind the ears and new on the job. It's a cold cruel world out there. And another thing, don't believe this crap about company loyalty. They can fire you at will, and if you're lucky, it might be with two week's notice."

I flew back to Denver, rented a car and drove to Boulder. I needed to read some theses at the University of Colorado that were relevant to my project. I also had dinner with Don Eicher and his wife. He had been at Colorado for two years teaching paleontology.

I flew back to Tulsa and returned to work.

By this time, I became better acquainted with Glenn Visher. Glenn was articulate and loved putting people down and arguing with them. I noticed that others in the lab were not that appreciative of his style. I didn't mind the arguments, but did mind some of his acerbic comments. While at Shell, he completed their training program and was trying to build research on one aspect, namely the vertical sequence concept of changes in grain size, sedimentary structures, well log shape and trend to characterize depositional environments. I reviewed his work and felt he was on to something, but it was a hard sell to the company. Moreover, there was concern that he used Shell proprietary data and this could lead to difficulties.

As I went over Glenn's work, I realized I needed to review the literature of the present day Gulf Coastal Plain of the Gulf of Mexico. It was the incubator of predictive models of sedimentology of the 1960's.

Around Thanksgiving, I reassessed my situation. I missed the academic environment and the freedom to pursue research of one's own choice. I found Tulsa a disappointing place to live particularly because I was single and dating opportunities were few. Most of my age group or younger were married with families. All eligible women I met were divorced with at least two young children requiring a major adjustment. Tulsa was a great place to live if one had a wife and three or more children. It was also heavily church oriented, and although I joined the local Unitarian church with a young charismatic minister, single ladies were few. There weren't many other attractions.

I met two great couples at the Unitarian church, Ken and Marge Ackley and Hugh and Grace Hay-Roe. Ken was regional exploration manager for Humble and grew up in west Texas, graduating from the University of Texas. Early in his career, Ken was assigned to the district office in Mattoon, IL. He met Marge there at church. She was teaching school in Mattoon after graduating from the University of Illinois. His knick-name for me was 'Rocknocker.' Ken explained that we were fellow rocknockers pounding on rocks to find oil and natural gas. Meeting the Ackley's was one of the bright spots during my entire stay in Tulsa.

Hugh Hay-Roe (BSc, Univ. of Alberta; MS, PhD, Univ. of Texas @ Austin; Jersey Production Research, International Petroleum – Peru; Belco; private consultant; BPZ Energy) became a life-long friend. A Canadian, we developed a good friendship. After I left Tulsa, we saw less of each other but reconnected when I moved to Houston and I completed some spec work for BPZ Energy, and later when BPZ Energy asked me to help them with a book on Peruvian basins. Hugh is best known for writing columns and a book on proper geological writing.

Sinclair Research, in many respects, was a great training ground, but I kept looking for teaching positions. I applied for a two-year temporary position at the University of Wisconsin and had a memorable breakfast interview at the Denver GSA meeting with three people, including Louis M. Cline who was now chairman. They hired someone even younger and in their view, cheaper. They were unconvinced I'd take a 30% pay cut if the job was offered.

At Christmas, I arranged to take time off and make up the time working Saturdays for six weeks. During that vacation, I interviewed for a faculty appointment at Villanova University but knew it was a poor match. A year later, Ed Belt, who was Catholic, accepted a job there. I also drove to New Haven and met with John Sanders to assess the situation and my options.

During January, there were changes on the job. I was expected to put out "fire-drill" requests from the operating company and did so. The first involved a granite wash problem in the Permian basin. Bernie and I worked together and completed our report which was well received.

Robbie then told me that management wanted Glenn and I go in the field with Allen P. Bennison, Sinclair's regional geologist, to get exposed to some of his ideas and places he had worked in southern Oklahoma and Arkansas. Bennison published a definitive paper on the Potato Hills of Southern Oklahoma and correctly recognized it as a window within an overthrust belt. I already spent some weekends in that area looking at the Stanley-Jackfork (Mississippian) and Atoka (Pennsylvania) turbidites which Lewis Cline described at the 1959 Pittsburgh GSA meeting. It was an opportunity to obtain background information and also thought it advisable that I work with the person who Sinclair's management considered their regional tectonic guru.

Glenn was incensed. He said because he worked with great professors at Northwestern (Krumbein, Sloss, and Dapples) and because I completed coursework with John Rodgers at Yale, we did not need to be taught anything by Bennison. I told him I didn't know enough about Bennison and welcomed a chance to go in the field.

I went on the trip. I did not agree completely with Bennison's approach, but learned new geology and at least saw it through his eyes. Geology is a science based on experience and observation. This trip with Bennison was an opportunity to gain needed additional experience.

Moreover, I also used the trip as an opportunity to propose a project to Sinclair management and it was approved. It dealt with a regional sandstone petrographic study of the Stanley-Jackfork boundary, utilizing field samples and well samples. That gave me reasons to go into the field more often at company expense.

In February, Gerry Friedman told me he accepted a position at the University of Pittsburgh as an Associate Professor. Frederickson offered him a job. I congratulated him.

Also that month, I received an invitation from the Yale Alumni Association to attend a function at a local country club. I went more out of curiosity. I met a young couple, Don and Ruth Nelson. He graduated from Yale and the Wharton School of the University of Pennsylvania. She graduated from Bryn Mawr and met Don while there.

Ruth's father, a German immigrant, owned the Kaiser Oil company. It sold drill pipe and got into oil exploration using a unique approach. He noticed where his clients drilled from pipe deliveries. He determined from total pipe length how deep they drilled, which wells were successful, and which failed. He reasoned that companies spent a lot of money putting a prospect together so they knew oil might be recovered. He drilled offset wells from dry holes, discovered oil, and made millions doing so. I spent many a weekend with the Nelsons and their children.

During the middle of March, Jimmy Johnson called unexpectedly to let me know I passed my six month's probationary period, had done well, and all colleagues were pleased with my work and interactions with them. My salary was raised from $9,000 (approximately $57,000 in 2009 dollars) per year to $9,500 (Approximately $60,500 in 2009 dollars) per year. He said he looked forward to a continued long-term working relationship with me at the lab.

As the spring continued, more "fire drill" requests were made. I wondered if I could ever finish any projects. Bernie called a meeting which Glenn, Robbie, John Rodgers, Bill Jacobsen, Nat Sage and I attended. Nat completed an eighth grade education but was a drinking buddy of Harry Sinclair, the company founder. When Sinclair passed away, Nat was assigned to the lab as an oil field water chemist.

Bernie showed a map of an area in west Texas. It displayed the organic content of soil samples and presumably the high organic content was correlated to possible oil seeps. When Nat looked at the map, he smiled and said, "Oh, I remember that ranch. All the highs are around the water troughs for cattle. Their anomalies correlate with bull shit." We recommended the proposal be declined.

In April, Apache, a small independent in 1961, made a major gas discovery in the Arkoma basin. Other companies leased offset acreage. Sinclair requested me to assess their opportunities with a report due in two weeks. I completed some field work, scoured all reports I could find, made an assessment, and wrote a report. Basically, Apache's discovery was a stratigraphic play exploiting a strandplain/barrier island system. I mapped the extent of this play and made my presentation. When Sinclair's exploration management looked at my maps, the landman at the meeting reported that all the good acreage in the play was leased by competitors.

The chief geologist asked me, "George, while you were assessing data, did you see any new opportunities that hadn't been leased or discovered?"

I explained that on the north side of the Arkoma basin, I observed Atoka outcrops which looked identical to the Oriskanie Sandstone (Devonian) of the Appalachians, a well-known gas reservoir. I interpreted them as fluvial channel sands.

He looked at me and said "Don't people at the research lab know that oil and gas only occur in marine rocks?" The meeting ended.

Eight months later, Apache found natural gas exactly where I told Sinclair to look for it. By then, I had left the company.

Gerry Freidman decided to stay at Pan Am's research lab and declined Pitt's offer. I wrote Frederickson applying for the job. He called two weeks later and invited me for an interview. He wanted a reference from Bernie so I called and explained the situation. Bernie gave a good reference. The interview went well, although I only met with Frederickson. Two weeks later, I accepted an offer to start teaching at Pitt that fall.

I gave Bernie my resignation letter and the first thing he said was "George, we invested a lot in you. Where's your company loyalty?" I replied, "Bernie, remember when you, Chuck Tenney, and I rode in the car going from Beulah, WY, to Casper back in November and you said that company loyalty was a bunch of crap?" He looked at me and said, "You're right. Good luck with it. But I hate to see you go to work for Frederickson"

A week later, Jimmy Johnson called me into his office. Bernie Rolfe was there as was the Vice President of the lab. They asked why I would leave. I explained I wanted freedom to pick my own research, that the inability to complete research with all the fire drills slowed me down, and there was a particular project in the UK on the Jurassic Great Oolite Series I wanted to undertake. They made a counter offer. They offered three months off during the summer during the next two summers to do field work in the UK, and they would pay me to do the lab work at Sinclair. I turned it down knowing they could reverse their offer any time.

I completed my reports, turned them in and said my good-byes at the end of July, 1961. I spent two extra days in Tulsa to visit the Carter Oil Research Lab and the Pan American Research lab representing the University of Pittsburgh which meant they could show me more of their work. I then packed everything I owned into my car and headed east.

LESSONS LEARNED:

1) When working as an employee of an oil company, one is an "at will" employee, meaning that the company can fire you any time. Therefore, it is best to build up and keep a cash reserve.

2) Risk adverse companies ultimately will fail or are merged into a more aggressive company. Sinclair Oil was merged with ARCO for their assets in Alaska.

3) When employed in a less-than-desirable location, use weekends to go out of town for R&R. While in Tulsa, I should have travelled on weekends to Dallas to counter the negatives of the Tulsa region. I learned later to do so when living in east-central Illinois.

CHAPTER 10
University of Pittsburgh (1961-1963)

The University of Pittsburgh was chartered in 1819 to serve the higher education needs of western Pennsylvania. It struggled because of lack of funds. During the 1920's, the campus moved to its present site near the old Forbes Field, home of the Pittsburgh Pirates baseball team. It was mostly a commuter school.

As a candidate the only thing I knew about Pitt was the discovery of the polio vaccine by Dr. Jonas Salk at Pitt's medical school. Salk was an immediate university icon because of the fame he brought them, including the royalty income from the patent.

In 1955 the Pitt Board of Trustees appointed Edward Litchfield to be president with a mandate to upgrade the university. He was Dean of the Business School at Cornell and negotiated with a member of the Scaife Family who was president of Pitts's board. The Scaife family owned Mellon Bank and Gulf Oil. Litchfield negotiated a side deal with Scaife which was never approved by the Board, and before he arrived, that Scaife family member died. Litchfield always believed he had resources to transform Pitt into the 'Harvard of the Alleghenies." Later events proved he didn't.

I arrived at the University of Pittsburgh in latest July, 1961. A. F. Frederickson let me stay at his home until I found an apartment. He lived in a large house with a very nice wife who was half Caucasian and half Native American. They had five daughters ranging in age from 6 to 17.

After unloading my rocks and books into my office, I found a furnished one-bedroom apartment in two days and moved in. I then drove east to visit my parents and made a brief visit to Yale. Sanders told me that Clark Burchfiel, who accepted a faculty appointment at Rice University, and I, were appointed to the two best geology academic positions in the USA that year.

John Sanders also told me he was chairing the 1963 SEPM (Society of Economic Paleontologists and Mineralogists; renamed Society of Sedimentary Research in 1993) Research Symposium on cyclic sedimentation and asked if I had any recommendations. I nominated Glenn Visher to talk about vertical sequences and explained what Glenn had developed. I also said that Glenn learned some of this at Shell so John needed to be careful how this was handled. Eventually, John invited Glenn to that symposium.

I also visited the headquarters of the Geological Society of America in New York City. I submitted my thesis to them as a possible Memoir, but it was returned in February with reviewer's suggestions to break it up into several papers. I resubmitted one paper and wanted to know what happened. I met with the editor, Agnes Creagh who reviewed everything with me because two reviewers recommended further breakup and publication. A general paper on environments and sandstone petrology appeared in the *GSA Bulletin* in September, 1962, and one on sandstone classification in May, 1963. My paper on the Keuper Marl, submitted from Tulsa, was also accepted and appeared in March 1962 in *Geology Magazine*, a journal published by Cambridge University.

I owe Agnes Creagh an eternal debt of gratitude for spending three hours showing me how to put a manuscript in good order. When I thanked her as I left, she said "George, you will be training many PhD's. I spent the time with you so you can show them what they must do to save me and my successor's lots of time." She was exactly right and if I did nothing else for my PhD students, it was to help them turn their standard theses into publishable prose which appeared in major geological journals.

The department of geology at Pitt, like the university, was undergoing massive change. The geology department had been at best, average. It was headed during the 1950's by Chip Prouty who left in 1958 to head the department of geology at Michigan State. Norm Flint (BS, Univ. of New Hampshire, PhD, Ohio State) a Carboniferous coal stratigrapher, served as acting head until Frederickson arrived in 1960. Frederickson was an international authority on clay mineralogy, and I recalled reading his widely-cited paper on weathering at Yale.

Frederickson's goal was to get rid of so-called 'deadwood' and rebuild the department with new people. He brought in Takesi Nagata (PhD Tokyo, paleomagnetics; Univ. of Tokyo) as a permanent visitor to spear-head a program in geomagnetism, hired Kazuo Kobayashi (PhD, Tokyo) one of Nagata's students, as an Assistant Professor of Geomagnetism, Joe Lipson (PhD, U. Cal, Berkeley) as an associate professor of geochronology, and me. Staying on were Norm Flint, Tracy Buckwalter (PhD Michigan, Petrology;

Pitt), and Martin Bender (MS Pitt, U.S. Steel exploration geologist; Pitt) who taught physical and historical geology. Flint taught stratigraphy and structural geology.

During the next month, I held several conversations with Frederickson. He grew up in a working class neighborhood in Seattle doing odd jobs to help the family survive. He had a summer job at age 16 on a floating fish cannery working off Alaska. One day they were in a bay near port. Fred took the day off and climbed a nearby hill to view the scenery. As he looked at the boat, it suddenly exploded, sank and killed everyone on board. He was one of three survivors. Fred returned to Seattle, ran errands for the Teamsters Union and described graphically how the labor bosses kept the rank-in-file in line. He graduated in mining engineering from the University of Washington and fought in World War II. On returning, he earned a Master's in Geology and went to M.I.T to earn a PhD in Mineralogy. Part of his PhD preliminary exam consisted of identifying 40 white mineral specimens.

After M.I.T, he taught at Washington University, St. Louis, for seven years reaching the rank of professor, went to Pan Am Research as director of exploration research and joined Pitt in 1960. I discovered later that Washington University fired him, even though he had tenure, over a charge of financial mismanagement of research grant funds.

Returning from my trip east, I met again with Frederickson. He told me that during the fall term I would teach a graduate course in sedimentology and an undergraduate course in mineralogy. In the spring, I would teach a graduate course in sedimentary environments. My salary was $8,500 (Approximately $54,000 in 2009 Dollars) for an eleven-month academic year. The goal was that all of us would raise grants to reimburse the university for the 2/9 summer supplement.

I began course preparation. The sedimentology course was straight forward. I asked Norm Flint to take me on a local field trip and selected several great outcrops for a Saturday trip. Some of them were textbook cases for Glenn Visher's Vertical Sequence concept. Glenn Visher mentioned that at Northwestern Krumbein took his sedimentology class on a field trip to sample a beach on Lake Michigan next to the Northwestern Campus. The class used those samples to learn laboratory techniques and write an integrative report. I adopted this approach too and used a beach on Lake Erie. I also ran a bedrock trip illustrating the concepts covered in class and I could tell the students understood the linkages between modern and ancient sediments.

The mineralogy course was a struggle, so I asked Frederickson for advice. He had one suggestion. It was to review my undergraduate and graduate mineralogy course notes and pick what was important and the rest would follow. I did and quickly discovered that if I covered crystal chemistry, crystallography, silicates and carbonates, I could teach a useful course. Later, I discovered Don Peacor at Michigan developed a similar outline and it became the new way undergraduate mineralogy was taught nationwide (See Chapter 15). In the past, the focus was on sulfides, oxides, and native elements. Silicates and carbonates were covered in petrology.

Before the semester started, Joe Lipson arrived and because he and I were "Frederickson's boys," we hung around a lot. Fred met with just the two of us to decide departmental matters, excluding the others. Kobayashi arrived in Mid-October.

The department had one secretary, Mrs. Kinch, who was there at least seven years. Before Frederickson arrived, she was almost a defacto department head, and adjusting to Frederickson, Lipson and me was a major change. Socially, she interacted with Buckwalter, Bender and Norm Flint. I recall handing her an NSF proposal to be typed and her comment was, "Assistant professors shouldn't be applying for research funds." When nothing happened for a week, I informed Frederickson and she got it done and did a very good job. I realized she was capable of doing good work, but was too involved in what clearly was a split department.

Buckwalter, Bender, Mrs. Kinch and some of the graduate students often met for lunch in Buckwalter's and Bender's shared office complex. I walked by one day and they were playing parlor games. I realized then that upgrading the program was that much more difficult.

In mid-September, I attended a coastal marine geology conference at the Oceanography program at Johns Hopkins University. I arrived the afternoon before and stopped by the Department of Geology and visited with Pettijohn. I then paid a visit with Aaron Waters. Our conversation went as follows:

Waters: It's good to see you Klein. Why are you here today?

Klein: I'm attending a coastal conference in the department of oceanography which starts tomorrow.

Waters: Well, they never notified the department here. Oh well. What are you doing now?

Klein: I started a faculty appointment at the University of Pittsburgh and have been teaching now for a month. Before then, I worked for a year at Sinclair Research.

Waters: OK, Klein, since you are now teaching, I'm going to give you some advice. First, try not to stay in one university for more than ten years. Second, always take a sabbatical; even go into debt to do it because it will always pay off later. Third, always buy a used car, about one or two years old. You can get them for half the original price, and just run them into the ground.

I thanked him for his time. His advice proved to be golden and some of the best I heard, particularly the one about sabbatical leaves.

The GSA meeting that fall was in Cincinnati, OH. I joined a pre-meeting field trip that started in Chicago and went to the meteorite impact structure at Kentland, IN, the Silurian Reefs of Indiana described by Shrock, and the Lower Ordovician McMicken Hill Section in Cincinnati. The last stop at McMicken Hill was next to a public housing development. While digging into soft shale to extract fossils, we were joined by about 12 African American children who really knew their fossils. One found an *Olonellus* (Trilobite) and correctly identified it.

During the first seven years of my career, I went on a pre-meeting field trip before every GSA meeting to broaden my experience and see new geology. The slides I took were helpful for both research and teaching.

I presented my paper on the Keuper Marl to a sedimentology session which I co-chaired with the eminent carbonate geologist, Albert V. Carozzi (BS, MS, PhD, University of Geneva; carbonate petrology; Univ. of Geneva; Illinois) of the University of Illinois. Carozzi knew John Sanders so we had common ground to get acquainted. The session was attended by Phillip H. Kuenen who received GSA's highest research award, the Penrose Medal. During a mid-session break, he talked with me about my paper and was pleased I identified lacustrine turbidites. I previously met him in Copenhagen and he recalled meeting me.

On arrival in Pittsburgh, I joined the Pittsburgh Geological Society. It was smaller than the Tulsa Geological Society. Both were affiliated societies of the American Association of Petroleum Geologists (AAPG).

One individual I met was Vint Gwinn, a geologist working for the Pittsburgh exploration office of Mobil. Vint graduated from Rutgers in geology where he received a baseball scholarship. On graduation, he was offered a contract by the New York Giants baseball team, but declined to attend Princeton where he earned a PhD in 1960. Vint developed a thin-skinned hypothesis for the origin of the Appalachian overthrust belt and also had done thesis work in sedimentology. He and his wife and I socialized and became good friends. He was curious to know how I moved from Sinclair to a faculty appointment at Pitt because he had similar goals and complained he was not getting help from Princeton.

Early after my arrival, I arranged to visit the Gulf Research Lab via one of my Yale fellow graduate students, Bob Hodgson (BS, MS. Wyoming, PhD, Yale) who completed a definitive thesis on jointing in the Colorado Plateau and was offered a job as a research structural geologist. On arriving at the lab, I first met Mel Hill, the director of exploration research. Mel handed me an organizational chart and then turned me over to their geological oceanographer, Jack Ludwick (PhD, Scripps Institution of Oceanography; coastal sedimentation processes; Gulf Research, Old Dominion University) who worked with coastal sediment models. Jack later introduced me to Wayne A. Pryor (BS Centenary College, Louisiana, MS, Illinois, PhD. Rutgers, sedimentology; Illinois Geological Survey, Gulf Research, Univ. of Cincinnati). I heard Wayne present a paper at the 1960 AAPG meeting which was just published. It integrated the paleocurrents work by Paul Potter, Ray Siever and, later, Wayne. Wayne and I immediately established common ground and good rapport and developed a lifelong friendship.

I asked why he left Illinois to go to Rutgers and he explained the quirks of that department. The department head, George White, ran a tight ship, and White's wife, Mildred, who had no children, developed a social group of all the wives of married students to teach them to become traditional ladies. How the wives got on with Mildred influenced the outcome of their husband's fortunes in the department. Pryor's wife was a nurse from Germany. He met her while on active duty immediately after the Allied occupation. She had many good qualities but also was a 'hard case," and Mildred was not pleased.

Wayne spoke fluent German and to help his wife adjust to America, they spoke both German and English at home. Moreover, Wayne had an independent streak that did not impress White. When Wayne took his German exam administered by the German department, he failed, and in fact, he failed three times. Normally, White went to bat for students but chose not to do so in Wayne's case. Wayne finished his PhD at Rutgers in 18 months, passed their German exam, and returned to the Illinois Geological Survey (IGS).

He left in 1960 because of changes in research management at IGS. When he left, he presented a colloquium about how "clods rose to the top." (It's based on the deflation principle forming desert pavement). He later accepted a teaching Fulbright professorship at the University of Heidelberg in 1968-69 where he presented his lectures "auf Deutsch."

I had heard similar stories about George White and Illinois from Stuart Grossman, Terry Offield, and Dick Benson, an Illinois PhD who taught *Ostracoda* micropaleontology at Kansas while I was there.

The second semester started and I was teaching only one course. By now, Brock Powers finished his PhD at Yale and returned to Saudi Arabia. He sent me a bound copy of his thesis which I read. I found it routine and underwhelming. I decided to ask my graduate class to evaluate it and we discussed it. They were as critical of it as I was without my prompting.

One evening two weeks later, Frederickson met the graduate students and gave them a talk outlining his goals for the department's future. Afterwards, he invited them for beer at a nearby pub and asked about the courses they were taking. The students in my class told him about their experience with the Brock Powers thesis, and told him good things about my teaching. Frederickson visited me the next day and revealed his discussion with those students. He said he was extremely pleased I gave them a chance to evaluate Brock's thesis because as Fred put it, "They didn't think much of it." He then said that he appreciated my approach because it taught the students they were as capable as anyone and their being at Pitt should not be taken as a reason to assume they were second class geologists. Fred said my teaching approach built up their professional confidence and encouraged me to keep it up.

I required a field project in the Ames Limestone as part of that course. Students were assigned areas to sample and examine and I joined them at least once with the rest of the class visiting each area and helping the students draw comparisons to where they were working. All data was pooled and when the semester ended, they were to complete an integrated report. We later published a small version of it for the Pennsylvania Association of Geologists Fall field trip.

During mid-February, I met Vint Gwinn again and mentioned that McMaster University was looking for a sedimentologist. I recommended he apply. Because he earned his PhD at Princeton, he contacted his advisers and they told him they had already nominated one of their final-year PhD's. Princeton traditionally nominated only one candidate for faculty appointments and told the university doing the hiring that the person they nominated was their best graduate. Other Princeton graduates were told to stand aside.

Prior to the early 1970's with passage of the affirmative action laws, academic jobs were not advertised and hiring was done by word of mouth, personal contact, or mailing announcements. That was the traditional way.

I saw Vint again three weeks later and asked if he applied. Vint told me why he couldn't. I told him that I thought the Princeton approach was unfair to the universities doing the hiring because they, not Princeton, could determine their real needs, and the practice was worse than medieval.

I then said, "Vint, I have an idea. Why don't I write to McMaster nominating you and you will be the 'Pittsburgh' candidate. Do you want the job?" Vint assured me he was confident enough in his abilities to win the job if he earned an interview. I wrote a letter to Gerard V. (Gerry) Middletown (PhD, Imperial College, University of London; sedimentology; Imperial Oil, McMaster University) who was chairman of the department. Six weeks later Vint was interviewed, and was offered the job which he accepted.

Vint also told me that Gerry Middleton was chairing the annual SEPM Research Symposium on sedimentary structures in 1964. I filed that information away depending on how well the summer research went.

Because I was teaching only one course, I wrote papers remaining from my thesis, one on Bay of Fundy tidal flats, one on the Quaco Conglomerate of New Brunswick, and one critiquing Sandstone Classification. John Sanders and I agreed to write a joint paper comparing the Bay of Fundy Intertidal zone with the Dutch Wadden Zee described by Van Straaten and which Sanders visited. I completed them and sent them off by the semester's end. All appeared during 1963 and 1964.

When I left Sinclair, I received permission to continue my work on the sandstone petrology of the Stanley-Jackfork Boundary and they let me take their samples and thin sections with me. After classes ended, I returned to Arkansas and Oklahoma to collect more material to complete my regional study. I visited Tulsa and reconnected with friends and returned to Pittsburgh in early June.

While collecting samples in Arkansas and Oklahoma, I drove off the main roads into the interior of the Ouachita Mountains. One day, while working an outcrop, a steady stream of locals drove by in pick-up trucks and stared at me. I noticed a big pot of water boiling while driving to the outcrop and suspected it was a backwoods still. As I finished and turned the car around, I noticed five pick-up trucks behind me riding shot gun. They let me leave. I discovered later that the favorite disguise for a "revenuer" was a geologist and I was in the heart of Ouachita moonshine country.

I then flew to London, connecting to Oxford. Stuart McKerrow arranged for me to 'house sit" for one of their younger faculty who was doing field work in Northern Ireland.

First, I had to arrange transportation. Stuart told me his family needed a second car and suggested we go 50-50 and buy a used car. I would use it during the summer and he used it until I came back. I would use it again the second summer and he would buy me out. I agreed and together we bought a medium-sized Austin four-door sedan.

After visiting outcrops together and Stuart explaining the current stratigraphic framework, I went to the library, reviewed literature to make a selection of outcrops to visit and commenced work. I experienced great difficulty in resolving their stratigraphic terminology. The Great Oolite was subdivided according to a series of three "fossil beds", the first, second, third, and so forth. Recognizing these in the field was not easy. The fossils were a mixtures of shallow-marine and coastal pelecypods, brachiopods, and gastropods. I began field work at a large quarry where all three were observed and noticed that contrary to past paradigms, the fossil beds pinched out. Moreover, the shells were concentrated and concave up, reminiscent of intertidal tidal channel fills reported by Van Straaten from the Wadden Zee and which I observed in the Bay of Fundy.

As field work continued, I also observed that in other outcrops, the shell lags were overlain by a cross-bedded oolite with bipolar orientation, and graded up into a ripple-bedded oolite. They were capped with a limey mudstone. In short, it replicated the fining-upward meandering channel model of Visher, but in a carbonate system, and in modern terms represented a parasequence. It clearly was a tidal channel fill. I checked the library again and reread parts of Arkell's book on 'The Jurassic System of Britain" to see if he and others reported anything like I observed and none had.

During the middle of the summer, I went to a small active quarry worked by one man. I first went to the farm house to get permission to cross their land and open and close gates.

Driving into the quarry, I immediately observed three such stacked vertical sequences in the quarry face. I introduced myself to the quarryman and began working, measuring a section and taking samples. After a half-hour, the quarryman came to me and said (spelled phonetically):

"Ya know, Englund was coovered by the sea three toimes."

I asked him why he thought so and he pointed to the base and top of each of one of my vertical sequences!!

I then asked him why three events. Couldn't it have happened during a single episode of marine drowning?

He said, "Well, ya know, the lite professor Arkell told me that England was coovered by the sea three toimes."

He had a contract to quarry limestone for restoration of the college buildings at Oxford. He explained that to quarry the building stone, he scraped off the limy mudstone at the top, and cut down to just above the fossil bed to get good quality stone.

I made many trips back to the quarry, and brought geologists from all over the UK to show these sequences.

That experience was a good reminder that people lacking formal education can make accurate scientific observations and ask intelligent questions about them. It was a sobering reminder that although I made a sedimentological discovery, there were forebears who had seen these features too. The difference in understanding was timing and training.

The summer ended and on my return to Pitt, I immediately wrote a short paper about the three tidal channel sequences in the Great Oolite, submitted it to *Nature*, and they published it in mid-February, 1963. I also wrote Gerry Middleton inquiring if he wanted a carbonate talk in his 1964 SEPM Research symposium and enclosed a preprint of my *Nature* paper. He invited me to present a paper at the 1984 SEPM symposium.

The fall began with nothing unusual on the horizon. In Late September, the Pennsylvania Association of Geologists held its field trip in Somerset, PA, and I brought my class. John Rodgers wrote earlier that he was attending the trip, so I invited him to stay with me the Saturday the trip ended, and to join me and Bob Hodgen and Bob's wife for dinner. He accepted. Bob Hodgen read my GSA paper on Triassic sedimentation, because it was published the week before, and congratulated me. Rodgers had not seen it yet and complimented me on getting it published so quickly.

I served as department colloquium chairman that year and discovered that Robert S. Dietz of the Naval Electronics Research lab was to be an AAPG Distinguished Lecturer at the October evening meeting of the Pittsburgh Geological Society (PGS) discussing plate tectonics. By this time, I was elected to PGS's council. Because Frederickson just hired Alvin Cohen, an Illinois PhD glass chemist who was conducting tectite research, I called Dietz to see if he could also present a colloquium about meteorites to the department. He told me to contact AAPG to get approval and it was granted.

I met Dietz at 7:00 am at the Pittsburgh airport and checked him into the Pittsburgh Hilton. He hung his clothes, left his brief case in the room, and took his slides with him. He visited different faculty in the department, gave his talk, and we returned to the Pittsburgh Hilton. I dropped him off to park my car, walked back, and waited in the lobby. He didn't come. I picked up the house phone and called. Bob asked me to come to his room. When I arrived he explained he returned to the room, took his clothes off to take a shower, and after drying off, opened the closet to pick a different suit. His clothes and luggage were missing. He was on the hotel's case, but luckily, he kept his slides.

I called the restaurant where PGS had its meeting and asked to speak to the PGS President, Scotty Affleck, to let him know we were running late. Eventually, Dietz gave his talk, and met everyone. I connected with Dietz two weeks later at the GSA meeting in Houston and he told me everything was found. The hotel mistakenly switched keys giving him keys to the room next to the one where his luggage was.

GSA that fall was in Houston and I went on a field trip led by H. A. Bernard (BS, PhD, LSU, sedimentology; Shell Research), Rufus LeBlanc (BS, PhD, LSU, sedimentology; Shell Research) and Charles Major (BS, MS, Illinois, geology; Shell, Pennzoil) to the Holocene sediments of Galveston Island and the Brazos River Valley. It was the same trip used by Shell for its training program. It was the best part of the meeting. I took numerous slides of both sedimentary features, and all their color diagrams. These were used for many years in my courses.

On my return, I noticed the atmosphere changed. First, Mrs. Kinch was fired. Freddy hired a Mrs. Orso to replace her and she was barely up to the job. One day, Joe Lipson chewed me out about a trivial matter I can't recall. There was tension between me and Buckwalter over the way I taught mineralogy because he had to revise his petrology course. Frederickson suddenly cooled towards me. A friend in Tulsa called to alert me to some news I did not know. Pan Am Research fired Harry Werner and he was looking for a job. I was warned that at Pan Am Research, Harry was Frederickson's lap dog.

Originally, Fred tried to hire Werner to teach petroleum geology, but the administration declined because of the recent hiring of Alvin Cohen. Because I was untenured, my position became vulnerable.

In December, Martin Bender died. Frederickson asked me to teach his historical geology for one semester on an emergency basis and I agreed. When I heard Martin died, I anticipated such a request so was prepared to help out.

In early February, Frederickson called me in and told me my faculty appointment would not be renewed. The tenure committee of Frederickson, Buckwalter, Lipson and Flint chose to turn me down. This came as a surprise because I was never asked for an up-dated CV, reprints, or other supporting documents. I asked why and he told me people found me difficult, students complained about my teaching, high standards, and my demanding workload in courses.

I visited my colleagues to get feedback and got little help.

Some of the graduate students were shocked by this turn of events. They were satisfied with their course experiences with me. One with whom I became friends over time was Dick Gray (BS, Engineering, Carnegie Tech, MS, Geology, Pittsburgh, various engineering consulting firms, past member of GSA Council). With an engineering background, he keyed in on sedimentary processes quickly and led both my graduate classes. He had an exemplary career with various engineering firms including his own, served on the GSA Council, and was North American President of an international engineering geology association. From my perspective, he could have made it anywhere, something I found true of many students every place I taught as a permanent and visiting faculty member, or when teaching industry short courses.

I contacted people around the country to get job leads so I could apply for other positions. I contacted John Sanders, and Carl Dunbar who was spending a year at KU as a visiting professor. In his letter back, Dunbar wrote "once emeritus, stay emeritus." I also contacted Moore, Foley and Hambleton. Soon, I had a list of leads, but they did not interest me. I applied anyway.

A week later, Frederickson wrote a memo ordering me to cancel my summer plans because I was to teach summer school. I went to David Halliday, Dean the College of Science to object. Halliday explained course staffing was the responsibility of the department head and he could not intervene. He asked me to keep him informed as developments occurred.

I was also advised by friends to move key papers, rock collections, and personal items from my office to my apartment, which I did nights and weekends.

During late-February, a box was delivered to the entrance door of Freddie's office but left in the Hall. One night I read the visible packing slip and it was a tumbler for polishing jewelry. The packing slip showed that the tumbler was charged to Frederickson's PRF (Petroleum Research Fund) research grant. Two Sundays later, while packing things in my office, I heard a noise in the hall from Freddie's office. As I left the building and passed his office, I noticed the box was missing. Frederickson was loading it into the trunk of his car while I entered the parking garage. I said, "Good Morning Dr Frederickson." He jerked up, hit his head on the trunk lid and turned around red in the face, but said nothing. Clearly, this item was bought for personal use and I let Dean Halliday know.

I made arrangements to attend AAPG in Houston. I interviewed for a position at San Diego State University, but it was a poor match.

On my return, Frederickson sent a memo about my trip to AAPG. I decided there was a simple solution. I wrote a letter to the Dean with a copy to Frederickson resigning my appointment a week after the semester ended, and attached a copy of Frederickson's latest memorandum. I did so even though I did not yet have a job offer. The Dean wrote back accepting the resignation.

Frederickson, however, was opposed and the dean reminded him one doesn't ask someone who was just denied tenure to give up their summer research plans, and because a termination notice was given to me, the resignation was accepted automatically. Halliday told me later he also informed Frederickson that he would have to investigate one of his equipment purchases because there was a complaint. Halliday disclosed to me earlier that Frederickson had become a problem because he became close to Litchfield and was going past Deans and Vice Chancellors to get things outside normal protocol.

When I returned from AAPG, I received a phone call from Howard A. Meyerhoff (BS, Illinois, PhD, Columbia, geomorphology; Smith College, U.S Manpower Commission, Pennsylvania) about my application for a faculty appointment at the University of Pennsylvania. Howard was appointed chairman in January, 1963, cleared the decks, and fired all but one person. He invited me to interview. I flew to Philadelphia early on a Thursday in mid-April.

I arrived and we talked for two hours. Meyerhoff taught for many years at Smith College, mapped Puerto Rico for the U.S.G.S. and during World War II, moved to Washington to head the US Manpower Commission where he remained until December, 1962. He was recruited by Penn's president, Gaylord Harnwell, to return to academe that winter. He explained that because he had no faculty for me to meet, I would meet with a special search committee consisting of the Provost, David R. Goddard (PhD, UCB; botanist; Univ. of Rochester, Penn, Member of National Academy of Science), Ray Nichols (Dean of Graduate School, Pulitzer Prize Historian), and Otto Springer (Dean, College of Arts and Sciences, PhD. Gottingen; Germanic literature).

We walked to College Hall where the offices of the College of Arts and Sciences were housed and went to the Dean's conference room. Springer was already there and we were soon joined by Nichols. Goddard arrived five minutes later.

The provost did the interviewing and after explaining he knew a number of paleobotanists and learned about geology from them, he asked me to explain sedimentology. I talked about four components of sedimentology: clastics and carbonates, and ancient and Recent sedimentology. He understood.

He then pulled out my publication list and read each title. Goddard asked me to identify where I did the research, what was thesis related, what work was done after leaving Yale, and when I had done the work. I gave him answers (the first publication was work completed after leaving Yale). Clearly, he wanted to know whether I was coasting, or whether after leaving Yale I continued new research every place I worked and was contributing new science to my field.

After discussing other things, the Provost said, "George, if we hire you, we may not be able to hire a second sedimentologist for many years. What kind of background should we look for when making our selection?"

I explained that they should consider a person who either completed research and published about BOTH ancient and modern clastics, as well as EITHER ancient or modern carbonates, or someone who had worked on BOTH ancient and modern carbonates, and EITHER ancient or modern clastics.

Goddard picked up my publication list again and asked me to identify which of the four topics the papers were about. The first five he named were either ancient or modern clastics. Then he read the title of my *Nature* paper and I said "Ancient Carbonates." He stopped, looked at his watch, mumbled about having to leave for another appointment, said good bye to me, and asked Meyerhoff to come with him. I stayed with the two deans.

Howard returned shortly afterwards, and after finishing the interview with the deans, Howard and I left. Once out the door Howard told me, "George the Provost wants me to hire you. His comment was 'Klein is the first Yale PhD I met in my entire career (about 25 years) who has published several papers besides his thesis within the first three years of completing his degree. Make sure he leaves town accepting our offer'."

I knew accepting an offer at Penn had its risks. Space was at a premium and being renovated. I had not met Lehman, the mineralogist who was staying, and had no idea who else was coming. There were many loose ends. I accepted the offer strictly because of the strength of reputation of the university. I believed that with Penn's overall reputation behind me, I should be able to advance my career in geology and either make the place into something better, geology-wise, or find another place in time that would provide a better opportunity. I was only 30 years old, had a long career ahead of me, and was confident that the future would turn out positively for me. The nationwide trend in Higher Education was one of expansion and consequently, there would be other opportunities. All I needed to do was work hard, keep doing research, keep publishing, and do the things I was good at.

I returned to Pitt, closed out my teaching obligations, collected my last pay check and left.

LESSONS LEARNED:

1) When travelling, be sure to keep one's lecture slides with you at all times.

2) In academe, one can be a hero and quickly turn into a goat and then again into a hero.

3) Never assume, particularly as an assistant professor, that everything is in order. Events can shift and an assistant professor is vulnerable to termination.

4) When a new head is charged to upgrade a department, it is a mistake to hold meetings and make decisions only with the people s/he hires and not the entire faculty.

5) Always take a sabbatical leave, and even go into debt to take one. It will pay off very quickly.

6) If wishing to undertake a career in a Tier I Research university, always continue to publish to remain a viable member of the scientific community.

7) When it is time to retire from academe, remember Carl Dunbar's comment "Once emeritus, stay emeritus."

POSTSCRIPT 1. Why did Pan Am Research fire Frederickson? During the GSA meeting in New York in 1963, I met John Hower. Hower earned his PhD at Washington University, St Louis, in 1955, when Frederickson accepted the job of Vice President, Exploration Research at Pan Am Research. Freddie offered Hower a job there. Frederickson was fired from a tenured appointment at Washington University, St. Louis.

In February, 1960, Hower accepted a faculty appointment at the University of Montana. He notified Frederickson who told him to continue working at Pan Am until the late summer. However, during the evenings, Hower wrote a proposal for the University of Montana to submit to the National Science Foundation (NSF) with himself as PI (Principal Investigator). The proposal was submitted in mid-March.

One of the NSF reviewers forwarded a copy of the proposal to Frederickson suggesting there may be a conflict of interest. Hower started a project like he proposed, but Pan American terminated it after two years.

Frederickson told Hower he was fired, and furthermore, if NSF funded that research, he would instruct the corporate lawyer to sue the University of Montana. Frederickson contacted corporate legal counsel in Chicago concerning this issue in the middle of April, 1960.

Pan American Petroleum was both a major producer of oil and natural gas in Montana and a major tax payer and benefactor in the state. The corporate lawyer read Frederickson's letter and contacted the corporate president stating that Frederickson was placing Pan American Petroleum at risk.

Frederickson was fired the next day.

POSTSCRIPT 2. The Rise of Mrs. Kinch and the fall of A. F. Frederickson. When Frederickson fired Mrs. Kinch in 1962, she went to the university personnel office to explain her termination and filed a grievance. They contacted Norm Flint (who told me this at the 1965 GSA meeting). Norm discretely supported her. By this time, Mrs. Orso was hired, and the personnel office told Mrs. Kinch that because she was a loyal and reliable employee for 18 years, had accumulated vacation time, and was considered an asset to the university, she should take a vacation, remain on payroll and wait till another vacancy opened. A month later, she was offered and accepted a position in the office of Vice-Chancellor for Academic Affairs Peake.

During the summer of 1964, Litchfield was fired from the presidency of Pitt for bankrupting the university and the state of Pennsylvania took it over. Dean Halliday, Dean Jones (Graduate Dean), and Peake met because Frederickson by-passed them getting things from Litchfield and thus undercut the three of them. The geology faculty members also were upset with Frederickson's management, Lipson moved to the School of Education, and Alvin Cohen was ready to quit. An investigative audit turned up some financial irregularities.

On the day Peake was to meet with Frederickson to let him know that he was fired, Freddie sat in the reception area. Suddenly, the door to the office wing opened and a voice familiar to Frederickson said, "Dr. Frederickson, Vice Chancellor Peake will see you now." That familiar voice was Mrs. Kinch.

CHAPTER 11
University of Pennsylvania: The Meyerhoff Years (1963-1966)

After leaving Pittsburgh, I briefly visited my parents and went to Europe. My first stop was The Netherlands to attend the International Association of Sedimentologists' (IAS) meeting. IAS accepted and published one of my papers on the Bay of Fundy in their symposium volume, but only selected contributors presented papers and I was not on the speaker list. Before the meeting started, I went on a two-day field trip along creeks and canal exposures in the reclaimed Polder area and also a tunnel excavation in Amsterdam showing the Quaternary succession. I met many people from all over the world and met Van Straaten again. I arranged to visit Groningen to see the Wadden Zee tidal flats after the post-meeting field trips were complete.

One individual who I met but can't recall his name, was the deputy ministry of petroleum from the Ukraine. He was a typical Soviet scientific bureaucrat. Secretary of Defense McNamara had negotiated a scientific exchange program with the USSR that spring, and I inquired about a possible joint research program. We came up with some possible projects.

When we were done, I asked "Look, architectural history is a hobby of mine. If we do this program, is there any chance I can make a side trip to Samarkand and Tamarlane to look at the classic 13th and 14th century Moslem architecture there?"

The reaction was unexpected and it took nearly 20 years to get an explanation (See Chapter 21, Postscript #2). His faced turned every color of the rainbow. He replied (phonetics to replicate accent) "Vell George, it vould be deefficult but not eempossible."

Needless to say, the exchange program never happened. But that conversation reappeared in an unusual way 15 years later (See Chapter 21).

After the Dutch part of the meeting, we took a field trip through the Delta Works in the Rhine and Schelde Estuary looking at Holocene estuarine sediments, and ended in Brussels for the Belgian part of the meeting. I participated on a field trip examining the Psammites du Condroz displaying pseudonodules and the ilots (erosional remnants) in ancient tidal sediments. The trip ended in Liege. From there, I participated on a field trip examining Devonian reefs led by Marius LeCompte. Ray Moore discussed these reefs in Geological Development of the World in 1957 which piqued my interest.

I recall a couple of interesting incidents. We were to spend the night at Bastogne after a bus trip across Belgium. First, however, we were taken to the Bastogne War Memorial with loud music and combat sound effects. Second, the following day, we passed a U.S. Military cemetery and the bus came to a stop. I was the only American on the bus and the field trip leader came back to me, snapped to attention, saluted, and thanked my countryman for their supreme sacrifice. These people had not forgotten.

One of the participants on LeCompte's trip was Eugen Seibold of the marine institute at Kiel University. We sat together at breakfast and I asked questions about German universities. He complained people could switch majors at will. He then said (phoneticized), "It vas better during zee nazi regime. Ve still had standards zen." There was shocked silence and from that point on until the end of the trip, no one talked, walked, or had anything to do with Seibold.

I spent the summer of 1963 in England completing my field work on the Great Oolite. At the end, I ran a field trip for interested British geologists and all were surprised that the Great Oolite displayed so much sedimentological diversity.

I took a vacation trip to Norway, first going to Oslo and looked at museums, and then on a Sunday, took the 'Oslo-Bergen Express' which provided a scenic view of the Caledonide Mountains of Norway. It was a clear day and spectacular. At Vik, the highest train station in Europe, I briefly left the train to take pictures and a German couple who were constantly trying to fake an English accent to hide their Germanic heritage, followed me. I got back on the train and the rail car was filled with seaman heading for the Monday morning boat shifts in Bergen.

I was travelling with another person who held my seat, but the German couple lost theirs. One of the Norwegians explained that it was open seating from Vik to Bergen, about a three hour train ride, and the German couple had to stand the rest of the trip. First, they tried to reclaim their seats and got stone-walled. Then they asked me to help. I told him I didn't know what happened and couldn't help them. The Norwegians were chuckling, winking at me, and laughing all the way to Bergen.

After a day of sightseeing in Bergen, I rented a cabin on the main ferry boat that went up and down the Sognefjord, planning to spend the night on board and return the next day. I boarded early to claim my cabin. It was a pristine clear day and the scenery was outstanding.

On deck there was an amalgamation of tourist groups from the UK, Israel and France. The Israelis were the most unpleasant. Two British ladies kept throwing food at passing seagulls which dived in my direction.

Lunch was serve buffet style with many Norwegian delicacies. The Israelis adjourned to a private dining room and when they returned, were subdued and better behaved. I surmised that some of their senior cohorts explained how the Norwegians had helped many of their compatriots survive World War II.

Late in the afternoon, the buses off-loaded and I had the ship to myself. I ate dinner with the officers. Around 10:00 PM, we arrived at a small city which had an aluminum smelting plant and I slept. I arranged with the Captain to be woken up when the ship was ready to cast off, and enjoyed a good breakfast. I was the only passenger on board. Four hours later, we picked up three tour buses and things got crowded. Again, the weather was perfect.

From Bergen I flew to Stavanger, and after a day there, flew to Copenhagen where I met Kai and Dora Van Der Mandele who was now the Netherlands Ambassador to Denmark. I went to the Embassy Chancery where he welcomed me with his entire family; two children appeared some time during the past 25 years. Their son and I partied around Copenhagen that evening and I flew to the US the next day to start my career at the University of Pennsylvania.

The University of Pennsylvania was founded in 1740 by Benjamin Franklin. Although the fourth oldest university in the USA, it was the first to be established as a university. Harvard, Yale, Columbia and Princeton were originally founded as colleges.

The Department of Geology was founded in 1833 by Henry Darwin Rogers, a pioneering American Geologist. It was for some time a distinguished and reputable department especially in paleontology. It awarded PhD's; the last one awarded before my arrival was earned by Horace Richards (PhD, Pennsylvania; coastal plain stratigraphy; Curator of Paleontology, Philadelphia Academy of Sciences) in 1933. Horace was a part-time faculty member affiliated with the Philadelphia Academic of Sciences. The department went into decline during the 1920's, and Paul Storm, another Penn PhD, became department chairman. Paul never was promoted to a full professorship, a bad sign for the longevity of the department.

Over the years, Paul hired a series of short-term (three- to-six year) assistant professors who moved on. Storm also left the department after lunch every day and returned around 4:30 PM before going home. Howard Meyerhoff explained that one day, two assistant professors followed Storm to see where he went. Apparently he spent every afternoon at the local burlesque house.

An attempt was made to revitalize the department in 1961. John Moss (BS, Franklin and Marshall (F&M), PhD Princeton) was chairman of the F&M Geology Department and was offered the job. He interviewed people and for reasons no one knew, declined. I surmised that Penn's administration likely weren't prepared to commit the sums of money to do it and later events (Chapters 12 and 13) confirmed their parsimony.

Penn's University President, Gaylord Harnwell, a physicist, was active in Washington scientific circles, knew Howard Meyerhoff and persuaded him to come in January, 1963. On arrival, Howard terminated two assistant professors, but kept Lehman, who arrived the previous fall, but still had not completed his PhD thesis.

I arrived at Penn two days after returning to the USA and rented a studio apartment in a high-rise. Once moved in, I went to the geology department in Hayden Hall. My faculty appointment was a three-year assistant professorship, but it was to be a second-stage three year contract and either it was to be followed by a promotion to a tenured Associate professor rank, or I was out in 1966.

The first phase of renovation was completed and I was given a basement office next to the stairwell. I unloaded my books and files, and found an empty lab which was to be converted next semester to a combined sedimentology and wet chemical lab and stored my rocks there. Meyerhoff then introduced me to someone he hired over the summer, Peter Fenner (BS CCNY; PhD Illinois, clay mineralogy and geochemistry; Penn, AGI, Governor's State University, Health Care consulting). Fenner was put in charge of Freshman geology and to teach geochemistry. I met with Peter and he loaned me his thesis. I was troubled because it was underwhelming and I could not understand why he did a clay mineralogy thesis with Arthur Hagner, an economic geologist, rather than with Ralph Grim, the pioneering, eminent founder of clay mineralogy. Fenner, I discovered in time, had good political skills and ingratiated himself with Meyerhoff.

We were basically a department of four faculty members, Meyerhoff, Fenner, Lehman (ABT Harvard) and me, offering the basic undergraduate curriculum. Horace Richards taught part-time. I was assigned a year's Physical and Historical Geology course, including one laboratory section, and asked to teach Geomorphology during the fall. I used Flint's course notes to teach it. I also chaired the colloquium and library committees. We had about a dozen Junior and Senior majors.

The goal was to rebuild the department and develop it into a nationally ranked program. A month after arriving, Howard asked me to become Graduate Advisor, serve as the departmental representative on the science quadrant of the graduate school with other department chairs in the sciences, and prepare a statement about a Masters program. The Master's program was approved three months later.

The department was housed in the basement and one first floor office in Hayden Hall, a late 19th century brownstone period piece. The remainder of the building was used to house the school of architecture. That school had new construction slated and finally moved out in December 1966. Geology then had the building to itself and slowly it was to be converted into a modern geology building.

The annual GSA meeting in 1963 was in New York. I gave a paper on the Stanley-Jackfork boundary. My results showed that sand was dispersed laterally into the basin, whereas paleocurrents mapped by Louis Cline's Wisconsin students indicated dispersal and transport parallel to the basin axis. This contradiction became known amongst sedimentologists as "Pettijohn's paradox." Other instances of it were reported as well.

I finished my paper on the Stanley-Jackfork and sent it to the *AAPG Bulletin*. After revision, it was published in 1966. I proposed a solution to Pettijohn's paradox, namely that the cross-slope paleocurrents formed by reworking contourite bottom currents. I also completed my lab work on the Great Oolite and began preparing a preliminary draft of a paper for the 1964 SEPM Research Symposium in Toronto, Canada.

The semester continued and there were no major changes. Additional renovations were approved to be done during the summer and Fenner was delegated the responsibility of seeing them through.

I recall running an afternoon field trip for my beginning geology class and told them to wear their oldest clothes. Two co-eds arrived in sweaters, skirts, nylons and heels as if they were off to a Philadelphia Main Line afternoon tea. When re-boarding the bus after last stop, I noticed their nylons were in tatters, and there were brambles and burs on their skirts. I reminded them I had said "old clothes." During the spring field trip, they wore them.

Howard also announced at a faculty meeting that the administration approved the appointment of a visiting Brazilian economic geologist, Elisario Tavora, for the following fall and his line was to be converted to a new faculty appointee for 1965.

I received a letter early during the semester from a Yale sedimentology graduate student informing me that John Sanders was denied tenure. The letter asked me to petition the president of Yale, Kingman Brewster, to reverse the ruling. The letter explained Sanders had the full backing of the entire geology faculty and the college dean. A higher-level committee turned him down.

I knew I had to be careful because one doesn't meddle in the affairs of other departments on campus or in other universities. I tried to phone Richard Flint to find out if all appeal procedures were exhausted. Flint was out of town and his secretary told me where I might send a letter. I wrote and asked for information. He replied later that all appeal efforts were exhausted, but if I wanted to write Brewster, I could. He asked me to write in such a way so that it did not appear as if I was "electioneering."

I wrote Brewster stressing that the growth of any university was dependent on a steady cadre of younger faculty who were contributing to their fields. Brewster replied and agreed with my assessment but also said something else. Because a tenure decision is a lifetime **financial** commitment by the university approximating $1 million (in 1964 dollars; approximately $3 to $4 Million in 2009 dollars), other factors had to be considered too. It was a sobering, but accurate assessment

The second semester was uneventful. I recall two incidents. First, Lehmann needed photomicrograph's for his PhD thesis and I offered to take them with a Polaroid slide system in my office that could be attached to a microscope. To do everything, Lehmann took twice as long as anyone else.

He was a stickler for perfection and I wondered if he could ever finish his thesis on time, much less make it. He eventually finished that summer but never published. I said nothing but felt his appointment was headed for trouble.

The other incident involved a sophomore student in my physical and historical geology course, Mark Cohen. Mark was an amiable student and on both field trips, he brought a high-end Nikon camera and took excellent photos of his classmates. However, his grades steadily dropped and by the end of the second term, he was flunking. I kept track of student class attendance by requiring everyone to sign an attendance sheet at each class meeting and noticed he missed many classes. I did not know that Mark was on probation and if he failed a class, he was out.

At the end of the semester, I received a phone call from the Dean's office. The associate dean explained Mark's dad, a Boston lawyer, was flying in next day to find out why his son flunked geology. I offered to meet with Mark's father and was told if I explained it to them, I didn't need to meet him. I said I'd talk with him.

Next morning, I prepared for the meeting and the associate dean called and said that Mark's father was in his office and they could explain everything. My response, with bravado, was, "Send him over."

Mark's father arrived and asked why Mark had failed Geology. I remember the dialog went like this:

Mr. Cohen: I would be grateful if you could explain why Mark flunked geology this year.

Klein: You know, I wondered about this too. Mark seemed reasonably bright. Let's see how the year went.

I then showed Mark's dad a set of take home preparation questions for the first exam, and let him read the text of the first exam. I explained 1/2 of the questions on the first exam came VERBATIM from the take home preparation questions. For the second exam, it was about 1/4 and the last exam, zero percent but it should have been easy to pass the exam with proper preparation of the take home questions. That practice continued during the second term.

Mr. Cohen: I must say, Dr. Klein, you've been very fair, providing guidance to give everyone a chance to pass. I don't understand why Mark didn't do better.

Klein: I don't understand it either. Do you suppose Mark wasn't coming to class?

I laid out the attendance sheet fanning them out like a Los Vegas card dealer.

Mr. Cohen (completely changing his demeanor): You know you've been very fair. Mark is such a baby. His mother died four years ago and I remarried and he never adjusted. Thank you very much for your time."

And he left.

The Dean's office called and explained Mark's dad was satisfied and then asked, "What did you tell him?" I replied, "The truth. Mark stopped showing up regularly for class two months ago."

I thought that was the end of this incident. It wasn't. In 1975, Dean Robert Rogers of the College of Liberal Arts and Sciences at the University of Illinois at Urbana-Champaign (See Chapter 18) nominated me to be invited to apply for a Guggenheim Fellowship. I applied and mine was declined (no surprise; it's hard for a scientist to get one). In their rejection letter, they enclosed a list of recipients and as I read the list I found a familiar name: *Mark Cohen, photographer, New York City.* Apparently, Mark found his true calling and was a successful photographer. The field trip pictures were an indication of his true avocation.

I often tell this story because it is a good reminder that the college experience is not for everyone, faculty can often get it wrong about the students they teach and evaluate, and parents often have inflated ideas of what their children should become and which 'name' college they should attend. Moreover, when people have a clear idea of what they truly want to do, they should be given help initially to try and achieve such a goal even if it does not fulfill one's own wishes for one's children. It worked in Mark's case, but I am fully aware of cases where it didn't.

I taught a spring course in sedimentology for seniors. We went on an overnight field trip to Western Pennsylvania where I knew many outcrops and added some in central Pennsylvania.

We also received applications for our Master's program. As graduate advisor, I received, read and filed all of them. The most disturbing thing was the uninformative letters of recommendation. I complained to Howard and he said, "Why don't you prepare an outline of what you would have liked to see in such letters and show it to me in a couple of days?"

Two days later, I completed such an outline and gave it to him. Howard said, "Now George, whenever you write a letter of recommendation for one of our students, whether it is for graduate school, or a job, I expect you to follow that outline." Howard liked to smoke a pipe from which smoke signals were always rising to the ceiling, and they did while we talked.

I followed his advice and can honestly say it helped not only place Penn students into good graduate programs and jobs but also, later, my students at Illinois. In doing so, I realized as a professor, the only tangible thing I really offered my students was the credibility of my recommendation letters. I used that outline regularly and also to decide whether or not I would even write a recommendation.

We admitted only one graduate student for the fall term, Neil Petersen. Neil graduated from Hofstra with a double degree in geology and chemistry. He ended up working with Fenner, but took both my graduate courses. Other applicants we accepted chose to go elsewhere.

I also received a card from Gerry Friedman announcing that he was leaving Pan Am Research during the summer to go to RPI as Professor of Geology.

The semester ended and I flew to Toronto for the Annual AAPG-SEPM meeting to present my Great Oolite paper in Gerry Middleton's SEPM Research Symposium. The most convenient route was to fly to Cleveland, OH, and change planes for a flight to Toronto with customs clearance at an intermediate stop in Guelph, Ontario. As I boarded the flight to Guelph, I noticed two younger men looking at me and smiling and had no idea who they were. Half the passengers in the waiting area were geologists I recognized or knew.

Preferring to sit in the aisle seat behind the bulkhead seat on an airplane, I bounded off the plane at Guelph to get through customs immediately. The customs agent went through my belongings, asking why I was coming to Canada. The minute he heard "AAPG-Meeting in Toronto," he asked, "Sir, did you bring any slides with you?" I explained I did and they were for my technical talk.

He demanded to examine them. I showed them one at a time (to keep them in order). After the third slide he said that I was cleared and to go on. I asked, "Why did you need to see my slides?" He explained, "Sir, we are required to check for pornographic slides which are not allowed into the country." I replied in a booming voice for all the geologists to hear, "Well sir, there is NOTHING pornographic about a limestone." All the geologists roared with laughter, and the red-faced agent waved me on.

While in the waiting room to reboard the plane, most of the geologist came by and said that my answer probably saved them time going through customs. The younger people who smiled at me in Cleveland introduced themselves, "Jim Coleman at LSU," "Woody Gagliano, LSU". They were giving a paper in the same symposium and we developed a long-term friendship. They heard me give papers at the previous two GSA meetings.

On my return, I received a letter from the Geological Society of America. That winter, I applied for promotion to "Fellowship" and in those days, candidate names were circulated to current fellows. A.F. Frederickson wrote recommending that I be turned down alleging a variety of derelictions.

I showed the letter to Howard who immediately offered to write a rebuttal. I asked if I should contact Dean Halliday and Dean Jones (Graduate School) at Pitt and get their help and he approved. I contacted both. They both wanted a copy of Frederickson's letter (probably for their own dossiers on him), and wrote GSA that Frederickson's letter did not represent the views of the University of Pittsburgh.

GSA wrote a month later saying that the letters from Meyerhoff, Halliday and Jones satisfied them and approved the elevation at their next annual meeting. A GSA Council member told me afterwards that Frederickson hurt his credibility because by that time, I published two major papers in the *GSA Bulletin* and both were well-received.

That summer, I completed field research on directional properties of beach and barrier island sands on the New Jersey shore. The project started as a reaction to a book by Paul Potter and Francis J. Pettijohn on "Paleocurrents and basin analysis." That book stressed that by mapping paleocurrents, one identified paleoslope, paleoshorelines, sediment dispersal patterns, and sediment source areas. Their emphasis was on fluvial sediments and turbidites. The concept was, in fact, originally derived from outcrop work in the Pennsylvanian of the Illinois basin (fluvial), and validated by several of Pettijohn's PhD students in the Appalachian basin (mostly fluvial and turbidites).

However, the concept lacked validation from marine and coastal geology. I discovered on the New Jersey shore, directional features were controlled by up-slope swash, down-slope backswash, and shore-parallel longshore current systems. I compiled information from the literature and wrote a paper, which was published in 1967 in the *AAPG Bulletin*.

One day that summer, on returning to Philadelphia for a brief period, I met Arthur A. Meyerhoff, Howard's son. He earned a BS from Yale and a PhD (both in geology) from Stanford and was working for the California Company (now Chevron) in oil exploration. We established immediate rapport. Several months later, he became Editor and Science Director of the AAPG.

That fall, I taught a year of freshman (physical; historical) geology, and at the last minute, Howard asked me to teach structural geology. I told him I would on one condition. Instead of a traditional lab, I would divide the class into two-man teams and each of the three teams geologically mapped 1/9 of a 7.5 minute quadrangle within driving distance from campus. I provided field supervision every third time the student teams were in the field, and spent an afternoon going over what they mapped while watching and instructing as they continued mapping. Howard agreed. By Thanksgiving, they were done and had until the end of the semester to write a report. For lectures, I extracted materials from John Rodger's course notes.

Elisario Tavora arrived and I was forced to share my office with him. It was a poor arrangement. I was busy with professional society committees and was on the phone a lot. Space was tight and it made forward progress on my research difficult. Private meetings with students were virtually impossible and students complained. I found out later that one evening, I left the room focused on my own thoughts and turned out the light not realizing he was still there. He complained to Howard.

Howard basically said that I should have been more deferential to a senior geology faculty member. I explained that perhaps such a senior colleague deserved a private office. Howard finally arranged one, and we were both happier with the arrangement.

Students also complained about Horace Richard's stratigraphy course. We finally agreed as a faculty that I should offer a course on "Physical Stratigraphy" and it be required, but Horace could still teach his regular course as an elective.

Early in September, Howard called me in. He just ended a phone call with Art Socolow (BS, CCNY, PhD, Columbia, field geology; State Geologist of Pennsylvania). Art was part of an organizing committee to form the Northeastern Section of the GSA and invited Howard to join the effort. Howard agreed. However, Howard was not attending the annual GSA meeting in October and suggested I take his place at an organizing committee meeting. Art knew me and agreed.

Howard and I talked about when and where the first meeting should be. The target date was 1966. Howard suggested we should offer to host one if no one else did, but not to push for it.

That fall, GSA was in Miami and I gave a paper. Before the meeting, I went on a field trip led by Ed Purdy (BS, Rutgers, PhD. Columbia; Carbonate sedimentology; Rice University, Exxon; consultant) and John Imbrie (PhD, Yale, paleontology; Kansas, Columbia, Brown). The trip lasted two days and was based on a catamaran schooner. We reached Joulter's Cay, stopped the boat, offloaded into skiffs, and went onshore to collect samples and snorkeled to make observations. We showered, changed clothes and had dinner after our return. During the meal, Imbrie appeared and said, "Would the gentlemen on board please get up and move to the bow."

We did. We were told to jump in unison. The captain told Imbrie nothing had changed and we were still stuck on an oolitic shoal. We finally were afloat at midnight, six hours behind schedule.

The next day, the wind died down. The schooner supposedly could make eight knots per hour, but half was by motor and half by sail. We finally reached our next stop at 5:00 PM and drew lots to determine who would go on shore to examine and collect Holocene dolomite. Four people, including Vint Gwinn, went and they promised to bring us samples, which they did. We then steamed back to Nassau to board planes to Miami. In short, the trip was an utter bust.

At the meeting, word of our misfortune spread. One of the people at the University of Miami told me that while Imbrie and Purdy did their field work years earlier, they sank a Grumman Goose float plane. They forgot to latch the entrance door near the rear of the plane. During takeoff, water poured into the rear door.

On my return, I shared the story with Howard. He looked at me poker-faced, smoking his pipe, while smoke signals rose steadily. When I mentioned the sinking of the airplane, he asked, "Are you through?" I replied, "Yes." Howard responded, "George, if I ever hear any similar logistical fiasco on any of your field trips, I'll know you failed to carry out the proper advance preparation to assure everyone's safety. If that happens, there will be consequences." He then smiled and added, "But I admit, it was a good story. Just don't let it happen to you."

It was good advice and during the next 29 years when I ran student field trips, only two things went amiss. One involved a student who had a bee sting (he was allergic to bees), and a second one had an upset stomach and we gave him Pepto-Bismol.

During that GSA meeting, I met with Art Socolow and his organizing group. By this time, a petition to form a Northeastern Section was circulated. I signed and also signed for Howard. We were told that it would come up at the Council meeting for approval.

Then we discussed where to hold the first meeting. A few suggestions were made but they weren't solid. I then explained that Howard authorized me to give them a conditional invitation provided a better one was not available. The offer was immediately accepted and Art said this would clinch the deal with GSA Council. When I returned to Philadelphia and gave Howard the news, we agreed he should be the general chairman of the meeting and that I should be the Technical Program co-chairman, and form a committee.

In December, all the assistant professors (Fenner, Lehmann, me) were asked to submit revised CV's because we all were in the middle of our three year contracts. A decision for continuation, termination, or promotion had to be made by February. Howard said that he wanted to promote me to a tenured associate professorship and therefore, he needed a list of prominent sedimentologists in the USA and internationally to obtain independent reference reports on my worthiness for promotion. I provided that too.

In early March, the decisions were made. Fenner and Lehmann received new three-year contracts. I was told the Dean said I qualified for promotion except I had not taught long enough. I was offered a two-year contract with a review to come early my second year. I asked if I could be paid at the highest level an assistant professor or what a first year associate professor earned and Meyerhoff said he would do what he could.

Howard, at the beginning of the year, asked me to prepare a proposal to start a PhD program. I wrote a variety of major PhD programs and requested their graduate catalogs and internal department requirements. I also wrote Brown University where I interviewed four years before because they recently had completed a similar exercise. All but one department chairman sent what I requested.

The one person who refused to send me information was Harry Hess, the eminent Princeton geologist who was involved with the plate tectonic revolution, marine topography, and ultrabasic rock petrology. I met Hess (Chapter 5) when interviewing to go to graduate school there in 1954 and 1956, again when he gave colloquia at Yale, and at Yale Alumni Geology parties at GSA. He suggested visiting him on a Friday afternoon. Because Penn was also looking for a structural geologist, Howard suggested I find out if they had any candidates.

Hess was very cordial when we met and he said that catalog statements are nothing more than boilerplate. He stressed that to rebuild the Penn department, we needed funds to give people starter grants, funding for fully equipped laboratories, and hire people who were active in research, or likely to be active in research if they were younger, and give them light teaching loads so that they will produce research papers. A strong record of research will help grow the department. He was very candid and said of the all the people at Penn, I was the only one who was actively publishing, and Horace Richards was still publishing but because he worked for two organizations, his impact for the department's benefit was minimal. Howard was not publishing. Hess hadn't heard of Fenner or Lehmann. Hess explained it could be done if Penn followed some form of his plan.

I then brought up our need for a structural geologist. He pulled out a three-ring binder which had a list of the annual graduating PhD's, their field of interest, and their career history. He identified one person, Reginald Shagam (BSc, Witwatersrand, PhD Princeton, structural geology; Venezuela Geological Survey, Univ. of Rochester, Penn, Ben Gurion University of the Negev, Rider College) who worked in Venezuela mapping the Northern Andes and was on a temporary appointment at the University of Rochester.

I asked if there were other candidates, and he said "no." I challenged him politely and then he said that their structure graduates were happy where they were. I asked if he would give me names and contact data so I could make that determination. He declined. His approach confirmed what Vint Gwinn shared with me several years earlier.

Returning to Penn, I met privately with Meyerhoff and reviewed my conversation with Hess. Meyerhoff listened, said nothing but smoked his pipe and as usual, smoke signals were rising. Finally, he took the pipe out of his mouth and said that he had been working to get more resources for space renovation, equipment and staff. He appreciated what Hess said about hiring research-oriented people and that was his goal. But, a department also needed people to offer basic service courses to recruit majors and increase enrollments, particularly as the department moved from a "honeymoon" phase to a formula-funded one based on total enrollments.

Howard instructed me to write up catalog descriptions and forward them to the graduate school for approval. Three months later, the Penn Department, with 5.5 FTE's, had a PhD program on its books. My proposal was approved.

I next brought up Shagam. Howard was keenly interested and wrote Shagam inviting him to apply. Shagam did. We interviewed him. Shagam was that rare individual with whom one established instant and lifelong rapport after talking with him for five minutes. We offered him the job to start in the fall of 1965.

We also were given permission to hire a second person, a petrologist. We hired Patrick Butler (PhD. Harvard; metamorphic petrology; Penn, later career with NASA) and he also arrived in the fall of 1965. We were gaining critical mass and the future looked better as we added and renovated space.

During the spring, we evaluated graduate applications, having sent out a marketing flyer nationwide and admitted six new students. Two were PhD candidates, Allen Ludman (BS Queens College, BY; MS, Indiana, petrology and structural geology, Queens College of CUNY) who later spent a career teaching at Queens College in New York City, and Leons Kovisars (BS, CCNY) who worked with Shagam in Venezuela and with whom I lost contact.

John Moss, Chairman at Franklin and Marshall College (F&M) called to inquire if Penn had room for a student in sedimentology. I encouraged an application to see if he could be waitlisted, and late in the spring, we admitted him. He was John H. Way who later went on to RPI to work with Gerry Friedman, earning a PhD, and spent the rest of his career as a geology professor at Lock Haven State College, PA.

Early in July, 1964, John Sanders accepted a job with Hudson

Laboratories of Columbia University, housed in the physics department. It was a marine physics laboratory contracted by the U.S. Navy and they were actively involved in marine acoustics and Anti-Submarine Warfare (ASW). They were responsible for all the U.S. Navy's east coast marine physics. Sanders was hired to help geologically calibrate their sonar images so as to improve image resolution and interpretation. The goal was to separate different hard marine objects, Soviet submarines, oceanic basalts, large marine mammals and other oceanic surprises.

He called to inquire if I was available during the summer of 1965 to work with their lab to undertake detailed mapping of tidal areas in the Bay of Fundy. I was to select two areas. The plan was to run the sonar prototype instrument over these areas at high tide whereas I would map sedimentary and other features at low tide. They would pay me 2/9 of my annual salary as a consultant and salaries for two Penn undergraduate student assistants. All living expenses, supplies, boat charter, and other related costs were covered.

I visited their lab in Nyack, NY, and advised they map two different areas, Parrsboro Harbor, because it had a combination of low-tide river gravel bars and tidal flats, and the Five Islands area where there were two distinct sand bodies. That way, the sonar gear could test and evaluate different geological features. I also requested permission to complete additional related science, namely mapping bottom current velocities and trying to correlate them to bedform evolution. Hudson Lab's director approved my proposal and funded it.

We left in mid-June and arrived in Parrsboro, NS. I stayed in the local hotel (now a museum) and the assistants found room and board in a local home. I then scouted for a fisherman to charter a boat and went to the dock at Five Islands, NS. The first one I met told me it was lobster season and he couldn't handle his fishing needs during a short season and mine at the same time.

It was low tide and I did not see other boats, but my conversation was overheard by Adrian LeBlanc who owned a bigger boat and was repairing it out-of-sight at the end of the dock. In the world of small, closely knit communities, word of my whereabouts spread quickly and Adrian contacted me. He gave me times he was available. We agreed that during lobster season, he would drop us off at areas of interest so we could deploy buoys, map exposed seabed geology and other features, sample sediment and take box cores. He would pick us up as the tide rose. Some of the time we mapped

Parrsboro Harbor which didn't require a boat because we walked out to our area of interest at medium and low tide.

In late July, Sanders and a team of people arrived from Hudson labs with their sonar system. They came in two trucks marked "U.S. Navy." Adrian LeBlanc now was available full-time. During periods of high tide, we criss-crossed Parrsboro Harbor and the two sand bars at Five Islands recording sonar images. The two people helping Sanders were a marine electronics tech and a mechanical tech. The recorded images were marginal and the electronics tech adjusted the sonar equipment, but did not achieve improvement. They returned to Hudson Labs to rebuild the equipment and to repeat everything next summer. Sanders contacted the director of Hudson Labs who assured funding for the following summer and asked me to give them first refusal rights on my time. I agreed.

After they left, I completed current meter surveys at selected buoy stations. Slowly a pattern emerged that suggested time-velocity asymmetry of tidal currents controlled facies patterns, bedform orientation, grain size distribution, and bar topography. I knew I needed to extend our survey areas to a bar complex at Economy Point east of Five Islands to validate these findings. To complete the project, I would need two more summers.

As I planned my return to Penn, I contacted Sanders about returning Hudson Lab's two station wagons and all the supplies we brought. I was instructed to dispose of supplies and not return anything except the station wagons. I asked if I could take them to Penn to restock the sedimentology lab and he assured me I could. In the process, we overstocked our teaching laboratory supplies for three years.

After settling in and going through my mail, I discovered my salary raise was not what I expected. A week into the new semester, I held a frank discussion with Howard. Because he could not increase my salary then, I requested I be nominated for promotion to associate rank and tenure again because I would have an extra year of teaching. He agreed. I assembled a dossier and gave it to him.

Next, I prepared a NSF proposal for funding my part of the research program in the Bay of Fundy. When I took the signature pages to Howard, he reluctantly signed but also asked if I couldn't get those funds from any other source but the government. He said he did not approve of raising research funds from the U.S. Government. I explained this was my best hope. He signed.

Planning for the GSA regional meeting in February 1966 was underway. I assembled a technical program committee of Fenner, Butler, Lehman, Shagam, Peter Goodwin (PhD Iowa, stratigraphy; Temple University), and Bill Crawford (PhD, UCB, petrology; Bryn Mawr College). I also chose to run a symposium on the history of non-marine sedimentation in northeastern North America and invited speakers. Speakers had to bring manuscripts ready for review.

GSA handled everything. They received the abstracts, helped arrange facilities and programs, and the job was easy. I received all the abstracts, parceled them out by specialty, and the program committee met.

Except for one session which was over-subscribed, all were undersubscribed by one or two papers. We then identified the abstract in each remaining sessions with the most scientific impact and designated them as a session keynote speaker giving each 10 minutes of extra presentation time. I wrote the individual keynoters accordingly and encouraged them to give it their best effort in exchange for our confidence in them.

The meeting was a success. At the end, we held a small party in the department and showed people around. I recall John Rodgers, Gerry Friedman, Art Socolow and Bob Jordan (PhD Bryn Mawr, coastal processes; Delaware State Geologist) attended. It was great for the students to meet these senior people.

My symposium went well and all but two authors gave me manuscripts. One was Peter Buttner, a PhD student at Rochester who requested more time. I gave him one month. The other, to my disappointment, was John Sanders. I gave him a month also. Buttner strung me out till early September at which time I sent a letter saying if he didn't have a manuscript on my desk by September 15, I would publish without him. Sanders sent an incomplete manuscript after two months. I sent it back with suggestions and he returned it later. It passed a review. The collected papers appeared in the fall of 1968 as *GSA Special Paper 106*.

A week after the northeastern GSA meeting, I walked out of my office and saw an unscheduled visitor, Dr. Henry Faul (BS, M.I.T., MS., Michigan State, PhD, M.I.T., nuclear geology; USGS, Southwest Center for Research, Penn). I met him three years before when he visited Alvin Cohen in Pittsburgh. Faul was a Czech immigrant, had worked at the U.S. Geological Survey and was at the Southwest Center for Research in Dallas, TX. When meeting Henry earlier, I found him entertaining, although jocular. He edited a book, *Nuclear Geology,* which was well reviewed. I asked what he was doing in town and he said he was visiting the area and dropped by to take a look at the department. He then excused himself for an appointment.

I immediately went to see Howard who told me that the administration decided this was his last year as chairman and they were interviewing Henry Faul to replace him. We would meet individually with Henry during the next one-and-half days.

When I met with Faul, he already knew I applied for NSF funding, was publishing many papers, and that Goddard told him his geological friends at the National Academy of Science spoke highly of my research. Henry also disclosed he knew I was up for tenure and told the administration that this was a decision they should make before he came. He explained that to delay it with a chairmanship change would not sit well with the geological community at large. When asking about his visions for the future of the department, he implied he had none.

I then asked how he heard the chairmanship might be vacant. Henry disclosed that Goddard headed a team of scientists organized by Secretary of Defense McNamara to negotiate scientific exchange agreements with the Soviet Union, and the Warsaw Pact countries. The US government kept a list of all scientists including their language skills and his name came up because he knew Czech, German, Russian and Polish. McNamara assigned Faul to join the Goddard task force as a translator. Faul told me he and Goddard enjoyed each other's company the previous summer while they barnstormed Eastern Europe and the USSR. Goddard inquired if Henry would be interested in replacing Meyerhoff. Faul agreed to visit.

I was unsure how this would work. Faul had no experience as a faculty member in a university geology department. He worked in a government agency and a forward-looking research institution (which eventually failed and was taken over by the state of Texas to become the University of Texas at Dallas). In Texas, he was a section head supervising five or six scientists and technicians. His specialty was isotope geology, mostly K/Ar age determination. I knew to get him to join the department at Penn would cost money to establish his isotope geology lab (In 1966 dollars it was close to $150K).

While we talked, we seemed on cordial terms and I assured him I would help to the best of my abilities if he was offered the job and accepted. I also assured him, when he asked, that I had no interest in the chairmanship. I was too young and too much into my research and needed a lot more seasoning before choosing an administrative assignment. I sensed, however, he was not totally comfortable with me.

A month later, NSF notified me that my grant application was approved. A week later, Dean Springer wrote saying my promotion to tenure and Associate Professor was approved, the first geologist to achieve such promotion since 1921.

Henry Faul accepted the chairmanship two weeks later.

That summer, I returned to Nova Scotia. We had a successful summer. The Hudson Lab technicians arrived with John Sanders and tested their upgraded Clay sonar system. They achieved 95% replication of the geology of the seabed at Parrsboro Harbor and the Five Islands sands bars.

I also expanded the project to Economy Point. Ken Aalto, a graduating senior who went on to earn a PhD at Wisconsin with Bob Dott, and John Way were field Assistants. I gave John time to do field work on a master's thesis at the famous Joggins Carboniferous section where both fossil upright tree-trunks and preserved fossil reptiles were discovered inside the trunks, probably a flash flood burial preservation phenomenon.

We returned to Philadelphia, off-loaded our gear and returned the station wagons to Hudson Labs. A new year under different leadership was about to start.

LESSONS LEARNED:

1) When a young scientist enters the International scientific circle, they are often not given the opportunity to present papers because they are unknown. However, they should attend to get recognition for the future.

2) A careful and thoughtful recommendation letter carries far more weight than the research reputation or experience of the letter writer.

3) Student field trips require considerable preparation, not only for instruction content, but also for safety procedures and, if necessary implementation of emergency procedures.

I followed Howard Meyerhoff's advice and as a result, I had only two incidents during 33.5 years of teaching and running student field trips. In one case, a student on a field trip got stung by a bee (he was allergic to them). We got him to a hospital fast and saved his life. The second time, a student got an upset stomach and we gave him Pepto-Bismol.

4) An academic environment is an unstable system. There is constant change of personnel, departmental and administrative leadership, and intellectual ferment.

5) When proposing joint international research with countries that are potential adversaries, or dictatorial regimes, proceed cautiously.

6) Reading every scientific journal paper or book will provide a new research opportunity.

7) Not everyone should go to college. Not everyone should go a so-called "prestigious" or "name" institution.

POSTSCRIPT #1. Parents often ask me what is required for their children to be admitted to prestigious schools like Harvard, Yale, M.I.T, and Stanford. If their children want to enter a field of science or engineering, I tell them to enroll at the state flagship university (cheaper option) and study like hell. Then when they want to go to graduate school, those prestigious name institutions will fully fund their graduate education via fellowships and assistantships.

POSTSCRIPT #2. When those same parents ask about student work opportunities, I advise parents to tell their children to go the main office of the department where they wish to major and ask if hourly work is available in the laboratories, research programs, or other sectors of the department. That way they gain experience and meet faculty and other students at all levels and learn not only about the field, but whether they really like the work.

CHAPTER 12
University of Pennsylvania: The Faul Years (1966-1969)

"It was the best of times; it was the worst of times."
- Charles Dickens

"Those who cannot remember the past are condemned to repeat it."
- George Santayana, *The Life of Reason*, Volume 1, 1905

The above two quotes best characterize the Faul Years during my service at Penn.

Although Faul and I got along at the beginning of his service, I sensed there was something about him I could not place. He appeared shifty and I realized I had to be careful with him, particularly about money and grant management.

Because Faul was new, an office was constructed for him next to the stairwell between the first and second floor and it was visible from my office. At the entrance to his office was a reception area where the department secretary, Mrs. Fannock, worked. I could always see when both were there at the same time.

When the September monthly statement for my grant expenditures arrived, I went through it and decided to see if something would work. I picked an item and went to Mrs. Fannock's office when she and Henry Faul were both in their offices. She kept track of the departmental books, invoices, receipts and bills. When I walked in I said, "Mrs. Fannock, I just got my grants account statement for this month and noticed an item I don't recall ordering. When you have a moment, would you be kind enough to bring me a copy of the invoice at your convenience". I then explained the item and showed her the code number on the spread sheet. Two hours later, she brought me the invoice.

When the October invoice arrived and I saw that both Faul and Mrs.

Fannock were in their offices, I went through the same exercise. The outcome was the same. I knew that the message to Faul was received and the message I wanted to send was: 'Don't mess with my grant money.' I told this to no one. Two years later in a memorable conversation with him I discovered it registered more than expected.

Our faculty meetings were somewhat disorganized. Faul decided that because the School of Architecture was vacating the building in six months, we should apply for NSF funds to renovate the building into a modern geology facility. After several meetings with the university architect, we developed a proposal, submitted it, met with an NSF site-review committee and it was turned down.

Henry had space renovated to build a K/AR isotope lab shortly after his arrival. I was surprised that he brought a mass spectrometer with him from the Southwest Center for Research. Normally, that was not permitted because although the equipment was bought on NSF funds or other grant funds, title belongs to the university or institute to which the grant went. It seemed unusual but I said nothing.

I received a notice from the graduate school about a new government graduate fellowship program for scientists and engineers, the NDEA (National Defense Education Act) fellowship. With Henry's concurrence, we applied and were awarded two. That meant a massive letter writing campaign to get applicants. We awarded one to Leons Kovisars, and the other to a new applicant from Bryn Mawr College, Julia Badal. She wanted to work with me in sedimentology.

That fall's annual GSA was in San Francisco and I presented a paper on Nova Scotia tidal sand bars. I enjoyed the many amenities of the city. From there, I went to a working group of "Friends of the Microscope" of 12 sedimentary petrologists in Boulder Colorado. We brought and exchanged samples and shared thin-sections from our collections. I was informed while waiting for my plane in San Francisco that my father suffered a heart attack but was recovering well. My mother asked me to come after I completed my trip. From Boulder, I went to the University of Nebraska to present a colloquium before returning to Philadelphia.

To travel to Lincoln Nebraska, I flew from Denver, CO, on a twin engine, puddle-jumping plane. I noticed a tall, older man talking to a middle-aged woman in a fur coat in the waiting room. He looked very subdued.

On boarding, the subdued-looking man took the aisle seat in the first row, and I took the aisle seat right behind him. After take-off, he reclined his seat all the way back and hit my knees. I stuck my knee into his seat-back to discourage his behavior. The plane made two intermediate stops and he reclined his seat too far back on each take off and again hit my knees. I put my knee into his back again and he got the message. He turned around once and gave me a dirty look.

On arrival at Lincoln, I jumped up to leave the plane. The man in front of me glared again as I passed ahead of him, and he followed me down the stairs that the airport staff rolled out for deplaning. I then noticed a Nebraska state trooper at the foot of the stairs and just as I passed him, he snapped to attention and said "Good morning Governor."

I realized immediately I had been sticking my knee into the back of the Governor of Nebraska all across his state.

I was met by M. Dane ("Duke") Picard (BS Wyoming, PhD Princeton, sedimentary petrology; Nebraska, Utah) from the University of Nebraska, who said, "I see you flew here with our Governor".

I told Picard about the flight and he explained to me that the Governor had just lost his re-election bid a week before. One of Duke's colleagues later suggested that they ought to get me on television to tell my story because I would be the most popular person in Nebraska that day.

On my return to Penn, I immediately went to my parent's home. My dad was doing well and fully recovered several months later.

The year bumped along. Nothing seemed amiss or ominous during the first semester. Faul wanted to hire some new people and had at least two positions to fill. I recall we interviewed Charlie Gilbert (BS, Univ. of Oklahoma; PhD, UCLA; experimental petrology; Geophysical Lab, VPI, Texas A&M, DOE, Univ. of Oklahoma), Stefan Gartner (PhD, Illinois, micropaleontology; Texas A&M) and Norm Sohl (PhD Illinois, paleontology; USGS). All accepted a better opportunity or stayed where they were. We also met Henry's contacts from all over the world who shared his interest in tectites, including two from Czechoslovakia.

It might be worth commenting on what is involved in obtaining a research reputation in geology and likely in any scholarly field. As my career advanced during the 1960's, certain things came to my attention (The numbers that follow may differ now). In the USA, if someone earns a PhD, perhaps 50% or less will publish one or two papers or a book (in the Humanities) from their thesis. Approximately 10% will later publish one or two papers dealing with new research after completing the PhD. During the 1960's, likely about 500 geologists were regularly publishing scientific papers in refereed international journals. The same numbers applied to presentation of papers at annual meetings of GSA, AAPG and SEPM. A smaller amount did so at international meetings. Likely, the numbers are greater now.

Because this number was relatively small compared to the total membership of a professional society (today, GSA has about 22,000 members and AAPG about 33,000 members), the names of these 500 people are known, or quickly recognized as they emerge as practicing professionals. Moreover, they become acquainted with each other at meetings and are selected to serve on society committees. Of published papers, about 40% of the titles are read, 25% of both the abstracts and titles are read, and anywhere from five to seven percent of geoscientists read the entire paper. The active group of geologists who are publishing regularly and being cited is relatively small. As their work is cited and followed, their standing grows. When David Goddard, Penn's provost, asked me where I had done the work I published, he was probably trying to determine if I would become part of that group. It can happen quickly but must be sustained throughout one's career. Stop publishing, or reduce the amount published, and one's career is viewed as in decline. That's what publish or perish is all about in academe.

Although I may be omitting some people, in 1966, of the PhD students with whom I interfaced at Yale, Chuck Ross, Lee McAlester, Clark Burchfiel, Pierre Biscaye, Steve Porter, Brian Norford, Pete Robinson, and I were part of the geological 500 active publishers of research papers. The nine of us came from a group of 40 people who were admitted to the Yale PhD program. By 1980 it was down to five people.

During the second semester, Faul met me in the halls and said that we had four assistant professors, Fenner, Lehman, Shagam and Butler. He decided we could only keep two and wanted to know from me (as a tenured member of a non-existent tenure committee) which two I would support. I told him "I could support Shagam and Butler." I did not however say anything about terminating Lehman or Fenner. Faul's approach was outside the rules. Normally, dossiers and CV's should be prepared, reviewed by the tenure committee, a vote taken, and a recommendation and report sent to the Dean.

Faul terminated Lehman and Fenner. Lehman found a job in the New Hampshire state college system and I lost track of him. Fenner accepted a job as Director of Education at the American Geological Institute, and from there became Dean of Science at Governor's State University in Illinois. He and his second wife (a nurse) formed a health care consulting firm during the late 1970's. They did very well financially.

I attended the second Northeastern GSA meeting in Boston. I met John Rodgers briefly who said he read I received an NSF grant. Instead of congratulating me he said, "Well George, you benefited from being in a smaller department. They ration out NSF grants to departments of different size as government policy." I told him I was unaware of this, and in fact, it was incorrect. But it was his perception.

The geology faculty at Yale voted Flint out of the chairmanship in 1964 after their failure to get Sanders tenure and a variety of other issues regarding their new building. Rodgers was now department chairman. Turekian became the number #2 person as Director of the Geology Laboratory. Sanders was replaced by Bob Berner (MS Michigan, PhD Harvard, geochemistry; Chicago, Yale), a low-temperature geochemist who made a distinguished career there.

During 1967, Brown University asked me to serve as an external PhD examiner for one of their candidates, Norman D. Smith (BS, St. Lawrence, PhD Brown, sedimentology; Univ. of Illinois at Chicago including service as department head; editor Journal of Sedimentary Petrology; Chairman, University of Nebraska at Lincoln). I agreed and the thesis draft was sent to me. It dealt with the Clinton Group (Silurian) in Pennsylvania and because some key outcrops were within a day's driving distance, I visited these outcrops to confirm his work. A week later, I went to Brown, met Smith, discussed the thesis, and then held the exam. He passed.

Some of my friends made career changes around this time. Vint Gwinn left McMaster to go on sabbatical at Penn State. He was offered the opportunity to stay, go to Oregon State or to LSU. LSU received a large 'Centers of Excellence' grant from NSF so he accepted their offer. He called me to discuss it and I told him his choices were between a very competitive environment, a department with potential in a beautiful state, or a place with potential but a troubling history. I said, "Vint, it's very difficult for universities to change from their traditional mold."

Wayne Pryor accepted a faculty appointment to the University of

Cincinnati. I saw it as a great fit for Wayne and knew he could attract students to develop a strong program and he did.

When the semester ended, I returned to Nova Scotia. By this time, the Navy closed Hudson labs and moved all personnel to a different location on a naval base. Sanders resigned and was appointed department chairman at Barnard College. I contacted NSF and they awarded me a supplemental grant for my research. We had a short season with miserable weather.

Faul visited me two weeks before I left for Nova Scotia to discuss an item in my NSF budget about renting a station wagon. He suggested we use it as a partial payment for a departmental carryall, add funds from a similar line item in his NSF grant, and have the university pay the rest. I agreed with his proposal.

Henry had to leave town for three weeks and asked me to buy one. The Penn procurement office found one in Baltimore, so I paid for it with a cashier's check, picked the carryall up and brought it home, got it registered, had the University name and department name painted on the side doors, and attached two University Decals. It also became my responsible to arrange servicing, assign it to people for field trips, and park it at my house because there was no secure campus parking spot. We used it on all student field trips and when attending the New England Geological Field Conference and the Pennsylvanian Geological Association Field Conference. People noticed the carryall and it conveyed the impression that Penn was on the move. I credit Faul with a smart marketing ploy.

As I loaded the new carryall with our supplies and equipment and two field assistants for a short six-week season, I reflected on the year. I saw signs of concern but felt I had come out reasonably well. I was left with impression that Faul and I could work things out if trouble occurred along the way. However, I was not 100 percent sure.

The short field season produced more good results and validated earlier work done in 1965 and 1966. It also opened up some new possibilities. It was a cold and wet field season. When we arrived, I observed the last iceberg moving past Parrsboro Harbor.

The season was shortened because I had to return to the UK. IAS asked me to run a one-day field trip in the Great Oolite Series for their 1967 Congress. I went early to finalize trip arrangements and also give a paper. The meeting was held at the Sedimentology laboratory of the University of

Reading.

The University of Reading developed an internationally-recognized sedimentology program through the leadership of its department head, Professor Percy Allen. He knew that to succeed he needed a brand new building but the university did not support his plans. He decided to raise the money himself. After two years of presentations, a Quarry Association put up an initial pledge which he then parlayed into matches and more from BP, Mobil, Humble (now ExxonMobil), Amoco, and Shell. As the building was built, he added John R. L. Allen to the faculty. Roland Goldring was added in trace fossils, and Allen Lees was their carbonate geologist.

Professor Percy Allen, it turned out, was a unique character and a bit out-of-sorts with the establishment. The University of Reading had a spring tradition called Rally Day, the final campus party before exams. It usually started with a hoax.

Sometime in May, 1958, early on a foggy morning, a grizzled, unshaven individual came out of the mist, appeared on campus and asked to speak to the head of the geology department. When asked why, he said in a cockney accent. "Oi think Oi've found some doimonds in the Thames". (The City of Reading is located on The Thames River). After 15 minutes, Professor Allen appeared, took the grizzly man to his office, chatted over tea, and came out and said to a suddenly assembled crowd of reporters, "Ladies and Gentlemen, these do appear to be real diamonds, pending a detailed assay." Before he finished with the disclaimer, the reporters ran from the scene and telephoned their stories to their papers. Had they stayed, they would have heard Professor Allen add "Let Rally Day begin!!!"

Nearly all major newspapers in the UK published stories with headlines about a diamond discovery in the Thames. The *London Times* was the first to print the story. Only one paper, *The Manchester Guardian* was unconvinced and called the University of Reading to discover it was closed for Rally Day. They didn't print the story. The other papers, red-faced, retracted it. Percy Allen was harshly criticized for using his position of expertise and authority to perpetrate a student hoax, and it delayed many honors he eventually earned in the UK. The IAS meeting turned it around for him.

I remember one incident during the IAS meeting at breakfast while talking with K.O. Emery (BS, MS, PhD, Illinois, Univ. of Southern California; Woods Hole Oceanographic Institute). Emery commented how the Russian delegation always moved as a group and it was difficult to talk alone with any of them. He pointed to a stocky person who was about 6'4, with bushy eyebrows, looking similar to a cross-breed between Joseph Stalin and Leonid Brezhnev. Emery said that man was the political officer and if you talked to any of the Russians, he'd come over.

I thought Emery was off-base, so when the Russians left the dining hall, I caught up with one of their stragglers and introduced myself. 'Brezhnev-Stalin' immediately came over, shouted something to the straggler who turned white as a sheet, and left me standing there. Not only was Emery correct, but when I shared this incident with Henry Faul later, he said I may have damaged that straggler's career.

The 1967-68 academic year started with Henry teaching freshman geology, Shagam (Structure), Butler (Mineralogy, Petrology), me (Sedimentology and Physical Stratigraphy) and Horace Richards (Paleontology) and our graduate specialty courses to graduate students. The departure of Fenner and Lehman reduced our size, so it would be a tight year, duty-wise.

Things started to sour between Henry and me when I balked at doing a public service talk at a local high school about geology during their annual career day. It seemed a minor disagreement to me but not to Henry. There was an issue with an undergraduate, Henry J. B. Dick, about the use of the sedimentology lab. I had no problem with people using the lab or the equipment as long as I checked out their capability to use and care for the equipment. Henry Dick was not checked out and damaged the fine mesh sieve, which was expensive to replace. I had set rules of access to the lab and Faul felt this was inappropriate. Henry Faul also was constantly going around university procedures and rules and I cautioned him, suggesting better ways to avoid hurting the department. He resented this.

Early in October, we interviewed Robert F. Giegengack (BS. Yale, MS. Colorado, PhD, Yale), a geomorphologist. Bob took Benson's field methods course in 1959 (which I TA'd) so we reminisced a bit during his interview. Bob's father was Yale's track coach. He gave a very well-prepared, excellent candidate talk.

Bob also disclosed he would be Dick Flint's last PhD student because Flint was retiring in two years and stopped accepting students.

Faul offered Bob a job starting in the fall of 1968, and he accepted.

Two weeks later, Faul called a faculty meeting. He had finished a phone call with Art Boucot who moved from MIT to Cal Tech in 1962. Boucot was coming for an interview because he was interested in returning to Philadelphia where he was born and raised. Boucot also thought Penn's geology department had great potential.

Faul explained that to hire Art required a major commitment from the administration not only for Art, but for a disabled junior colleague, Jess Johnson, his lab associate, Peggy Losey, and two graduate students. He asked if we would be onboard so he could go to the administration with the full support of the faculty.

I told everyone at the meeting that Art's appointment would have my fullest support and encouraged everyone else who did not know him to support the appointment. I explained I knew Art from my Nova Scotia thesis days, saw him periodically at GSA meetings, got along with him, and was aware of his international reputation as a world-class Silurian-Devonian brachiopod paleontologist. I stressed it would give the department world-wide visibility, enhance our reputation off-campus, and attract good graduate students. Moreover, I also suggested that perhaps Henry might want to come up with some other things we would need that could be bundled into any package the university could fund to get Boucot. Shagam also knew Boucot and echoed everything I said.

Henry then pointedly asked me, "George, you're developing a growing track record that is well thought of too. Does it worry you that Art might overshadow you?" I was surprised by the question but replied, "Henry, I would welcome Art here. Anyone who enhances the department's reputation is a benefit to the entire department. This will be win-win for Penn."

Art visited two weeks later. A month later, an offer was made that met all his demands and he accepted. I felt it was a good move for Penn but I also wondered how Henry would be able to work with Art who had a very strong personality. I saw the possibility of a major conflict between them at some time.

I kept working on my own research, staying in touch with my network, keeping my ear to the ground for better opportunities should things get worse, and avoided Henry as much as the job allowed.

During the spring, Henry invited a Russian scientist, Dr. M.A. Semikhatov to give a colloquium on his work on Precambrian stromatolites of Siberia. The lecture was given a week after I reviewed carbonate depositional environments in my Physical Stratigraphy class. The stratigraphy students were required to read the definitive work of Bob Ginsburg, Brian Logan, Ray Murray and Gene Shinn. I sat through the lecture and was dumbfounded with Semikhatov's conclusion that the stromatolites he described were interpreted as "deep water" in origin. He showed many features on his slides that replicated observations from the Bahamas and the Persian Gulf. My class was required to attend and they were joking amongst themselves in their immature undergraduate way having heard my class lectures and read the critical papers.

When Faul asked for questions, I raised my hand and was recognized. The dialog went like this (including phoneticized text).

Klein: Dr. Semikhatov, I was puzzled about your interpretation that the stromatolites were deep water. Have you considered the work by Ginsburg, Shinn, Murray and Logan that they might be intertidal and supratidal?

Semikhatov: Intertidal! Supratidal! Efer since I arrifed in America, efryvere I have given zis lecture people tell me ze stromatolites are intertidal and supratidal. Zay are wrong. Zay are deep water!! (At this point my class started to laugh). I don't believe it! (More student laughter).

Klein: Did you see any examples of exposure like mudcracks and birds-eye structure, or diagenetic dolomite, such as described from many intertidal and supratidal areas in the Bahamas, Persian Gulf, and Shark Bay, Australia, that were interbedded or adjacent to your stromatolites?

Semikhatov: Eempossible!! Zese are deep water!! (My class laughed again).

At this point, I was irritated with the dogmatic replies so I thought a minute and said, "Dr. Semikhatov, you have an interesting hypothesis that may be easy to refute, but I don't wish to discuss it further. I tell you that if I had been alive when these stromatolites were growing on the Siberian craton, I could have walked from one end of that craton to the other without getting my navel wet." (My class then laughed and applauded).

Faul interjected himself into the discussion to close the meeting and said, "Dr. Semikhatov, I suggest you listen to what Dr. Klein is saying. I also have seen some of the features he mentioned in many places associated with stromatolites. I strongly recommend you reconsider. I do want to thank you for presenting an interesting talk that led to such lively discussion."

Next day, Faul told me my questions were legitimate and polite but to be careful how I challenged international experts. I said if I hadn't, I would have lost credibility with my class and they would conclude that dogmatic views take precedence over scientific facts. I reminded him that is not how western science is conducted and that Semikhatov may know stromatolites but was totally out-of-touch with work by others. Faul explained that most of Soviet geology is that way because of their political isolation.

In early March, I received a telephone call from Vint Gwinn. He left LSU to accept the Headship of the Department of Geology at the University of South Carolina and asked me to refer undergraduate majors to him for graduate work. I agreed to see what I could do. When asked why he left, he said, "George, it's a long story, but I will tell you this. Your comment about universities going back to their traditional mold when things go wrong was so true." That turned out to be our last conversation.

Because I planned the summer of 1968 to be my last in Nova Scotia and only needed a month to get everything finished, I decided to apply to NSF for one of their new program grants, the Advanced Science Seminar program. It was designed for graduate students and I thought a five week field program for graduate students in North America would be a good way to end the project. The Bay of Fundy was not easy to reach in those days and because of its unique setting, I thought I could get funding. Fortunately, I succeeded. That meant screening applicants and advertising the program.

I offered one slot to Julie Badal because she needed field experience. We admitted a new graduate student, Paul R. Schluger (BS Temple, MS Penn, PhD, Illinois, AGAT, Mobil)) in January and I suggested he do a Master's thesis on the Perry Formation (Devonian) of New Brunswick with the idea of expanding it into a PhD thesis later. I offered him a summer TA in the seminar as a way to finance his thesis work and he agreed. I also offered a visiting appointment to Gerry Middleton.

Eventually, we had 12 students who came from Saskatchewan. McMaster, Rutgers, Wyoming, Oregon, Texas at Austin, Cincinnati, Oregon State and other institutions I can't recall. It was a success. I arranged for a field trip into a salt mine, and scheduled colloquium speakers including Raymond C. Murray (BS Tufts, PhD Wisconsin; carbonate sedimentology; Shell Research, Dept Head at Rutgers, later Dean of Graduate Studies and Vice President of Research, University of Montana) and others about whom I received word that they were visiting the area.

We rented the science lab in the local high school and brought sieves, sieve shakers, and basic equipment so everyone could complete a project, write it up and present it. Students rented rooms and obtained board in Parrsboro. Gerry Middletown brought a movie camera and made a movie of the entire activity during the seminar.

When Henry heard the news I received the grant he wished me well but was concerned about safety issues. I reviewed them and he seemed satisfied.

I was also invited to interview for the department chairmanship at Duke University that spring. I made two visits. During my first visit, I met with the faculty and developed a plan for the department. I refined it slightly and presented it to the Dean. I was offered the job with a promotion to a full professorship but the administration would not endorse my plan. I turned it down.

I also received a phone call in mid-April from Adrian Richards (PhD Scripps Institute of Oceanography, sedimentology; US Office of Naval Research, Illinois, Lehigh, Fugro, Consultant) at the University of Illinois. He inquired if I would consider moving there. I assured him I'd be interested, sent him my CV and heard no more.

The academic year ended and I was disturbed that Henry recommended my salary be unchanged, in other words, no raise. I knew it was time to get out but to do it, if possible, on my terms. In another year, I was eligible for a sabbatical and I wanted to take one before moving.

The summer season of 1968 went well. That summer, the International Geological Congress was held in Prague, Czechoslovakia, and was interrupted and cancelled by an invasion of Russian tanks. Henry Faul and his wife attended and he told me that to get out, the American Embassy arranged transport to West Germany for all US citizens in the country. He was most subdued about the experience.

There was one unhappy event during the summer. I rented the Parrsboro school superintendent's summer cabin on the Bay of Fundy shore and it had a telephone. After dinner one evening, the phone rang and the caller was Gayle Billings at LSU. He explained that Vint Gwinn died in an accident that afternoon. I suddenly heard a shrieking woman's voice in the background saying "Is that George? Give me the phone." It was Janet, Vint's widow. She was clearly shattered, incoherent and frightened. It was, up to that time in my life, the most tragic event I experienced and left me fumbling around trying to calm her down, expressing the right words, while calibrating my own shock. Vint was only 35, too young for anyone to die.

On my return, I applied for a sabbatical leave for the 1969-70 academic year. I proposed a program to go to Oxford University where Stuart McKerrow had made arrangements for me to be a Visiting Fellow of Wolfson College, a new graduate college. The request was approved within two weeks.

Art Boucot was still overseas and was expected to arrive in late September. His wife, Bobbie, bought a house, moved the family, and Peggy Losey, Art's assistant, who moved Art's fossil collection, Jess Johnson's family, and the students.

A week after the fall semester started, Henry came to my office, which was something he rarely did. He looked ashen and was very agitated. I invited him to sit down. The dialog, I recall, went like this (and a conversation like this is seldom forgotten):

Faul: George, we have a very serious problem with Art Boucot. I talked with people at Cal Tech because we're transferring his grants and there are things I didn't know.

Klein: Like what?

Faul: Well, since arriving there, he overspent his budgets, never re-budgeted when changes are made, never received clearance from anyone to shift funds around budget categories, and when asked to follow grant funding rules, never cooperated. He will be difficult to manage and control.

Klein: Well, Henry, I didn't know any of this. What do you plan to do?

Faul: I'm going to have to impose strict financial controls and will do so by managing his grants. I have to do it quickly. I shall insist on approving every expenditure he wants to make.

Klein: Have you discussed this with Art, or plan to when he returns?

Faul: I discussed it briefly with his wife. Art is careless with his grant funds. He's not like you. **You watch over your budgets like a hawk. I can't get away with a thing with you.** So I must move quickly now with Art. (NB Bold type added).

Klein: Have you discussed this with Shagam and Butler?

Faul: There's no point discussing this with Reg because he has Venezuelan government money and no NSF funds. Pat got an NSF grant six months ago and I went over procedures with him because he is new. I'm leaving him alone, but will check his monthly statement to see all is in order.

And one other thing, I will not approve Art's appointment of Peggy Losey, or her moving expenses.

Klein: What! Gee Henry, she just got here. I met her briefly and she doesn't look like someone who has a lot of money to stand a loss like that.

Faul: Well, I won't have it. You know when faculty move from place to place and bring a woman assistant with them, there's always some fooling around going on.

Klein: Oh come on Henry! I can't imagine anything going on between Art and Peggy Losey, especially because Bobbie is so much part of his career and is a terrific lady. She does much of his drafting, edits his papers, and helps out in other ways. He wouldn't do anything stupid to jeopardize that. In fact, I can't imagine any guy wanting to fool around with Peggy. She's very ordinary, immature, and not that appealing.

Faul: George, you are wrong. That's always the way it is.

Klein: Well Henry, why are you bringing this up with me? This is a matter between you, Art and the Grant's Office. You know how I try to operate, namely to stay out of things where I lack authority.

Faul: Well, I may need your help later and was hoping you will do so.

Klein: Henry, all I can promise is to stay out of the way. Art's a long term friend, and I have no direct knowledge of what happened at Cal Tech. If you want advice on an informal basis where you think I can help, you can ask me, but I'm not sure I know enough to deal with a situation like this. I haven't encountered it before.

Faul: Well, didn't you report some irregularities at Pitt by Frederickson?

Klein: Guess you talked with Freddy and knew him at M.I.T. Anyway, I was aware of a problem and referred it to the Dean. I didn't have the authority to do more.

Faul: Well, I hope I can count on you if things get ugly between Art and me.

Henry then stood up and left.

When he left I was speechless. Henry doesn't give me a raise, gives me all the signals he wants me to leave, and when someone with Art's stature comes along, he runs around like a frightened child and wants my help! It made no sense. Moreover, he gives more leeway on grant management to an assistant professor than a senior professor. I foresaw a big split and after my experience at Pitt, I decided to be careful and figure out a way to come out ahead without getting hurt too badly, if at all.

I called Bobbie to let her know I was back and she invited me to the house for dinner. After the children went upstairs to study or go to bed she said she had been to the office several times since arriving and had some difficult meetings with Henry Faul about Art's grant management and his refusal to sign Peggy Losey's appointment forms. She knew Peggy and loaned her some money in the interim. Because Art was inaccessible, she had no way to reach him. Did I have any suggestions?

I thought for a few minutes and advised she keep a diary of every conversation she had with Faul, when, where, time of day. I also recommended she sit tight and when Art returned, go over everything with him before he set foot in the office. I emphasized it was critical she spell out everything she knew so when Art arrived at the department he wouldn't be blind-sided.

I also asked her to tell Art to call me so I knew when to expect him. Art also was aware, she told me, that Faul did not like me around because he felt threatened. Henry was concerned that I had more experience in academe than he did and knew my way around the campus administrative structure better than Henry as to what any of us could or could not do. She added that Henry thinks that one day I shall go after his job. I also told her to keep our meeting quiet and not even mention it to Jess Johnson, Art's students, or Peggy, and she agreed.

Art called the last Sunday in September. I welcomed him back and agreed to see him next day. On Monday, I taught a class, worked on research and around 10:30 am Art came to my office and closed the door. He looked more somber than I ever remember. He came straight to the point, "George, I just had a two hour meeting with Henry and we had a big fight. I'm on my way to the provost now to get him to resolve this. It might be best if we weren't seen talking in the department for now and called each other in the evening." I replied, "That bad, huh." Art said, "Yes."

I said, "Are you going to call Dick Ray at NSF and put them in the picture?" Art replied "Yes, and also the people at Cal Tech to find out what went on between them and Henry." I then said, "Gee Art, this is your first day here." Art replied "Yes." I shared Henry's comment about Peggy Losey with him and Art laughed and replied, "Henry should credit me with better taste" and left.

That afternoon, Shagam and Butler came to see me. Art talked with them too and tried to enlist their support. They wanted to know my views. I told them I wanted to think long and hard about this not knowing all the facts. I then disclosed I was not totally surprised and repeated Henry's comment about my watching my grants like a hawk. I explained that I sensed Henry was capable of taking control of other people's grants and that he reminded me of Frederickson but wanted to give him a fair chance from the very beginning. Now my worst suspicions were confirmed.

I also said that we needed to think of our own futures. If we weren't careful we could get hurt and we needed to work together as much as we could to avoid that happening.

Art met with the provost and within a week, after Art, Henry, and the Dean of the Graduate College met again with the provost, grant expenditure authority was restored to Boucot and Goddard immediately approved his hiring of Peggy Losey on the spot, back-dated to the day she arrived.

It was a momentary victory for Art but I knew it wasn't over. Art came by my office to let me know the outcome and then said, "George, I'm going to try and get Henry fired." I suggested he take it easy.

The economy was slowing, the Penn endowment was not doing well, and money was tight on campus. I also said that it might be a good idea if he Jess, Pat, Reg and I, and eventually Giegengack, got together and did some forward planning about what kind of a department we should become. I told him that what troubled me about Henry was that his vision of geology was 1940-ish with traditional fields and using his K/Ar age determination to work within that framework. I added, "Much as I like Gieg, no one has hired a faculty member in geomorphology since 1965 and I'm unaware of anyone planning to do so now."

I assumed all would calm down soon enough and did nothing further. However, I was wrong and realized that the war between Art and Henry would continue until the administration intervened. I talked with Art one day in his office about graduate admissions when he took a call. He was telephoning everyone he knew on the continent and either left call-back messages, or told them what had happened his first day and how the administration settled it. He asked me to stay and over the phone he castigated Henry and then inquired if they knew any 'dirt' about Henry that Art could use. He was calling senior paleontologists, friends from his U.S.G.S. days, faculty at M.I.T. including Bob Shrock, Lee McAllister at Yale, past officers of societies, and members of the Academy of Sciences, amongst others. I was disturbed because this would hurt our efforts to recruit graduate students.

Soon word spread nationwide and overseas. Moreover, I received inquiries about my availability from LSU and Texas A&M. I sent them my CV and reprints. By this time, my GSA Spec. Paper was published and in those days, editors and authors received 50 complimentary copies. I included one with each packet. They wrote back and told me that they would know more in the spring.

In December, Art, Jess, Peggy Losey, Reg, Pat and I had lunch in Art's office. Art said he turned up some interesting things about Henry. First, he called someone at UCLA who told him that Henry's standing in the isotopic age determination community was very poor. Apparently, questions were raised about his analytical methods and about one third of his age determinations were wrong. Second, his U.S.G.S. contact told him that Henry Faul was the only person in its history who was fired from a civil service job there.

The third item was most disturbing. While at the Southwest Center for Research, the director, Anton Hales, barred Henry Faul from the isotope lab and the use of their mass spectrometer, even though Henry raised the grant funds to get one. There were complaints from other users and arguments occurred. The director settled it after an investigation.

During the middle of the summer of 1966, Anton Hales, a South African, was leaving for a trip late on a Friday afternoon. At 3 PM that day, Henry walked in and handed Hales a resignation letter effective the following Monday. Hales told him he would deal with it on his return.

Early next morning, a Saturday, Henry arranged for a moving truck staffed with chemical technicians to come to his office, he entered the lab, they dissembled the mass spectrometer, packed it, and drove it to Philadelphia. When people arrived for work in Texas on Monday morning and called the Institute's security division, the moving truck was long gone.

Henry had flown ahead to meet them.

We discussed these allegations. Henry's standing with isotope geochronologists wouldn't matter much with the Administration. The firing from the USGS needed to be explored. Art mentioned that Henry's branch chief at the time was Sam Goldich who was now at Northern Illinois University. It turned out I was the only one in the room who knew Sam. Sam was crusty but had a decent streak. Art asked me to call him.

We discussed Hales. Both Shagam and I knew him. Shagam had no desire to call. I was asked to call him too.

When I called Goldich he said that Henry had applied for a Senior NSF Postdoctoral fellowship to work in Switzerland without USGS approval, was awarded it, and left without notice. USGS policy was that after trying to trace someone for a month for failure to show for work, they were terminated automatically. Goldich explained that Henry was his most difficult geochemist, couldn't be trusted, never followed USGS rules and procedures, and was happy he could terminate him.

I then called Hales. Hales said that he couldn't tell me much because the matter was still (after more than two years) in dispute between his institution and Penn's administration. Southwest Center for Research had since acquired a new mass spectrometer so they were seeking damages and were in negotiation for a final settlement. He disclosed the mass spectrometer, a $60,000 instrument (1966 prices), was removed from the premises the day after he left for South Africa in 1966. Given the sensitive negotiations, I was surprised he confirmed that much.

I was struck by the parallelisms to Frederickson. Both earned PhD's from M.I.T. (probably a coincidence). Both were fired from tenured or civil service (equivalent to tenure) positions. Both removed equipment bought in the name of their home institutions from their premises. It confirmed my initial suspicions when Faul arrived.

When Art and the rest of us met again, I brought them up-to-date. I felt it was a stalemate. His firing from the USGS would likely matter little to the administration. Disclosing we knew about the mass spectrometer dispute would only tip our hand, and risk exposing them in a sensitive deal. I said as much and Art said, "No, we got him. They'll kill him for this." Reg and Pat agreed with me and three of us told him to let it ride. Art became angry and said, "Don't you want the son-of-a-bitch fired? Whose side are you on?" I finally said that this was not the time. We might find it helpful later.

Giegengack was invited to our meetings but told us he would stand down because he hadn't been at Penn long enough. We supported his position.

In December, I reviewed Paul Schluger's progress and he asked about the dispute between Art and Henry. Paul spent four years in the Air Force and was more mature so I spoke frankly in confidence. I told him the situation was not good, that I might be leaving, and it might be best if he wrote up his Master's thesis in a timely manner. He should plan to continue on for a PhD on the Perry Formation and apply for funds. If I stayed in academe, I would arrange for him to move with me and get him financial aid as part of any negotiations I had. He appreciated my frankness and assured me he would keep things confidential and plan accordingly.

In February, Faul told me he wanted Art to take over graduate advising because I would be on sabbatical next fall. I agreed. Henry also said he hoped by giving Art this opportunity it might improve relations between them. It appeared as if Henry was unaware either of Art's phone calls or our meetings.

I also received a phone call from Richard L. Hay (BS Northwestern, PhD Princeton; sedimentary petrology; LSU, UCB, Illinois). He was appointed to a Miller professorship at UCB and the department needed people to teach there during the year he was on-leave. He asked if I could come spring quarter, 1970, as a visiting professor. Bob Dott at Wisconsin was on board for the fall quarter. Because my sabbatical was for a year at half pay, I accepted to get extra income to cover me for the year, and reorganized the sabbatical to a semester at full pay and a one-semester unpaid leave-of-absence.

During one of our meetings with Art, we discussed future plans for the department and developed a general outline and cost estimates. It would be pricey. Shagam asked who should succeed Faul as department chairman if we got Henry removed. Art wanted me to do it and I made it clear I wasn't going to sacrifice my sabbatical under any circumstances because of this dispute. Someone would have to be acting chairman for at least a semester. Unfortunately, a month later, Peggy Losey saw Henry in the halls and he asked her to do something. She told him she didn't have to do anything for him because I would be the next chairman when he would be removed by year's end. I really gave Art hell about that and also explained it exposed our efforts.

Indeed ten days later, we were meeting in Art's office and in walks Henry. He said, "Ah-ha, I've caught all you conspirators red-handed. None of you will get a raise next year." Art and Henry started to slug it out verbally and I got up trying to separate them. Henry then said, "George, you're fired at the end of the semester" and walked out.

I knew, of course, that Henry couldn't fire me because I was tenured. He would have to present charges, arrange a hearing, allow me due process and rebuttal, and then have the matter settled by the Board of Trustees.

We discussed this and I said, "Guys, here is what I could do. Why don't I file a grievance with the local chapter of the AAUP (American Association of University Professors)? That way we get them involved and Henry has to explain his actions to the administration. This is one I think we can win, but it will only expose Henry to more scrutiny and that's all we'll get out of it."

We agreed, and I filed a grievance. It was referred to the Provost and I was asked to meet with him, the local AAUP Chapter president, Faul, the new dean of the college, Bill Stevens, former head of physics, and the Dean of the Graduate School. We were to meet in the College conference room, and Art, Reg, Pat and I arrived with me, as did Henry. The provost was shocked to see the others and Art took him aside and told him he wanted to use the opportunity to request a meeting of the Provost and the geology faculty to settle outstanding issues. Goddard finally agreed to such a meeting.

I met alone with the deans, Goddard and Faul. I explained my side and also shared with them Henry's comment about my watching my grants like a hawk. Henry claimed that was just my imagination. The Dean of the graduate school, a former Vice President of Metallurgical Research at Bell Labs, said, "Oh, I don't know Henry, I've heard you make similar outlandish and inappropriate statements many times in committee, usually in good fun, but not this time. You're quite capable of saying these things." Goddard finally said, "George, you know Henry can't fire you and you can stay on the faculty as long as you wish. I hope this ends this matter." I responded to Goddard by saying "I appreciate your supporting my continued work here. However, I want to go on record that the appeal was filed because Faul represents Penn's administration to the department and the geological profession and vice-versa. You clarification is much appreciated and I thank you for taking time to meet with me." I then got up to leave knowing the meeting was over. Then Henry said, "Gentlemen, I now recommend we leave Dr. Klein's salary next year as is." The provost and deans ignored him and walked out with me.

I then received a telephone call from the University of Ottawa in Canada asking me to serve as an external examiner for one of their doctoral candidates, Andrew D. Miall (BSc, Univ. of London, PhD Ottawa; J.C. Sproule and Associates, GSC, University of Toronto). I agreed and went during spring vacation. The thesis was long but well done and Andrew passed. Afterwards, I drove with some of their faculty to attend the fourth Northeastern GSA Meeting in Albany, NY.

Bob Shrock attended the Northeastern GSA section meeting also and asked, "George, I've been hearing things from both Henry Faul and Art Boucot. Who's right?" I replied, "Art is." Bob replied, "Then I'll back him, I know you're honest and I believe you. I don't trust either of them."

I reviewed my options again and recalled Adrian Richard's phone call. I was aware that Illinois interviewed five people for a sedimentology position, but none were offered an appointment. Two years before, they hired Fred A. Donath (BS Minnesota, MS, PhD Stanford; Structural Geology and Experiment Rock Deformation; Columbia, Illinois, CGS, Inc, Earthtec) from Columbia as a new head after a departmental revolt removed George White. I wrote Donath with the usual materials and in a week, received a polite acknowledgement.

The departmental meeting Boucot arranged with the Provost, the Deans and Henry was scheduled to be held ten days later. On the morning of that meeting, I received a telephone call from Donath asking if I was attending AAPG in Dallas in late April and could I route my trip back through Urbana for a two-day interview. I accepted the interview dates. He asked me to make airplane reservations and to let him know so he and I could fly together from Dallas to Urbana, IL, via St. Louis. Because I was attending an NSF conference on Deltas at LSU, followed by an interview at Texas A & M before that meeting, I changed my reservations and called back to give him my itinerary.

That afternoon, Art, Pat, Reg and I walked over to the Provost's office in the main administration building. We all wore suits. We got there early and were told to sit at the conference table. I sat at the end closest to Goddard's desk knowing I would sit next to him when he chaired the meeting. Goddard and the Deans came and sat down, Goddard next to me, and the two deans at the opposite end of the conference table. Henry arrived five minutes late looking disheveled and wore a dirty white shirt with a field jacket. He sat opposite me.

Goddard opened the meeting and said, "Gentlemen, I've been hearing about difficulties between Henry Faul and Art Boucot since October. It was only recently that I discovered there were other problems too. What the deans and I need to know is what it will take to build Penn's geology program into a nationally recognized program now and in the future."

Immediately, Art started arguing with Faul, and they went back and forth for 20 minutes. The Provost, the deans, Shagam, Butler and I sat and listened. Finally, Goddard said, "Gentlemen, unless you start telling me immediately what it will take to build Penn's geology program into a national program, I shall close the meeting." I immediately raised my hand and Goddard said, "George, go ahead."

I said, "Sir, if Penn really wants to become a leading national program, we need to build on existing strength represented by Shagam, Butler, Boucot, Giegengack, Johnson and myself. To do it, we need to add four people during the next year, four more the following year, and two each year after that until we reach a critical mass of 20 faculty. That's the size of leading departments like Cal Tech, M.I.T, UCLA, and Columbia. Areas we need to add are three people in geophysics, three in geochemistry, one in carbonate sedimentology, one or two in hydrogeology, and an igneous petrologist. At that point, the department should reassess its growth to determine what changes are occurring in the field and opportunistically select the next four people accordingly." As I talked, I looked at Goddard and had his attention, turned to face Henry who scowled, and ended my comments while looking at the two deans.

Goddard said, "Thank you George," and then pointed to Pat who sat next to me. Pat told the provost he agreed with my assessment but would add a mineralogist, and perhaps change some of the fields I proposed. Goddard then pointed to Shagam who said he agreed in principle with me and Pat but would add one or two people in tectonics.

Goddard then called on Art. Art started by saying, "I entirely agree with George," and then proceeded to castigate Henry again. Henry took a sheet of paper from his shirt and because he sat across from me, I could see it was on Penn letterhead. He proceeded to read it. The letter was a copy of one Art wrote to someone outside the university castigating Henry and asking both for information about him and leads to information. Art accused Henry of coming in at night and going through his private files and proceeded to attack his morals and ethics.

Goddard interrupted and said that he and the Deans heard what they needed to know, thanked us for coming, and adjourned the meeting. Henry quickly left. The rest of us slowly filed out to shake hands with Goddard and the deans.

As we left the building, Art was ecstatic. He said, "Did you see how I punched Henry out. We're going to win. They won't keep him."

Shagam stopped and we all did. Reg first said, "I think I'll go back to Israel and grow potatoes" (I forgot Reg was Jewish, his wife had been an airline hostess for El Al, and had many relatives there who were after Reg to immigrate and do his geology at an Israeli university). Reg then went on to say, "No Art, I think we lost. The deans are in Henry's corner and Goddard picked him. I don't think they really care what happens to geology at Penn." We then continued to walk back to Hayden Hall.

Pat said, "I really don't know how to read it, George, what do you think?"

We were now walking past College Hall and I said, "Guys, I haven't had a chance to see you until this afternoon so I need to tell you something. I got a call this morning from the University of Illinois inviting me to interview after the AAPG meeting. I'm also interviewing at Texas A&M before the AAPG meeting. If they make offers, what happened today may not matter. But I want to say that until I know if I'm leaving or staying, you guys have my support until we hear the outcome. We've played our last card and all we can do is wait. They'll take their time and let us know around final exams. From here on, they'll be watching us so we need to be very careful that anything we do becomes the slightest pretext for favoring Henry."

Art responded, "Naw George, we'll win this thing."

Reg said: "No, we've lost."

LESSONS LEARNED:

1) A university administration expects faculty to work together in a professionally civilized manner. They are uninterested if faculty members get along or not.

2) When arguments occur between a department head or chair and one or

more faculty members, the administration is only interested in the short- and long-term consequences for the future of the relevant academic program. Their options are to keep it intact, expand it, permit stagnation, or close it.

3) Academic programs are fundamentally free-standing units as part of the shared governance tradition of running a university. The administration allocates resources and fosters research, academic growth and accountability. It has no interest in running each department nor does it claim expertise to do so.

4) Department heads or chairs are rarely removed administratively. It only happens if the head takes actions that exposes the university outside its walls, a head/chair becomes mentally or physically incapacitated, or if the administrative structure of a university is threatened by actions taken by the head or chairman. If a department head or chairman loses the confidence or support of the departmental faculty, the university may exercise its option to remove the head or chairman, but it is not guaranteed.

POSTCRIPT #1. During the middle of July, 1969, The University of Pennsylvania compensated the Southwest Center for Research $27,000 to settle their dispute over Henry Faul's removal of a mass spectrometer without authorization.

CHAPTER 13

Closure: University of Pennsylvania, Interviews, Oxford (1969)

I left for an LSU Delta Seminar run by Jim Coleman, Woody Gagliano, and Jim Morgan several days later. Then I flew to Texas A&M for two days of meetings and interviews, including giving a colloquium on my Bay of Fundy work. It was invigorating to be away from Penn and in new settings.

Texas A&M had just become co-educational, but all the male students were still in the "Corps", wore cadet uniforms, and when walking across campus, there was a steady murmur of "Howdy" because every cadet was required to greet each other. The Geology Department was housed in the College of Geoscience that also included separate departments of geophysics, geography, and oceanography.

Bob Berg, the department head, told me he would like to have me join them but their budget was uncertain awaiting the Texas legislative session to end. He added that the Dean, a geographer, attended my talk, liked it, and was supportive of an appointment if funds came. We agreed to stay in touch. My own assessment was it was doable but it would be a major adjustment. The campus was in transition from being a partially research oriented military school to a Tier I coeducational research university.

I went to Dallas to attend the AAPG-SEPM annual meeting. Everyone I knew had heard about the Boucot-Faul war and friends were suggesting places to apply. On Monday evening, I dropped by the University of Illinois Alumni cocktail party to introduce myself to Donath and confirm our meeting at the airport. The meeting ended on Wednesday afternoon and my flight to St. Louis was scheduled to leave at 4:30 PM.

Donath and I met at the airport and after checking in together to get adjacent seats, we went to the waiting area. We started talking and fairly quickly, established good rapport. He was about two years older than me. Fred grew up in Minnesota, went to the University of Minnesota for his undergraduate degree, and earned a PhD at Stanford. After a year Post-Doc in experimental rock deformation at UCLA with David Griggs, he joined the faculty of Columbia University. He also was editor of the *GSA Bulletin* for two years.

Fred joined the faculty at Illinois as department head in January, 1967. He knew the Champaign-Urbana, IL area because his father was a flight instructor during World War II at the Rantoul Air Force base 15 miles north of Urbana. Fred explained that he also had a private pilot's license and rented airplanes from the University of Illinois's Aviation institute to make some of his trips.

We started talking about the department and his vision. Fred wanted to build on its existing strength in Paleontology, Sedimentary Petrology, Stratigraphy, Hydrogeology, Engineering Geology, and Quaternary Research by adding structural geology, experimental petrology, sedimentology, geochemistry and geophysics. His first new hire was Dan Blake, a paleontologist. He hired an experimental/metamorphic petrologist, Dave Anderson, a geohydrologist, Pat Domenico, and an isotope geochemist, Tom Anderson. Dennis Wood was a visiting professor in structural geologist during 1968 and became a regular appointee in 1969. He also hired Don Graf, a senior low-temperature geochemist, to come to Illinois in the fall of 1969.

We discussed criteria and time schedules for tenure and promotion, what they wanted me to do in terms of research, teaching and service, how he saw me fit into the department's programs, and what he saw down the road. By this time we were landing in St. Louis for a two-hour layover for the Ozark Airline flight to Champaign, IL.

After landing, Fred said, "Come with me", and I followed him to the American Airlines 'Admiral's Club' of which he was a member. By this time we established positive communication and were getting along well. We continued our conversation and I asked about his vision for the department. His goal was to make it one of the top ten in the USA, and believed it would be best if the faculty did both their individual research and interacted with people in other areas. He unfolded his cocktail napkin and drew a stick diagram of faculty research interactions he saw. Fred loved to develop diagrams to make his point, even if sometimes these diagrams were unintelligible.

We boarded the flight to Champaign, a 45 minute hop on a puddle-jumper. We were met by his wife, Mavis, who drove us to the Urbana Lincoln Hotel where I spent the night. I asked how many students the

University of Illinois had and he said "34,000. But George, Minnesota is bigger." I replied, "Yea Fred, everything's bigger in Minnesota."

At this point, Mavis said, "Gosh you guys just met this afternoon and sound like you've known each other all your life." When we got to the hotel which was attached to a new shopping mall, Lincoln Square, Fred showed me how to find the entrance, and gave me a note with an address and said "Give this to a taxi driver and when you get to the building, go to Room 252 and ask to see Dorothy Smith. She'll give you your schedule. I'll see you at your talk and at the end of Friday afternoon for an exit meeting. Dorothy is expecting you around 8:00 am."

The Urbana Lincoln Hotel was a renovated 19th century period piece with an excellent regional dining room. Next morning, I arrived at the department office to meet Dorothy Smith, a lady in her early sixties. She handed me my schedule. First up was Dan Blake (BS, Illinois, MS, Michigan State, PhD UC-Berkeley; Crinoid and Bryozoan paleontology). Dan and I talked and I had three questions for every candidate: Criteria for promotion and tenure, how were the changes Fred was making impacting the department, and how he felt everyone was being treated. I then asked him to tell me about his research. Next up was Al Carozzi (BSc, PhD Geneva) who I already knew. He then introduced me to Frank Patton (BS, Univ. of Toronto, PhD, Illinois, engineering geologist, later left to form his own company in Canada). When we were done, I was supposed to meet Dennis Wood (BS, PhD, Leeds; taught at Leeds, then Illinois, returned to the UK with Robertson's Research) but he was nowhere to be found. By chance David Anderson (BSC, PhD, Sydney, Post Doc at Harvard, metamorphic and experimental petrology; Illinois) walked by. He wasn't scheduled to meet me, but we talked. Dennis showed up late and was most apologetic. We were supposed to go to lunch, so Dave, Dennis and I went to the 'Bull and the Bear,' a pub attached to the Urbana-Lincoln Hotel.

Because I lived in Australia, and worked and lived in the UK, we established instant rapport. Dennis had his usual "pint", but I declined. I reminded him as friendly as all this was, it was an interview.

I don't recall the exact order of who I met because this happened 40 years ago, I recall meeting Art Hagner (PhD Columbia, Economic Geology; Illinois),Pat Domenico (BS. Syracuse, PhD Nevada, Reno; geohydrology; Desert Research Institute, Illinois, Name chair later at Texas A & M), Ralph L. Langenheim (BS Tulsa, MS, Colorado, PhD. Minnesota, Stratigraphy;

Grinnell, UCB, Illinois), Philip Sandberg (BS, MS, LSU; PhD Stockholm; Illinois, George Mason, South Dakota State – Dean of Science), C. John Mann (BS, MS, Kansas, PhD, Wisconsin, stratigraphy, mathematical geology; Chevron, Harza Engineering, Illinois; knew him at KU), Carleton Chapman (BS New Hampshire, PhD Harvard, Petrology; Illinois) and others. I met Tom Anderson (BS DePauw, PhD. Columbia, low-temperature isotope geochemistry, Illinois) at the end of the first day.

Prior to my trip to Urbana, I called Peter Fenner who provided a run-down on the entire faculty from his days as a graduate student and who were still there. His assessment was accurate and helpful. He described Langenheim as an "odd duck" and when Ralph and I met, his complaint about Donath was that Fred required him "to justify everything I do with my requests." He was the only one to complain openly about Fred. Dave Anderson told me, "Basically, Fred runs a fairly democratic show."

When I interviewed with Langenheim, I mentioned living in Tulsa and driving on 'Langenheim Boulevard." I asked if it was named after one of his relatives. Ralph explained his father was Dean of Engineering at the University of Tulsa and active in community affairs. They named the street in his honor. As I came to know Ralph later, it was obvious he learned the inner workings of academe from a ringside seat since childhood and knew how to push the system to its limits.

Pat Domenico and Tom Anderson hosted dinner at the Urbana-Lincoln Hotel dining room. Pat was a heavy smoker, Tom smoked a pipe, and we had a couple of drinks. Eventually, they asked why I would leave Penn. I mentioned I really wanted to be in a larger department like Illinois, that there was a dispute between Boucot and Faul, the entire department had met with the deans and the provost, and a decision was imminent, probably at the end of the semester. Langenheim, who knew Boucot, also asked me about it. I downplayed as much as I could because I didn't know what Pat and Tom knew, and I wanted the focus kept on Art and Henry. They seemed satisfied.

Interviews continued next morning. Lunch was with Chapman, Carozzi, Langenheim and Hagner. During the visit I did not meet with George White (BS, PhD, Ohio State, glacial geology; New Hampshire, Ohio State, Illinois), Hilt Johnson (BA, Earlham, PhD, Illinois, Quaternary geology; Illinois) or Don Henderson (BS Brown, PhD Harvard, Mineralogy; Illinois). Bill Hay (BS, SMU, MS, Illinois, PhD, Stanford, ocean micropaleontology; Illinois, Miami (FL), Colorado, Kiel) was away. I knew Hay through SEPM meetings.

After lunch, I presented my colloquium in 228 Natural History Building, a large (300 seat) lecture hall with which I would become most familiar in time. Donath introduced me. The talk on the Bay of Fundy tidal sand bar dynamics went very well. During the question period, two students asked about storms. I replied that during my field work, there never were any major storms. Before-and-after aerial photographs taken after the 1938 hurricane of the Big Bar at Five Islands showed no changes. I indicated awareness of the problem which is why I set aside funds in my NSF budget so if a major storm occurred, I could go there and make over-flights to check out changes. Those comments ended the discussion.

Afterwards, George White, who I had met at GSA meetings, said, "George, I hope you'll come here." I thanked him.

My second last meeting was with Lester Fruth (BA, Carleton College, PhD. Columbia, rock mechanics; Illinois, Earthtec). Lester was Donath's permanent Research Associate who kept his lab and research program going. After a tour of the impressive experimental structure lab, we talked. I found him a difficult person to communicate with. Moreover, Lester had no personality at all.

The afternoon closed with an exit meeting with Fred. He told me he heard nothing but favorable reports from people, but would need to review an appointment at a faculty meeting and hold a vote. He asked about equipment needs. If appointed, I would inherit Richard's lab because Adrian was leaving, and I noticed it lacked things. I estimated I needed an additional $50,000 in starter grant equipment funds.

I brought up Paul Schluger and Julie Badal asking if they could get assistantships for the fall. He asked about them. I had the foresight in late January to Xerox their Penn files and gave him a set. He glanced at them and said it would have to be decided by the Graduate Admissions Committee, but saw no problems.

Fred then said, "George, I see you are an associate professor now. I suppose you would want a full professorship, but we can only offer an associate professorship." I replied, "Well Fred, let me ask you this. Can you appoint me as an associate professorship but be paid at what a beginning full professorship would make. And if so what would that salary be?" He replied, "Yes, that could be done and you would get $15,000 (Approximately

$70,000 in 2009 dollars) for a nine-month year, but why would you want to take it?"

I replied, "Fred, right now everything looks good. However, we don't know what will happen in the future. If I take an associate professorship now, you can review me for a full professorship later and with it comes an extra raise. If I accept a full professorship now, I give up the chance to get another larger raise, and to review progress, renegotiate the position, and ask for extra funds for additional facility needs."

Fred looked at me, almost in shock because he never expected my response. He thought for a minute and said, "You know, that's not a bad idea, reasonable and fair. If the faculty vote in your favor, that's what we should do." He then smiled and looked pleased.

Then we discussed timing. I brought up the Oxford leave and going to UCB. He said, "George, we must have you onboard during 1969-70 or we'll lose the budget line." Then he thought a minute and said, "Tell you what. Go to Oxford for the fall. Come here for a couple of months in January and February to teach a graduate seminar with Dan Stanley at the Smithsonian, and then go to Berkeley. That way you accomplish your sabbatical goals and we sew up the budget line." I realized Fred enjoyed coming up with creative administrative solutions to problems and I told him that what he proposed was acceptable. We said our good byes.

Dinner was at John Mann's house with his wife, Diane, Al and Marguerite Carozzi and Don Deere (BS, Iowa, PhD. Illinois, Geological Engineering; Illinois, private consultant in FL, elected in 1971 to National Academy of Sciences) and his wife. It was all social and a fun evening. I left next morning.

I returned to Penn and nothing had changed. Not a word from the administration. While I was gone, Art interviewed for the chairmanships of geology at Oregon State and at Cincinnati, and Pat interviewed at NASA. Shagam decided to stay another year.

On Tuesday, Fred Donath called asking if I was attending the AGU meeting in Washington, DC, the following week. I told him no, but could make arrangements to meet him. He wanted to interview Paul Schluger and Julie Badal. I said I'd see if they were available and let him know. I met with both of them, explained my situation and Fred's request. Paul arranged to change his TA schedule so I notified Fred and he set a time and place a week later at the Washington Sheraton Hotel.

I then received a call from the acting head of geology at Cincinnati, Dr. Larson. He asked if I would consider applying for their headship. He explained I was nominated by the students because one of them, Jim Barr, completed my NSF seminar the previous summer and thought it well run. I told Larson I needed to discuss this first with Boucot because he interviewed there and I didn't want misunderstandings. Larson understood, and I talked with Art. Art said, "George, do the interview. It's their decision, not ours. Anyway, I'd rather go to Oregon State."

I called back and arranged to travel there the week after the AGU meeting.

Julie, Paul and I drove to Washington and met with Fred. He talked mostly with them and I listened. While sitting there, I heard someone say, "I thought I saw George Klein. We need to talk with him." I recognized the voice as belonging to Bill Phinney (BS, MS, PhD, M.I.T, petrology, Minnesota, NASA) who I met at M.I.T's field camp in 1956. When Fred, Julie and I were done, I looked for Phinney on the way out but couldn't find him. We drove back to Philadelphia.

Next day, Phinney called. He said he knew I interviewed at Illinois and that Minnesota was also looking for a sedimentologist. Was I interested? I told him yes and asked how he knew I interviewed at Illinois. He heard it from Don Graf who was leaving in the fall

I made the trip to Cincinnati and didn't like it. The place was in an uproar with student riots. It also reminded me of Pittsburgh. Wayne Pryor was on a Fulbright teaching fellowship in Germany at the University of Heidelberg, the undergraduate dean (a geologist) was a total pinhead, and the department was funded through the graduate school which had a great Dean. I knew he would move on because he was too good for the place (and that happened). The president and I met and he didn't want to provide necessary funds to the department for me to be successful.

I returned to Philadelphia and on my first day back at the office, Fred Donath called. The first thing he said was, "I hear you've been on the road and also applied to Minnesota." I said, "Yes, Fred, that's true."

Fred then said, "George, the faculty met yesterday and voted 100% to offer you the job. The salary and rank will be as we discussed, Julie and Paul will be sent letters offering TA's, the starter money is confirmed, and we'll have you start on February 1, 1970, at the beginning of the second semester so you can go to Oxford and Berkeley. Are you available to accept our offer?"

"Fred", I said, "Well, I'm no longer available to others because I definitely will accept your offer. Now I'll need appointment papers with a letter of offer so I can resign from Penn and prepare to move."

"Well George" Fred said, "That may take three months to be approved by the Board of Trustees, but I'll keep you posted every step of the way." I said, "OK."

Fred then asked "George, can I ask you something. Were you really thinking of going to Minnesota?" I laughed recalling he was a Gopher alum and replied, "Only if you turned me down!" He roared with laughter and said, "That's great. This is really going to work out for both of us."

I wrote Cincinnati immediately telling them I was going to Illinois. I decided to contact Minnesota, Texas A&M, and LSU after I received Fred's letter of offer and other formalities.

We were now in the last week of the semester and Art, Reg, Pat, Jess, Gieg, and I received individual letters from Dean Stephens about the outcome of our meeting with Goddard, the two Deans and Henry. The letter stated that the administration sustained Henry as Chairman but with constraints. They asked each of us individually to cooperate with him. The letter closed with a paragraph I'll always remember:

"If you feel you cannot work with, or cooperate with Dr. Faul, I ask that you please consider pursuing your career elsewhere."

Copies of the letter went to Faul, Goddard and the Dean of the Graduate School.

After receiving Stephen's letter and assessing my situation, I realized there were still some hurdles to overcome to keep my sabbatical plans intact and go to Illinois in February. The first hurdle was overcoming a universal stipulation in all universities about sabbatical leaves. Sabbatical leaves are given AFTER a stated period of service, but if approved, one must return to one's home institution for six months to a year when the sabbatical ends. When applying for sabbatical, I agreed to return for the additional year.

I met with my attorney, explained the problem, mentioned the Boucot-Faul war, and showed him copies of the university sabbatical policy, my letter of application, and Dean Stephen's letter. The attorney said such a university policy had no legal standing, was a violation of the bondage provision of the US Constitution, and I could ignore it. But he added that universities rely heavily on "collegiality" and recommended that once I had Illinois' offer in writing, I write Goddard asking for a release on the grounds of Dean Stephens' closing statement, with copies to Stephens and Faul.

That gave me a game plan that would keep me in good standing in the academic community.

The next problem was what and when to move. Fred Donath networked me to the geology department's business manager, Bill Latham who gave me an authorization number to ship my rock specimens, peels, Box Corers, books and files by Railway Express. I arranged shipping in mid-August. As for moving my house, I rented it to a visiting professor and could not break the lease. Selling it was deferred until I came through town on my way from Oxford to Urbana.

I spent the summer writing up my Bay of Fundy research and packing rocks, books and files to ship to Urbana. I visited Julia Badal in Connecticut where she started thesis field work in the Triassic looking at newly constructed road cuts on interstate highways as well as known outcrops described before, but not sedimentologically. Faul left for a summer's field work in New Mexico.

Fred Donath called before I left. He explained that my appointment was approved at all levels of the University except the Board of Trustees. Because they didn't meet during the summer, my appointment will come up at their September meet. He added that feedback from both the Dean's and Vice Chancellor of Academic Affairs' faculty review committee was extremely positive.

I flew in August to Oxford. It felt good to be away from Penn and its troubles.

Stuart McKerrow met me at the University Museum. He was 'Vice Gerent" of Wolfson College, which meant he was either second or third in charge. He took me to the college office which was in a building away from the main college. I filled out paper work registering as a Visiting Fellow, picked up keys to my apartment, was given a pass for their main building and left. Wolfson rented three buildings. The office building was three blocks away from a college building with a dining hall, meeting rooms and lounge.

A half block away was their apartment building. It was a mile walk to the University Museum. The apartment building was also a half block from new construction for the new Wolfson College building on a tributary of the Thames.

Wolfson College was a new college strictly for graduate students. It was originally founded as 'Iffley College' in the village of Iffley just outside of Oxford (where Roger Bannister broke the 4 minute mile world record in 1954). They wanted to expand and needed to endow the college. The Wolfson Trust agreed to endow the college if they changed their name.

I told Stuart that I wanted to find ancient rock equivalents of tidal sediments in the UK. Because no one had done such work, I asked what units were described that contained lots of sedimentary structures as a place to start. He referred me to some papers by E. B. Bailey who, after a sabbatical at Wisconsin with L. T. Meade during the 1930's, returned to Scotland and mapped way-up criteria using sedimentary structures, introducing the approach to the UK. I read those papers and prepared a list of places to visit.

I also wrote Hans Reineck at the Senckenberg Marine Lab in Wilhelmshaven, Germany about visiting him. He invited me to come in latest August and participate as his guest on a lecture and field short course for the German Geological Survey. I flew to Hamburg and spent three great days with Reineck looking at the work he and his colleagues completed.

On my return, I flew to Glasgow, rented a car and start going to Bailey's localities. The first stop was the Island of Islay. I reached it by ferry, drove across a flat elevated peat bog and got a room at the Bowmore Hotel in Bruichladdich. The next day, I started field work in the Lower Islay Quartzite (Precambrian) found Bailey's localities and observed they replicated what I saw the previous week in the North Sea at Wilhelsmhaven. I also observed several fining-upward tidal flat progradational sequences.

The Island of Islay is unique. It had 30 whiskey distilleries (now only eight), 10,000 sheep on the flat bogs, and a population of 1,000 people (now 3,200). Their whiskey production accounted in 1969 for 30 percent of Scotland's Scotch whiskey export trade. It now is a luxury tourist attraction. The movie "Tight Little Island" was set there.

I proceeded to Skye to examine the Eriboll Sandstone (Cambrian). Its features replicated the tidal sand bars in the Bay of Fundy. I went to Durness to examine more Eriboll Sandstone and observed more confirming similarities. I knew I had enough for two papers and returned to Oxford.

I settled in starting to write up my Scottish work. My manuscript on Fundy was reviewed and returned. I revised it and sent it back to the *Journal of Sedimentary Petrology* (JSP). It was later accepted, and published in their December, 1970 issue.

Fred Donath wrote during late September that the University of Illinois Trustees approved my appointment and two days later I received the official paper work. I wrote Goddard requesting a release stating that when I applied for a sabbatical I intended to return, but the turbulence of the previous academic year made it difficult. I quoted Dean Stephens closing statement and wrote I could not work with Dr. Faul. Therefore, I requested a release.

After six weeks, Goddard wrote back saying he would give one but he also wanted a resignation letter effective at the end of the fall semester and I sent one. I had also sent letters to Minnesota, Texas A&M and LSU telling them I was going to Illinois.

During that fall, I visited a British geology department each week during 'term-time' to present a colloquium. I attended a meeting of the British Sedimentological Research Group (BSRG) at the Geological Society of London (GSL). GSL also scheduled me to show Gerry Middleton's movie of the 1968 Advanced Science Seminar in the Bay of Fundy and an accompanying talk in November.

Although a GSL Fellow then, I found their meetings stultifying. The meeting room reminded me of the colonial parliament house in Nassau, Bahamas which I visited in 1964. Senior professors and geologists sat on the lower (front) benches whereas junior faculty, graduate students, and visitors sat in the back benches. At US meetings, where rooms are arranged in lecture hall style, senior and younger professionals intermingle thus facilitating not only meeting people, but also giving younger people easier access to senior colleagues.

In January, I returned to Philadelphia, arranged to sell my home, visited with Reg Shagam, and drove to Urbana, arriving on Friday, January 15, 1970 at noon. My career at Illinois was about to start. I was there for the next 23.5 years.

LESSONS LEARNED:

1) During academic interviews, come prepared with a realistic understanding of one's expectations in terms of rank and salary.

2) During academic interviews, ask the same questions of all people one meets to double check for consistency of answers. If answers differ substantially, be very careful about the place. That is particularly true about criteria for tenure and promotion.

3) Academe has many arcane rules and practices. Therefore it is best to get advice concerning their legitimacy and ways to deal with them to keep one's reputation and integrity intact.

POSTSCRIPT. In 1977, Shagam visited me at Illinois to give a colloquium. He said that the Penn administration sustained Henry because he was part of their social circle. I told Shagam, "No Reg, it was purely a matter of money. When they listened to our vision for the future of the department, they knew they couldn't afford it. It was cheaper to keep Henry, keep the department small, knowing you, Art, Pat, Jess, and I would find something better elsewhere and we did. Also, Reg, if Henry came back asking for money to replace people or hire new ones, the administration kept him in check by telling him 'Remember Henry, we saved your job.' He had no recourse." I couldn't convince him.

CHAPTER 14
Illinois and Cal-Berkeley (Winter-Spring 1970)

"Time flies when you're having fun"
– Anonymous

I parked next to the Natural History Building (NHB) in the Illini Union parking lot both of which were next door. Fortunately, I arranged a room for a few nights so that gave me a parking spot and time to get a permanent parking sticker. As I discovered later, getting a parking permit was a big deal. In fact, many things at Illinois and the local community that were routine elsewhere were a big deal.

Pat Domenico and another person came out as I started to enter the building. Pat told me that everyone had gone to lunch so why not join them. Pat introduced me to Vic Palciauskas (BS, MS, PhD, Theoretical Physics, Illinois; Illinois-geophysics; Chevron Research, DOE), a new geophysicist, and we headed for the student cafeteria in the basement of the Illini Union. We got caught up and Vic told me a bit about what he was planning to do. He and Pat already started some research. Vic also offered an applied math course which I required all my graduate students to take.

After lunch, I went to the front office. Fred Donath was unavailable so I met Dorothy Smith. She handed me a departmental manual and said, "Mr. Klein, you might find it very valuable if you read it over the weekend." She then introduced me to Bill Latham, business manager, who arranged keys to the building, my office space, the mail room, and all my labs and student office.

When I entered my office, which was on the basement level, all the boxes, bags and other packages I shipped from Penn were there. It was hard to move around. Fortunately, mail was stacked on a table on the south wall

My office had a front door to the hallway and a side door to the adjacent office for my graduate students. I opened the side-door and Paul Schluger and Julie Badal were there and we had a friendly reunion. They summarized their semester's work and activities and generally seemed happier than I recall them being at Penn. Soon, we were joined by a temporary occupant of the office, Joseph R. Hatch (MS, Minnesota, PhD, Illinois, geochemistry; USGS). Joe worked with Don Graf and used a corner of the adjoining core X-ray radiography lab, to continue his geochemical thesis work. Graf's new lab was far from ready so Fred made this temporary arrangement knowing I wasn't planning to do core x-ray radiography immediately. I assured Fred later, and also Don Graf, that I had no problem with it.

I then went through my first class mail and sorted the rest. It took most of the afternoon to read it. During a break for coffee in the student lounge on the second floor, The Wanless Room (named after the famed professor of stratigraphy, Harold R. Wanless), I saw Bill Hay in the halls and he invited me to a party for his marine geology class that evening at his apartment.

Fred Donath visited around mid-afternoon to welcome me. We talked briefly but he was clearly happy to see me as I was to see him. He mentioned he'd be at the party at Bill Hay's apartment that evening and we could visit more there.

Fred also arranged for a student to move my rock samples to the larger sedimentology lab next to the Radiography Lab. The larger lab was divided into a chemical hood area on the north, and the rest was split in two between a dry lab with a sink on the east side and a conference table/classroom on the west. A sliding blackboard separated the two. Adrian Richards did a good job of designing the space when he arrived four years previously.

Don Graf came by to see Joe Hatch and because the 'tween-door' to the student office was open, he stepped in and we introduced ourselves. Graf (BS, Iowa, PhD Columbia, low-temperature geochemistry, Post-doc, Chicago, Illinois Geological Survey, Minnesota, Illinois), and I chatted for a half-hour. He was more serious than many but had a dry humor that could trip you up. We talked about his research, his laboratory renovations, and mutual friends at the University of Minnesota. He thanked me for letting Joe use a carrel and some of the radiography lab space.

I asked Don how he found the department now that he returned to the community. He said he knew Chapman and Henderson from his Illinois Geological Survey (IGS) days and some of the other pre-Donath faculty. He was impressed with the new people Donath hired. He was concerned about the older soft-rock faculty (Langenheim, Carozzi) slowing momentum and hoped I could perhaps counter their stance.

Around 4:30 PM, I went to my room in the Illini Union, took a nap, a shower, and changed clothes. After a meal, I headed for Bill's apartment and arrived with about three students. Soon the place was full and I started meeting everyone. Most of the students knew who I was because as I discovered in the geology department at Illinois, word travels faster than the speed of light, sound, or laser beam. Around 8:00 PM, Fred and Mavis arrived.

I knew something was up because suddenly the party atmosphere changed for a moment. Fred, Mavis and I sat down and he asked about the sabbatical leave, and I asked about the fall. The highlight of the fall at Illinois was the 1969 GSA annual meeting held in Atlantic City. Fred discovered it was cheaper to rent the Institute of Aviation's university DC-3 to take the faculty and students attending the meeting, than everyone travelling separately. It generated a lot of positive publicity for the department. An over-flight of the barrier islands of the Jersey shore was arranged during the meeting.

One of the students, Sue Mahlberg, came by to me to talk during the middle of the party. She obviously had plenty to drink and started carping about the department. I finally said, "Sue, I just got here ten hours ago. I don't know about any of this." Then I saw Fred Donath at the bar and I said, "Sue, Fred Donath is over there. Why don't you talk with him? He's in charge." She replied that was a good idea and fortified with alcohol and my comment proceeded to march up to him and as they talked, Fred stepped back against a wall. She just stuck her right arm up to the wall next to him and gave him an earful for a half-hour.

Sue, as I discovered later, grew up in Rockford, IL, where her dad ran the local museum and didn't think his daughter needed a college education (I heard similar stories from many Illinois students over the years). She was awarded a scholarship at Illinois, worked part-time and earned both a BS and MS degree (Metamorphic Petrology) at Illinois and a PhD at Brown. She later became a Research Fellow at Cornell and then rose through the faculty ranks to become a full professor. She also was elected a member of the GSA Council during the 1990's. Sue married Bob Kay, also at Cornell, an igneous petrologist, the son of G. Marshall Kay.

I stayed until midnight. Around 10:00 PM, Bill Hay, a gourmet cook, served a Mexican chicken dish with a chocolate sauce. It was an unusual blend of flavors, although later, as a geological consultant in Mexico, I ate it many times.

I went to the office next morning and read the department's manual. It had a brief paragraph about its history. Later, George White told me more about the university so what follows is a summary of what I learned. Like many Midwestern state university, Illinois had a unique history with some major events that impacted its direction.

Founded in 1867 as a land-grant institution under the Morrill Act, the University of Illinois was located in the last area of the state to be settled. East-Central Illinois was swampland and before development, large drain tile systems were installed for agriculture. The university campus was on privately-donated land. Over time, the campus grew, but because of its isolation (140 miles from Chicago), development was slow. The university had the reputation of being the party school on the prairie. Geology was one of the founding departments. It was first headed by John Wesley Powell who left two months later to head the famous expedition to explore the Grand Canyon.

Of the two communities, Urbana was founded by settlers from New England who resisted progress. When the Illinois Central Railroad built its main line from Chicago to New Orleans, the Urbana civic leaders voted against a railroad station in Urbana, much less a right-of-way. The railroad established a train station at 'Champaign Junction' five miles to the west, to serve the agricultural interests of Champaign County. A second community, Champaign, IL, grew around the new railroad station. The two communities grew towards each other and their common boundary was Wright Street which passed through the middle of campus. Despite efforts to merge the communities, none succeeded.

During 1910, the university presidency was vacant and a national search for a new one began. The alumni in the Chicago University Club, including those on its Board of Trustees, were annoyed with the continued derision directed to them about the university's party image. Alumni from Wisconsin, Minnesota, and Michigan pointed with pride to the research accomplishments of their campuses including the patent on homogenized milk, mining engineering, and medicine, respectively. The alumni were determined to find a new president who would make Illinois a research university on par with Michigan, and Wisconsin.

The presidential search committee wrote each of the short-listed candidates a letter spelling out that during the interview, they would ask the candidates' views on how to make Illinois a competitive research university.

The man appointed to the presidency was a history professor at Harvard, Edward J. James. To travel to Urbana for an interview, he left Boston for Albany, NY, where he boarded the 'Commodore Vanderbilt', the overnight train to Chicago. He went to Chicago's Union Station to take the Illinois Central to Champaign, IL. That trip took three hours. He was taken to campus in a horse-drawn carriage and spent the rest of that day and the morning of the second day meeting people and touring the campus.

Professor James met with the Board of Trustees after lunch and they asked if he was prepared to answer their question. The first thing James said (according to George White) was, "Gentlemen, your location is against you. For Illinois to do as well or better than Michigan, Wisconsin and Minnesota, you need to build the best facilities available, the best library resources available, and people will come here and do research that will enable Illinois to succeed." James went on to say, "I also advise the university move into two new areas that will increase in importance. One is chemistry. The second is a new field emerging from Austria where a leading scholar, Jung, is developing a field called psychology."

The trustees, one of whom was president of the Illinois Central Railroad, understood and agreed. However, one trustee asked, "How will we know that the professors are doing the work after we give them all the facilities you requested?"

James responded, **"From their publication record. Illinois should establish a reward system for those professors who publish their research results in recognized scholarly journals."** (Bold type mine). That one sentence embodied the American system of scholarly accountability known as "Publish or Perish."

James became the 11th president of the University of Illinois and change was rapid. First, chemistry was expanded and became one of the top seven departments in the US. Psychology was established as a department and is ranked now in the top five in the US. The engineering school was expanded also and some of its departments rank in the top three to five nationwide today. In science and engineering, major discoveries were made including the invention of nylon in the chemistry department. The campus was home to several Nobel laureates including John Bardeen who invented the transistor and the semiconductor.

The library received the biggest largess. Because so many of the humanities and social science faculty went to Europe each summer to do research, each was given a blanket purchase order to buy rare and important books for the library. The effort started slowly but after World War I when many personal libraries from European estates were being sold, the University of Illinois bought a treasure trove of rare and significant books. That occurred again after World War II.

When George White arrived in 1948, that purchase-order system was still in place. Because he and his wife, Mildred, had no children, they took a European vacation each summer and bought significant geological collections for the University Library, including an original of William Smith's 1815 geological map of England and Wales. George was instrumental in making the Geology branch library at Illinois the third largest geology library in the USA. This purchasing arrangement ended in the mid 1960's.

The Department of Geology experienced slow growth until after World War I. During the 1920's Francis B. Shepard was appointed and developed a new field, marine geology. Harold R. Wanless also was appointed and developed a world class research program on Pennsylvanian stratigraphy. Harold Scott (BS, Illinois, PhD, Chicago, Conodont paleontology; Montana School of Mines, Illinois) arrived during the mid-1930's and Carleton Chapman (BS, New Hampshire, PhD, Harvard, petrology; Illinois), arrived in 1938.

PhD's that emerged from their programs include Robert S. Dietz (BS, MS, PhD, Illinois, astrogeology, plate tectonics, marine geology; US Navy Electronics Lab, NOAA, Arizona State), K.O. Emery (BS, MS, PhD, Illinois, marine geology; Univ. of Southern California; Woods Hole Oceanographic Institute), George V. Cohee (PhD, Illinois, stratigraphy, USGS), J. Marvin Weller (BS, Chicago, PhD, Illinois, stratigraphy; Univ. of Chicago (replaced his father, Stewart Weller)), and Raymond C. Gutchick (PhD, Illinois, stratigraphy; University of Notre Dame).

After World War II, the department expanded. George White was appointed department head in 1948 and hired Ralph Grim (BS, Yale, PhD Iowa, clay mineralogy; IGS, Illinois), George Maxey (PhD Princeton, hydrogeology, Illinois, Nevada), and Don Henderson. Later he added Michael Wahl (PhD, Illinois, clay mineralogy; Illinois, Florida, GSA Executive Director) and Jim Eades (PhD, Illinois, clay mineralogy; Illinois,

Florida), W. Hilton Johnson, and Albert V. Carozzi. White was removed from the department headship in 1966.

I read the manual and noticed that if one needed certain items, one contacted specific people. As I proceeded to unpack and sort through items shipped from Penn, I quickly noticed things I needed. I compiled my list and annotated who was to get a specific list and determined I needed about 20 items to get the office properly arranged, and typed three lists.

While at Oxford, I wrote all the major Midwestern sedimentologists to find out what projects they and their students were doing. They included Bob Dott (Wisconsin), Paul Potter (Indiana), Ed Dapples (Northwestern), Keene Swett (Iowa), and Wayne Prior (Cincinnati). I wrote because during my Kansas days, the region was replete with stories about turf wars about projects. With the exception of one person, all wrote monographic letters listing projects, many that were never published or finished. However, one letter, which I memorized, came from Paul Potter:

"Dear George,
Congratulations on the appointment at Illinois.
As for your question about research, ***just remember, there is nothing like looking at an old problem from a new point of view.*** (Bold type, mine)
Good Luck in your future endeavors.
Sincerely
Paul E. Potter"

The bold quote in the letter was added as a filter to my research agenda, and I often used it as an introductory slide when giving talks, papers and colloquia.

I read those letters (I had requested they be sent to Illinois) and filed them for five years. I kept Paul's letter until I left in 1993.

On Monday at 8:00 am, I handed each of my lists to the three people designated to receive them. I could tell it blew them away. The support staff consisted of Dorothy Smith as Administrative secretary, Jan Nicholson as Chief Clerk, and four younger secretaries who were married to students. In addition, there was a thin-section technician, Jack Pullen, a draftsman, and two machinists (Ed Bauerle, Don Dotson) in the departmental machine shop which was devoted primarily to working on parts and components for Donath's rock presses.

One of the letters I received came from Keene Swett (BS, Maine, PhD, Edinburgh, sedimentology, Iowa) who was also working with one of his students on the Eriboll Sandstone. I called and we agreed I should come to Iowa in ten days, give a talk and we should pool data and write a joint paper. It appeared in 1971 in the *Journal of Geology*.

I found a temporary place to live for two months. Bill Hay alerted me to a Champaign-Urbana phenomenon. Many people in construction and farming (who commuted from Champaign to their farms on a daily basis) had winter homes in Florida. I rented a house from the owner of a construction company.

I taught my graduate sedimentology course but because it wasn't in the catalog, it was taught as a graduate seminar and we met for three hours at night once per week. All went well except during a student riot, a rock was thrown through my office window. Bill Latham arranged a replacement window fast.

I also made a trip to see Bob Dott (BS, Michigan, PhD. Columbia; US Air Force Research Lab, Wisconsin) at Madison, WI, to give a colloquium at the University of Wisconsin. Bob and I became good friends since a GSA meeting in 1962. He was one of my references. Bob and his wife built a custom home on an old quarry floor on a street which ended in an abandoned quarry. The back yard had a quarry face of the Cambrian stratigraphic section. He mounted flood lights outside the back of the house to admire the outcrop at night through an enlarged picture window. It was impressive to be eating in their dining room and looking at the flood-lit rock face.

A week later, I was in the front office and Fred Donath came out with a big smile. He held a letter from Louis Cline. Louis wrote that he heard me present talks at meetings and thought that although I did good research, he questioned my ability to teach large beginning geology classes. He wrote after hearing my talk in Madison that he was convinced I could and congratulated Fred on hiring me. Fred explained the Big 10 Department heads meet at GSA annually and compare notes and administrative strategies. These people were networked through a common bond looking out for each other, and sharing good news about their faculty.

Before I knew it, it was the end of March and time to head for UCB.

The first Monday after my weekend arrival in Berkeley, I met the department chairman, Chuck Meyer (PhD, Harvard, Economic Geology;

Kennecott, UCB). He gave me keys to my office and showed me where it was. He mentioned that traditionally, the entire faculty had lunch in a conference room and suggested I attend and meet everyone. Meyer also introduced me to a supervisory lab technician who wanted to know what my teaching needs were for my one course. Again, no time had been scheduled, so I arranged to meet for three hours on a Monday morning.

At lunch, I met most of the faculty: Jean Verhoogen (PhD, Louvain, metamorphic petrology, UCB), Harold Williams (PhD UCB, Volcanology, UCB), Frank Turner (PhD, Otago, Metamorphic Petrology; UCB), Nick Christensen (PhD, UCB, Structural Geology; UCB), Fred Berry (Petroleum Geology; Chevron, UCB), Garnis Curtis (PhD, UCB, Geochronology; UCB), Ian Carmichael (PhD, Manchester, Experimental Petrology; Manchester, UCB), Mitch Reynolds (PhD, UCB, Field Geology; UCB, USGS), Clyde Wahraftig (PhD. Harvard, geomorphology; USGS, UCB), Bruce Bolt (PhD, Sydney, Seismology; Sydney, UCB), and two others. Dick Hay was out of town but returned a week later for a month. Bill Berry, who I met at Yale, was in the Department of Paleontology, and during the afternoon, I visited him. He enrolled all his graduate students in my course.

I then prepared to do field work in Nevada. I read papers on the Wood Canyon Formation (Late Precambrian) and the Zabriskie Quartzite (Cambrian) and concluded they were worth examining for tidal sedimentary features. I used one of UCB's carryalls for a field reconnaissance a week after classes began. I made three return trips and over time, completed a field program in these units and the Eureka Quartzite. All three were a good match with the Eriboll Sandstone.

I telephoned Dr. Charles Muscatine a month after arrival. I took bonehead English with him at Wesleyan 19 years before (Chapter 4). He remembered my name and we met in his office. We chatted and then he asked, "By the way, where did you earn your PhD?" I was waiting for this moment and with a smile I responded, "Yale." After a 30 second silence and a lock of shock on his face, his composure returned and we continued the conversation. The expression on his face was literally priceless. It was unfortunate that the members of the fall, 1951, bonehead English class at Wesleyan weren't there to witness it.

Overall, I found Berkeley very disappointing. First the campus was in an uproar with anti-war protests. Second, the department was oriented more towards geophysics, petrology and geochemistry, so I had little common ground with most of the faculty. In the Paleontology Department, it was a classical paleontology and stratigraphy program.

One afternoon, Dick Hay and I took a break at a coffee shop near the geology building. We were joined by Ian Carmichael and four of his graduate students. After some small talk, Ian turned to me and asked in a haughty voice (amplified by his Lancashire accent), "Well George, how are enjoying Berkeley with all the great people we have?" I replied, "Ian, do you really want to know?" He responded, "Yes," expecting accolades.

I explained my disappointment with UCB. I found it narrow and specialized and except for Dick Hay, I did not find interest in soft rock in the department. He became belligerent and I kept countering his arguments, rebutting them. After a pause, I turned to Dick Hay and suggested returning to the geology building, and said goodbye to Carmichael. On the way back, Dick said "You really put it to Carmichael. No one stands up to him."

Returning to my office, I barely sat down and there was loud pounding on my door. I opened it and there was Carmichael. He said, "Look buster, you have it all wrong." He entered my office and turned to a small blackboard on the wall. He proceeded to write down the names of the faculty and separated them into two columns. He conceded the names on the left (at least 2/3 of the faculty) were not in 'soft rock'. The conversation went like this as I recall:

Carmichael: Let's go down the right column, Wahraftig.

Klein: World class geomorphologist, but he deals with surface landforms. That's not soft rock.

Carmichael: Garnis Curtis. He's dating ash beds at Olduvai Gorge.

Klein: Yes, he is dating volcanic material. That's hard rock.

Carmichael: Dick Hay.

Klein: Agreed. He does soft rock research – sedimentary petrology.

Carmichael: Gilbert

Klein: Agreed. He's an eminent sedimentary petrologist.

Carmichael: Berry.

Klein: He's a hydrological modeler. He's not what we would consider soft rock. His work is similar to what my colleague at Illinois, Pat Domenico, does.

Carmichael thought a while and sat down. During the next two hours we went back and forth and his attitude changed. He discovered we were both argumentative, enjoyed being so, and holding forth on a range of issues about geology. We became long-term friends and while there, talked things over at lunch, had a few beers, and became better acquainted. He opened up about the department and many of his impressions coincided with mine.

On my return from my last trip to Nevada, Nick Christiansen asked what I did there. I summarized my research and he asked if I could give a colloquium and did. After being introduced, I went to the podium and said, "Before I start my talk, I wanted to share something with you. In 1955, I was looking for a graduate school to attend and I applied to UCB. I forget who the chairman was (at which point Gilbert covered his face with his hands) but I received a letter telling me that my application was denied and wished me well in my future endeavors. So when Dick Hay called inviting me here as a Visiting Professor, I said to myself, 'if I wasn't good enough to be a graduate student, I was eminently qualified to be a visiting professor here'." The audience erupted, laughed and applauded. I often used that introduction during the next ten years when presenting colloquia at other universities that denied me admission.

I was also asked to serve on a Prelim committee for one of Bill Berry's PhD candidates, Tom Yancey (PhD UCB, paleontology; Texas A&M) in the Department of Paleontology. A senior professor in the department chairs the committee, not the advisor. That prelim chairman was Dr. Robert Kleinpell (PhD, UCB, Unocal, prisoner of war in Philippines, UCB). To say Kleinpell was crusty is an understatement. During the next 90 minutes, his questions browbeat the candidate. The next 30 minutes Bill Berry asked questions and I could tell Yancey was having a difficult time. A 15 minute break was made and the exam questions resumed. I asked questions and because Yancey was taking my course, I asked about a topic discussed during the previous week. He knew the material but his answers gave everything in the wrong order. Kleinpell's browbeating impaired his effectiveness. I stopped. Pat Wilde (BS Yale, PhD. Harvard; sedimentology; Dept, of Mining UCB) finished the questioning and we excused Yancey.

Kleinpell asked what we thought. I made it plain that I thought that he (Kleinpell) behaved abhorrently, browbeat the candidate, showed a lack of respect, and this impaired Yancey's true effectiveness. All we proved was that Yancey could endure the experience. I recommended passing him.

Kleinpell was shocked anyone would stand up to him and castigated me. Wilde said this was the longest PhD prelim he sat in on since coming to UCB in 1958 and told Kleinpell my point was well taken. We argued and finally, in frustration I said, "Dr. Kleinpell, after your performance, if I were Yancey, I'd leave the building, go to a bar, have a few drinks, find a lady to have my way with over the weekend, and return on Monday to find out if I passed." Kleinpell froze. We voted to pass Yancey.

We left the room to give Yancey the good news, but he wasn't in the halls. Bill Berry told me later that Yancey left the building and returned as Bill left. He thanked me for standing up to Kleinpell.

When Nixon invaded Cambodia, the campus exploded. Governor Reagan closed the University of California system and I went on my last trip to Nevada, returning when the campus reopened. On my return, every campus department, including Geology, organized students to fan out through Berkeley to deliver leaflets against the Vietnam War. A classroom was designated as an operational headquarters.

During lunch a week before my departure, Chuck Meyer discussed a letter he received from the administration. They sent a questionnaire inquiring how the anti-war protest impacted instruction, space use, and if any changes in course content occurred. Chuck circulated the letter for everyone to comment. When it was handed to me, I read it and immediately passed it on to the person on my right, Nick Christiansen, who supported the anti-war movement. Nick yelled at me questioning why I forwarded it without comment. I responded, "Nick, I leave in a week. I'm here as a visiting professor. This is a Berkeley issue. I don't want to say anything that may influence the rest of you to make a decision that I won't be living with."

Frank Turner then chimed in, "George, you're absolutely right. It's our problem. But, is there anything you want to add?' I replied, "Actually, I did have one thought. A classroom was assigned to the student activity. If the administration finds out, you may lose jurisdiction over that space." Turning to Nick Christiansen, I added, "Nick, if you feel it would help, why not let the students use the garage at your home as a campaign office." Chuck Meyer said, "Yes, we need to think about how to protect that space."

I flew back to Urbana a week later.

LESSONS LEARNED:

1) When serving as a visiting professor, never get involved in or take a public position on an issue with which the host department or university is involved. Even if invited to participate be guarded. When the appointment ends, one returns to one's home institution and doesn't live with the consequences of what was decided.
2) When starting a new position, follow any and all guidelines that are given.
3) Learn as much as possible about the heritage and traditions of one's employer, particularly an academic institution. That heritage is the framework for its rules and by-laws and determines its traditional mold that guides institutional change.

POSTCRIPT#1. When George and Mildred White went annually to Europe, they often bought rare geological books in duplicate, one for Illinois, and one for themselves (from personal funds). After retiring in 1970, he progressively sold parts of his personal collection to supplement his retirement income.

During 1980, I invited Bob Dott at Wisconsin to present a colloquium, and he asked to meet with George White. When I met Bob again at George's office before his talk, White was putting away the last of the books he displayed for Bob in a locked cabinet. The University of Wisconsin received a bequest from the estate of Louis Weeks to build a new building and upgrade their geology library. Bob obtained White's list of books for sale and was negotiating purchasing most of White's collection.

CHAPTER 15
Illinois: The Good Donath Years (1970-Late January 1973)

"The University of Illinois is a community of scholars held for ransom by a bunch of swine herders"
– Ralph L. Langenheim, Geology Faculty Meeting, 1979.

I returned to Urbana, closed on a new house, and went back to Philadelphia to move. I saw Shagam briefly. The biggest shock, however, was that on the Sunday before moving the *Philadelphia Inquirer* had a headline screaming: "PENN CENTRAL BANKRUPT." The next morning's edition featured an interview with the chairman of the investment committee of the University of Pennsylvania's Board of Trustees. He disclosed that the previous Saturday night, he planned to sell Penn's entire endowment holdings in the Penn Central Railroad on Monday. The bankruptcy filing ended his plans. The University of Pennsylvania lost 20 percent of its endowment that day. Obviously, it would be a long time before Penn would recover, and it would have a negative impact on the geology department. It confirmed I left at a good time.

After moving, I settled in briefly to check mail, journals, and read all of the departmental memos (Donath memos were all on salmon-colored paper). I met Paul Schluger before he left for his last season of field work in New Brunswick, and with Julie Badal. Julie decided not to continue her project in Connecticut but instead to work on a Lake Michigan shore sediment study led by Charlie Collinson at the Illinois Geological Survey (IGS). Collinson provided funds for her field expenses and a Research Assistantship at the Illinois Geological Survey.

Charlie had a joint appointment in the department and I agreed he should be her co-PhD supervisor. I thought a research assistantship at IGS would help Julie mature and gain experience in the working world. This turned out to be one of the worst decisions I made as a graduate PhD supervisor and during my career.

NSF awarded me a grant to pinpoint time of bedform migration in the Bay of Fundy, using an underwater movie camera housed in a sealed chamber. The project achieved limited success because the camera housing seal kept breaking under the turbulence of the Bay of Fundy tidal regimen and water leaked into the camera chamber. However, we had very good weather with minimal wind, permitting us to define intervals of bedform migration from timing the presence of surface boils.

Al Carozzi wrote while I was still at Oxford. He returned from Brazil teaching a short course to PETROBRAS, the Brazilian National Oil Company, who then negotiated an arrangement whereby two Brazilians were assigned to him to work on a project. The Brazilians came to Urbana to complete the project, audit courses and return after a semester. Carozzi was paid for his time in Brazil. PETROBRAS was looking for a clastic sedimentologist to do something similar and Albert nominated me. Was I interested? I wrote back indicating interest, but also contacted Art Meyerhoff for advice.

During the Goulart regime in Brazil in the mid-1960's American geologists working for PETROBRAS were expelled and weren't fully paid. Payments were finally made three years after their expulsion. Art confirmed that information. I was advised by a retired Scottish business manager with a steel company who I met in the UK that PETROBRAS and I open an escrow account in New York into which advance payments would be made. Eventually we agreed on a 50 percent advance payment and 50 percent payable on my return to the USA.

Before I left for Brazil, I took advantage of a service provided by the University of Illinois McKinley Health Center. Immediately following World War II, many Illinois faculty went overseas on US Government and United Nations fact-finding missions. Most returned quite ill with strange diseases because in Champaign-Urbana, Illinois, obtaining proper inoculations was non existent for overseas travel. The health center began providing this service at minimal cost to all faculty. Over time, those of us who used the service came to know the nursing staff there quite well.

My colleague, Don Deere told me an interesting experience he had at the McKinley Health Center. Because he travelled overseas frequently, he used their inoculation service regularly. After election to the National Academy of Sciences (NAS), he grew a goatee. A month later, he went to get a set of inoculations and the nurses acted like they didn't know him. He was instructed to lie prone on a bench after removing his pants and undershorts. As he lay prone, the nurse said "Oh, I'm sorry, Dr. Deere, I didn't recognize you until now."

During July, I spent four weeks in Brazil. Carozzi told me that PETROBRAS was unhappy with my request for an escrow account, but I explained the reasons. He was aware of the situation and said that it wasn't going to happen again (and to the best of my knowledge, payment problems never arose with their US geological consultants and advisors since 1969).

I arrived in Rio de Janeiro and was met by a PETROBRAS representative who took me to a hotel on the Copa Cabana beach. Arriving on Sunday, it was a festive place with Brazilians playing soccer on the beach sand.

Monday I went to the PETROBRAS office, met Carlos Walter Marinho Campos, Chief Geologist, and their chief stratigrapher. They reviewed the project explaining that on the next day I would fly to Salvador in Bahia, to work at their Reconcavo basin facility because that was where the project data was housed. I reviewed past work. It looked like a slope to deep-water turbidite play, but controversy developed in the company. I knew this could spell trouble. Usually when a consultant is brought in, according to my AAPG sources, it either meant the problem was unsolvable, or a controversy arose. To keep good internal working relations, a decision is made to hire a consultant and the resulting recommendation/interpretation was adopted until proven wrong. Then the participants in the controversy could agree management was stupid to hire the consultant in the first place.

The chief stratigrapher brought up one matter I did not know. When Carozzi returned a month before I arrived, he complained he spent a lot of free time on weekends supervising the two PETROBRAS geologists during their semester at Illinois. He demanded extra compensation. It meant amending his contract, and obtaining approval by PETROBRAS management, the national governing council of generals, the PETROBRAS Board, and the Brazilian Treasury Department because it involved foreign exchange. They eventually got it for him, but were worried about being further exposed if I made a similar request. I explained I couldn't assess what my workload for PETROBRAS would be in Urbana until I completed my work in Salvador.

Before I left for Salvador, the chief stratigrapher asked, "Have you discussed your project with Nelson?" I said "No." He made a phone call and ushered me to Nelson's office. When the elevator stopped on Nelson's floor, I realized we were in 'carpet city', the executive offices, and when we reached Nelson's office it said 'Director of Personnel." We talked and he claimed the unit we were examining (Candeias Formation) was deltaic and showed me where he worked on a map. I was polite and left. The chief stratigrapher explained that because they thought his delta model was outlandish, they moved him into personnel management because of family connections.

After flying to Salvador and settling into an American-style hotel, I went to work. Glenn Visher was there teaching short courses, so we visited a lot. One evening, while talking in the hotel lobby, a younger man walked in with a PETOBRAS uniform asking to speak with us to learn more English. We talked for a while. Clearly, he was working hard and told us he earned the equivalent of $250 (US – 1970 dollars) per month as a pumper. The most disturbing thing he said when I asked about this life and family was, "every nine months my wife presents me with a child." He seemed neither to know the significance of his comment, nor why this happened, nor how he could control the situation.

I looked out the window of my hotel room after arrival and noticed a hill, a string of lights going up the hill and nothing else. Next morning, I looked again and the hillside was one large 'favella', ramshackle tin shacks cobbled together to house the poor. The string of lights was the only electricity into the settlement. It rained and one could see rivulets flowing between houses laden with sewage. These settlements occur all over Latin America and although this may not be political or polite to say, the effect of an anti-birth control policy by the Roman Catholic Church clearly hurt these people and the region, in my opinion.

I began work with two PETROBRAS geologists, Ubijarro de Melo and Jorge Della Favera. We examined cores, looked at previous reports, seismic lines, maps and cross-sections. In the field, it became clear we were looking at delta front troughs with all the key criteria. As we went up dip, I recognized deltaic facies where Nelson had them identified, and downdip, the cores showed the presence of turbidites. In short, the entire succession was a transition from delta mouth to a delta front with delta front troughs filled down slope with slumps, debris flows, slurry flows and turbidites. It meant a unifying model to connect all controversial arguments.

Because of PETROBRAS protocol, I could not go into the field with my colleagues until the district geologist and assistant district geologist took me there first. We went and at mid-day, arrived in a small town at lunchtime. We went into a dingy place with a central room for food and drink. In the hallways to the side were small rooms with beds. We were about to order when my compatriots said something to the district geologist in Portuguese and we left. On the way out they explained to me they thought it was bad form to take the American professor for lunch at a restaurant which was known locally as a bordello and told him so.

We went instead to a restaurant next to a police station. The food tasted fine but next morning, I got extremely ill. The hotel summoned a doctor and by Monday, I was ready to go in the field. Ubijarro came to the hotel with a different driver because both the driver from the first day and the assistant district geologist got sick too.

I needed to use a restroom an hour after leaving and we stopped at a gas station. As I went to the rest room, Ubijarro followed me and said "Here in Brazil, the shit houses don't have ass-wipe". Because I always take a supply of toilet paper before heading for the field for just such emergencies, I showed him my packet of toilet paper and went in. When I rejoined Ubijarro, I suggested that he never use words like that in the US. He explained he heard them from American drilling crews who obviously had fun with him.

One issue that arose was the in-house interpretation that the Candeias Formation was lacustrine whereas I assumed it was of marine origin. That interpretation was based on paleontology. I asked "Which fossil group?" and was told *Ostacoda*. I asked to see the specimens. A paleontologist was called and predictably, the paleontologist was a lady. I looked at her samples through a microscope and discovered that all the *Ostracoda* were abraded. I showed this to her and mentioned a British paper about Jurassic abraded non-marine *Ostracoda* being dispersed 150 miles from shorelines and mixed with marine fossils. She was not expecting my analysis, but agreed it did not rule out marine deposition.

One of the geologists I met in Salvador had just returned from the USA, after working with Gerry Friedman at RPI. His name was Hanfried Schlaffer, a German immigrant. He was poor but wanted an education to study geology and was admitted to the University of Bahia in Salvador. Hanfried played an accordion and went around the bars in Salvador and earned tips to pay his university expenses. He did well because as a tall, blue-eyed blonde, he was considered unusual and was an accomplished accordion player.

One day, as he walked on campus between classes, he was besieged by a group of people who, as he put it "acted crazy". They ran after him and said "He's the one". They were movie producers and needed a tall, blonde-blue-eyed individual to be the lead character in a movie called "Sunset in Bahia". The movie people heard about him through the local grapevine. He explained that he had no acting experience and they told him they would send him to acting class. They paid well. He took a year off, took the acting classes for six months, made the movie, and banked the money. He told amazing stories about his experiences, and he enjoyed his way with the leading ladies away from the movie shoots. The movie was a flop.

When he returned to campus to finish his degree, he didn't need to play the accordion to generate income, but he did. He missed his friends in the Salvador bars, so he played for free. He now had a decent car, those leading ladies were still after him, and showed up to hear his music much to the delight of the bar patrons.

I received a call from Carlos Walter Campos about four days before I was scheduled to leave. He was flying to Salvador to meet me next morning. I wrote a one-page summary about our findings and recommendations. He was very satisfied. He then asked about the extra compensation.

I realized that I should try a different tack because I realized in Latin America, contracts are difficult to change. Because my university raises were geared in part to my publication record, I proposed that he release the scientific findings for joint publication. I would manage publication in a US-based refereed journal. I explained the added raise would stay with me the rest of my career and more than make up the difference. He readily agreed and told me that was fair. He was relieved I spared him from dealing with the bureaucracy to get funds approved.

I returned to Urbana, flew to Nova Scotia, and returned for the fall semester.

Nothing unusual happened that fall, but there were a few incidents involving field trips. I took my Geology 437 (Sedimentary Processes) class to the Wabash River point bars near Grayville, IL. I had bought a coffee pot for the sedimentology student office, and discovered when my students returned in the evening, it was still on.

The course TA was recently married. He requested to join us next morning because he returned from his thesis field work three days earlier and wanted to be with his wife.

Paul Schluger joined the field trip to pilot a boat that came with the lab and to teach the students how to use the lab's current meter and fathometer.

After driving the lead vehicle for an hour, I stopped as did the carryall behind me. I asked Paul if he unplugged the coffee pot. After roars of laughter, he told me, "No." I stopped at the next pay phone and called the new graduate assistant, explained the problem, and asked him to check. He went ballistic and said, "Dr. Klein, you're just making this up because I wanted to stay here tonight with my wife" and refused to go. I then decided to call Julie Badal as a contingency and asked Paul where she might be. Everyone said in unison, "The Thunderbird Bar" for geology TGIF.

The owner of the Thunderbird bar was a German immigrant who fought with Rommel in North Africa. I called, identified myself and he said (phonetics mine) "Dr. Klein, vy aren't you here vis your friends?" I explained and asked him to find Julie. Three minutes later she took the call and agreed to unplug the coffee pot. Next morning the new TA confirmed he went to the student office and someone beat him to it because the pot was still warm. I explained I called Julie after his outburst.

During early October, Donath asked me to join Art Hagner and Dave Anderson on a weekend field trip. The course was Geology 110, an honors freshman course Hagner taught but the field stops were in sedimentary rocks with which Hagner said he had no familiarity. I agreed to go. We visited several spots and after looking at the rocks, I interpreted the outcrops.

We camped at night and two students, a guy and a girl, paired off, cooked their own meals, slept together in a tent and all seemed calm. The second night, she got in a fight with her boy friend, kicked him out of the tent (it was hers) and invited someone else in to share the meal and enjoy her charms. Art, Dave and I just watched and said nothing.

The trip ended in Wisconsin in the Baraboo Quartzite. When examining the outcrops, I saw all the key identifying sedimentary structures I saw in Bay of Fundy tidal sands, and the Eriboll Sandstone. Clearly this was a tidalite. On my return, I called Bob Dott because the annual GSA meeting was to be held in Milwaukee in two weeks and he was leading a Baraboo trip for which I signed up. He offered me the opportunity to discuss my observations with the participants.

With Don Graf and me on board, Donath wanted to integrate us into teaching our specialties within the undergraduate major. He asked me to meet with John Mann and Ralph Langenheim who team taught a stratigraphy course, suggesting we add sedimentology to it. He gave us no guidelines and had not done the necessary ground work ahead of time to provide a framework from which to develop a plan. The three of us met and I reviewed their course outline. I showed them mine from Penn. Both John and Ralph resisted and we reached a stalemate. The joint course did not materialize. I offered a new three hour elective in sedimentology (Geology 309, Sedimentology, 2 hr lecture, and 1 hr lab) instead at the undergraduate level. Between 1971 and 1992, it drew 85 percent or more of Illinois' geology majors, and students from geography, and paleobotany.

Fred asked me to meet also with what he called the "hard rock" group to see if they could find a way to incorporate a module on sedimentary petrology taught by me. Carozzi told Donath sedimentary petrology should not be taught at the undergraduate level whereas during my interview, I explained I incorporated it into my Physical Stratigraphy course at Penn. We agreed that to subdivide the existing petrology course into three equal components of igneous, sedimentary, and metamorphic petrology, with Chapman teaching igneous, me sedimentary, and Dave Anderson metamorphic petrology, and put it into effect immediately.

Henderson was teaching mineralogy for years and that course was poorly reviewed. Fred asked Dave Anderson to teach it and he develop a new course outline a month before the meeting. I sat through the discussion as Henderson and Chapman argued against Dave's outline, although it was similar to the course I taught in 1961 and 1962 at Pittsburgh (Chapter 10).

Fred turned to me and said, "George you've been silent. What do you think?" I replied, "Well Fred, you may remember that I taught mineralogy for two years at the University of Pittsburgh and used a course outline very close to what Dave is proposing. At that time, I inquired what other places were doing and in fact Don Peacor at Michigan developed a similar outline independently."

Henderson's facial expression was one of shock and his jaw literally dropped. Chapman, who usually controlled his non-verbal reaction also acted surprised. Dave taught mineralogy the way he wanted.

During November, I attended GSA and went on Bob Dott's Baraboo trip. Bob sat on the bus in the front row, and across from him were G. Marshall Kay (BS, Iowa, PhD. Columbia; stratigraphy; Columbia) and Gilbert Raasch

(Geological Survey of Canada). I sat behind Dott. Kay and Raasch talked non-stop the entire trip. When we arrived at several Baraboo Quartzite outcrops, Bob let me discuss my findings and show them to the group. However, Bob explained later that Marshall Kay told him, "You know, Klein really showed good stuff and is a good geologist. But he talked too much!"

The Milwaukee GSA was memorable. First I gave a talk on the Tidalite concept, sediments formed by tidal currents. It attracted a lot of attention. Second, I attended the Illinois cocktail party and discovered that many geologists I knew earned their degrees there making for a fun evening.

Third, Arnold Bouma (BSc, PhD, Groningen; Utrecht, Scripps, TAMU, USGS, Gulf/Chevron Research, LSU) and I co-chaired a session. The papers were poor. The last paper was challenged and we let the challenger tear the speaker apart. At the end of the session I asked Arnold, "Well what did you learn this afternoon that was new, significant and different?" Arnold looked at me with a startled and almost blank expression and said (phonetics mine): "Nossing."

The second semester was a busy one. Donath asked me to teach Geology 111 (Honors Historical Geology), one third of a team-taught Marine Geology, a trial version of my undergraduate sedimentology course, and four weeks of sedimentary petrology within the undergraduate petrology course. It was almost too much. I also was advisor to the Wanless Club, the undergraduate geology club and accompanied them on a spring vacation field trip to the Grand Canyon.

Early in February, the university budget picture became cloudy. Fred called a meeting of the tenure committee and asked us to individually write him a Memo listing the three assistant professors we should retain without question, and three who could be considered expendable. I recommended Dave Anderson, Dennis Wood and Vic Palciauskas be retained. In the end he terminated a younger economic geologist and a paleontological curator, and retained the rest.

In February, Fred asked me to become the departmental placement coordinator. He established an administrative system where Hilt Johnson was Educational coordinator handling details about the undergraduate and graduate program. I agreed provided I never had to teach Geology 111 again. He accepted. I also asked for additional resources to make it go well.

When I discussed this new responsibility with Paul Schluger, he told me the graduate students were unhappy with the existing arrangements and would welcome a new approach. He made several constructive suggestions including finding a way for advisors of all interview candidates to meet oil company recruiters. Students complained that recruiters only took soft-rock faculty to lunch.

Paul also mentioned his wife, Birgette, worked as a secretary in the Economics Department and every year, they prepared a brochure about their PhD candidates which was mailed throughout the country. It helped them get good jobs. I asked him to get me a copy and when I showed it to Fred Donath, I recommended we plan something similar and mail it out to all US and Canadian geology department a month before the annual GSA meeting. I explained it required a cost commitment of clerical help and postage. Fred agreed. I also suggested each candidate looking for a job should submit a CV and copies of transcripts into the placement files, and each of their faculty references should write a letter for that file.

Fred thought it was an excellent proposal and asked me to present it at a faculty meeting. During that meeting, I stressed that placement was the responsibility of a student's graduate advisor. My role was to facilitate job information only and advise students where needed. The faculty endorsed everything except the letters of recommendation. I suggested that at GSA, I could circulate a complete dossier with copies of transcripts to potential employers that might bring an offer. There was resistance, but in the end, Fred Donath instructed the faculty to do it.

I held a meeting with the students informing them about the changes and gave them general advice. I stressed that on days before and after a particular company recruiter was interviewing in the department, it was in their best interest to wear a decent sport shirt instead of a four-letter declarative T-shirt. The stranger they could meet in the hall before the interview date might also be the recruiter. Often recruiters stayed an extra morning to visit the departmental office to look at student files, especially transcripts.

Late in April, Fred called me in again. He explained the graduate program needed a complete revision and was terminating the existing members of the graduate study committee. Moreover, he wanted a stronger screening of petitions from students requesting delays or waiving of requirements. Some approved petitions were undermining departmental guidelines and standards. He wanted me to chair the new committee with Domenico and Graf as committee members, and Johnson as an ex officio member. Fred also wanted a proposal for a revision in a month for faculty review. I told him I needed clerical back-up and that in future years, I be excused from teaching any more parts of marine geology. He agreed. Thus, in one semester I renegotiated my course load to a manageable level.

I then checked the existing program guidelines again and found that each graduate student had to complete one graduate course in one of four areas: soft-rock geology, hard-rock geology, geomorphology, and structural geology/tectonics plus specialized courses relevant to a student's goals. It was similar to the Yale program I completed (Chapter 7). The Committee was to consider a new approach whereby students planned their program individually with their faculty thesis supervisor and it appeared to be more specialized.

Pat, Don, Hilt and I met twice weekly. We developed a plan and I parceled out various responsibilities. In three weeks, we had a draft and I gave it to Fred. He liked the new proposal, made some recommended changes, the committee reviewed them and he accepted our final proposal. He circulated it to the faculty but there was resistance.

The senior faculty (Hagner, Chapman, Henderson, Carozzi, Hay, and Langenheim) thought the program undermined the true training of a geologist. I countered that argument by suggesting they could still require their graduate student to complete the four area courses and those they chose for their specialty. I added I would do so to some extent. During that faculty meeting, Fred mentioned that breadth could include advanced courses in engineering, math, physics and chemistry. The new program won by one vote and Fred decided we should review it again as a committee, make changes, and bring it up for a second vote. The second time it passed by a 3 to 1 margin.

That spring, the department held elections for one of its two members to serve on the faculty senate. I was elected, but resigned after a year.

By this time I had three new graduate students. They were Roscoe G. Jackson (BS, KU, PhD, Illinois; fluvial sedimentology; Northwestern, Michigan, Jackson Brothers Oil and Gas), John J. Barnes (BS, MS, Michigan State, ABT, Illinois, stratigraphy; Amoco, independent consultant), and Gordon S. Fraser (BS, MS, PhD, Illinois, stratigraphy and sedimentology; Illinois Geological Survey, BP Alaska, Indiana Geological Survey, Director - Great Lakes Institute at Buffalo State University).

Gordon was the most interesting one. Departmental policy permitted people to earn two, but not three degrees from the department. Gordon countered that he did a BS thesis with Richards, a Masters with Langenheim and wanted to do a PhD with me. He said that changes in the department and my course requirements were no different from going to a different university. Donath accepted the argument.

Originally, Gordon planned to complete his PhD with Langenheim. He had a half-time research assistantship at the Illinois Survey and was already publishing. However, he wanted to do a thesis on the Ordovician Glenwood Formation which Langenheim discouraged.

One Friday, at the usual geology TGIF at the Thunderbird Bar, Gordon and his wife were sitting with me and he asked about the possibility of my supervising his thesis on the Glenwood. I asked what would be new, significant and different in such a thesis. He hedged because he hadn't thought about research this way. I then asked, "Gordon, what is the stratigraphic order of the formations below and above the Glenwood?" He replied, "St. Peter Sandstone, Glenwood Shale, and Platteville Limestone." I then asked if the lithologic succession or facies succession had any significance. He couldn't think of any.

Finally I said, "Gordon, do you remember your Historical Geology course where the main paradigm was that a succession of sand overlain by shale and capped by limestone represented a deepening water facies change across a shelf?" He replied, "Yes." I asked, "Do you believe it?" He thought and said that he wasn't sure. I told him, "Gordon, I'll supervise your thesis if you address that paradigm as a problem you wish to solve and will use the Glenwood to solve it." I added that if he met me Monday, I'd help him write proposals to GSA and Sigma Xi for funding. He said, "Won't the department fund it?" I responded, "Gordon, if you take the initiative to raise your own money and get it, the department will give you more money than if you applied only to them."

Gordon and I met. I helped him formulate a GSA and Sigma Xi proposal, which was funded by both societies two years in a row. The department gave him added funds. When he finished in 1973, he submitted a paper to the *GSA Bulletin* and it was published in 1975. The Glenwood Formation was a tidalite sandwiched between eolian sands and coastal, tidalite carbonates.

As an aside, in 1974, I visited UCLA to give a colloquium on "Tidal circulation on cratonic seaways." While there, the students were telling me that one of their faculty, Gary Ernst (BS Carleton, MS, Missouri, PhD Johns Hopkins, Metamorphic Petrology, UCLA, Stanford (Dean of Earth Sciences), member of National Academy of Sciences) discovered the reprints of his Master's thesis on clay mineralogy of the Glenwood Formation and gave them away to everyone. I said nothing.

After presenting my colloquium, UCLA's stratigrapher, Clem Nelson (PhD, Minnesota; UCLA) asked, "George how does your model of tidal circulation fit the Glenwood Formation?" The audience laughed. I replied, "Glad you asked. One of my PhD's, Gordon Fraser, just finished a thesis on the Glenwood." and I explained his findings. There was a stunned silence.

Five PhD's was a heavy load, but Paul would finish by January, and Roscoe wasn't starting a thesis for another year. I was also asked to serve as a co-chairman/thesis advisor for a geography PhD candidate, Dag Nummedal (BSc, Physics, MS Biophysics, Oslo, PhD, Illinois; sedimentology; Post Doc at South Carolina, LSU, Unocal, Wyoming, Colorado School of Mines).

Dag was an unusual case. When he finished his masters in biophysics at Oslo, he decided he wanted to earn a PhD in sedimentology in the USA. In Scandinavia, until the early 1970's, sedimentology was taught in geography departments. Thinking the US no different, he applied to the geography department at Illinois who were happy to get him. When arriving in August, 1969, he asked about sedimentology and was told that was done in the geology department. Dag visited Dorothy Smith who told him there was a new person coming in February named Klein. Dag waited to take my Geology 437 course until the fall of 1970.

He went on after his PhD to complete a Post-Doc with Miles Hayes at South Carolina, replaced Jim Morgan at LSU, then went to Unocal as manager of geological research, then to the University of Wyoming as Director, Energy Institute, and moved to Colorado School of Mines to head the Colorado Energy Institute. Looking back, Dag is my most distinguished and most widely known PhD student.

In April, I went to Houston for the annual AAPG-SEPM meeting. I met Donn Gorsline (BS Montana, PhD USC; Marine geology, Florida State, USC). He said, "Congratulations." I asked "For what?" He replied, "Oh you didn't know? Your paper on Intertidal Sand bodies in the December, 1970, JSP was voted the Outstanding paper for 1970. You'll get your award next year. Be sure to come to the business meeting when it will be announced."

Carozzi, Langenheim and Hay also attended the meeting. They returned to Urbana before me and let Fred Donath know. When I returned, Fred came to my office to congratulate me. He then asked me to make extra copies of the award letter and explained he would send a copy to the Dean, the Vice Chancellor for Academic Affairs and the President's office for transmittal to the board of Trustees. He added that this is how the major academic players are identified on campus and everyone in the upper levels of the administration kept track of their accomplishments. When I received the award letter, I did as asked.

The semester ended on a somewhat disappointing note. I taught my undergraduate sedimentology class as a senior seminar to six graduating seniors. It was strictly lecture. The format was more like I observed at Oxford, with only one exam at the end. The students seemed to enjoy the course, but there was one, Don Eggert, who had a reputation of being a loud-mouth, or as I liked to put in later years was 'part of the irrational disgruntled' of this world. Don grew up in a hard-scrabble Chicago neighborhood and definitely was rough around the edges. On the final exam, Don earned a 'D.' He filed a grade grievance alleging that the other five students in the class studied together, and enlisted Rick Forester, a graduate student to prepare them for the exam. Rick completed my graduate course the year before.

What was most disturbing was that Eggert's letter was written in flawless English. His exam was written in very poor English with multiple grammatical errors which is also how he spoke.

Chapman as chair of the grade grievance committee visited me, gave me a copy of the letter, and asked if I could respond in two days.

I wrote the committee stating the other five were not treated on a favorable basis, they could study together for the exam if they wished, and were free to get advice from whoever they wanted, as was Eggert. I also reviewed the seriousness of Eggert's mistakes on the exam, including a serious mislabeling of a diagram showing fields of flow regimes. I ended the letter stating, "If Mr. Eggert had written his exam questions with the same level of English proficiency as his grievance letter, this complaint would never have arisen."

I personally delivered the letter to Chapman who read it. When he read the last sentence, he smiled and said, "Yes, I wondered about that too. Where do you suppose he got help?" I replied, "Carlton, I really don't know, but because of the maturity of the prose, I suspect a faculty colleague." Carlton said, "Yes. I agree."

The outcome was that the committee asked me to give Don another exam. Dave Anderson who also was a committee member advised me privately that in the future I should give a mid-term and a final, although he too had attended a university where each course only had one final exam. It was good advice.

I also knew I could not prepare a substitute exam that was as intensive as the one I had given. I therefore gave Eggert the same exam and his grade improved about two percentage points. Chapman told me later that Eggert had only himself to blame. However, I realized that my successes during the year were arousing envy on the part of some colleagues and I needed to be a little careful.

I spent the summer taking a one-week trip to Brazil. Jorge and Ubijarro finished our project and we wrote a paper which appeared in the 1972 *GSA Bulletin*. It generated a lot of reprint requests and was widely cited. I completed a presentation to the Petrobras team in Salvador who were pleased with the unifying model. They realized they each had made a correct interpretation on a small part of the spectrum, including Nelson.

I returned to Urbana, field checked Gordon's work in the Glenwood and wrote several other papers I had in progress. I realized as I advanced in my career, there would be times to reassess during a summer and get caught up writing papers before generating more new data and field observations.

In August, I attended an IAS meeting in Heidelberg, Germany and presented both my general tidalite paper and a paper on my Bay of Fundy work. They were enthusiastically received. The meeting, however, was in the minds of many, not a success.

After the 1971 IAS meeting, I flew to Edinburgh to attend the first all British Geological Societies meeting, modeled after GSA. I was an invited speaker, along with Bob Dott at Wisconsin, and Keene Swett at Iowa to talk at a special session of the BRSG (British Sedimentological Research Group). The entire all-societies meeting was organized by Professor Gordon Craig (PhD. Edinburgh, General Geology, History of Geology; Univ. of Edinburgh) and it was held at the University of Edinburgh to commemorate the 100th anniversary of the founding of its department of geology.

After registering, I saw Craig and we chatted. He was very effusive about the celebration of the 100th anniversary of his home department and rambled on about it. Finally, I said, "Gordon, I wanted to talk about this. The University of Illinois' geology department was founded in 1867 so on behalf of that department I want to welcome the University of Edinburgh's geology department to the centenary club!" My comment was overheard by several people and Craig bristled. He replied, "Young man, I want you to know that we have had geology at Edinburgh for 200 years but it was first offered by a department of natural philosophy." I replied, "Yes, I understand that, but still, Illinois' geology department is officially older." He left in a huff but the interchanged reverberated through the whole meeting and generated much humor.

The meeting ended on a Saturday morning and I returned to Urbana. The fall semester started quickly. When new students arrived, I was expecting some based on accepted applications. One was Nancy Lee, but she did not work out. Another, John Nelson (BA, Williams College, MS, Illinois; Phillips Petroleum, Illinois Geological Survey) was clearly bright and did well in my course. He wanted to be a petroleum geologist and refused to take Vic Palciauskas' applied math course. Realizing he was unhappy, I privately spoke to Langenheim and suggested Nelson might be happy in his program. Ralph indicated interest and I referred John to Ralph and that worked out for everyone. There were two more students who wanted to work with me and I'll discuss them below.

I was teaching two courses, Geology 309, Sedimentology, and Geology 437, Sedimentary Processes. About 25 students were enrolled in Geology 309, and 15 in Geology 437. Paul Schluger and John Barnes were assigned as TA's in 309, and were available to assist with 437 on the fall field trip.

Because of fall commitments off campus, I decided to merge the 309 and 437 field trips. I would leave with Geology 309 on a Thursday, spend Friday with them on the Wabash Point bars at Grayville, IL, and rendezvous with the Geology 437 class on Friday night at a camp ground near Goreville, IL. Saturday, the two classes jointly examined several outcrops of ancient sedimentary rocks and the undergraduates were sent back around 3:00 PM. I took the 437 class back to Grayville and we spent Sunday sampling the Wabash River point bars and collecting current velocity data.

Planning meals for this group was difficult. Breakfast and lunch worked out. However, for Friday's dinner, I thought if we cooked canned beef stew all would be well. It wasn't. One of the students in the class, Margaret Leinen

(BS Illinois, MS, Oregon State, PhD, Rhode Island, marine sediments; Illinois Geological Survey, Univ. of Rhode Island as Prof, Dean of Oceanography, Provost; National Science Foundation, Associate Director, Geosciences; Climos, Inc., Chief Scientist) said, "This is the worst meal I've ever had on a field trip!" I replied, "Margaret, I'll admit that I agree with you. I made a mistake." That reply utterly shocked her and she didn't know what to say. In time, she acquired more confidence.

I began serving in the faculty senate and was appointed to the Calendar Committee. The committee was charged to change the calendar from the traditional start in September, ending in January, and starting again in February and ending in June. The goal was to start earlier and end the first semester by Christmas, and start the second semester in January and end it in May. The committee was chaired by the associate dean of the College of Liberal Arts and Sciences and included the associate dean of agriculture, an education professor and two student members, plus several others.

In those days, I still had a crew-cut, and wore a white shirt and tie with either a suit or coat and tie. When attending the first meeting, I wore a suit and the students inferred I was the 'right-winger' on the committee. As the discussion started, I mentioned I had come from Penn where the fall semester was 13 weeks and the spring semester was 14 weeks long to accommodate a schedule such as being proposed and advised it be considered at Illinois.

The students were shocked at my 'radicalism' and the rest of the committee pooh-poohed the idea. However, at the second meeting, the chairman explained he reviewed my ideas with the higher administration who told him to seriously consider it. The agriculture associate dean explained it would be difficult to implement in the College of Agriculture because most of their programs required accrediting and for graduates, licensing. He explained that accrediting and licensing boards watch such changes carefully and he couldn't place his college at risk. The representative from the School of Engineering concurred.

The faculty member in education, however, said what to me was (and still is) not only shocking, but rigid and in some ways utterly stupid, all simultaneously. He turned to me and said, "Traditionally and pedagogically, all courses are 15 weeks long. That's why textbooks have 15 chapters – one chapter per week." That comment aroused my combatative instincts and I responded, "I can't speak for the school of education, but I try to train my students to think and analyze critical data. I don't use textbooks. I assign both my undergraduate and graduate classes current and key journal papers which are placed on the reserve shelf of our branch library. That way, they are able to advance professionally and develop lifelong learning skills."

"But don't you teach any beginning geology classes where you use a textbook?" the education professor asked. I responded, "No, but as I recall, most have about 40 chapters."

The committee finally recommended starting the fall semester in mid-August (the height of ragweed season in Illinois) and ending in Christmas, and starting the spring semester in early January and ending in May. The committee was asked to sit in the front row during the senate meeting to be available to answer questions or participate in discussion, along with Deans and other administrators. On arrival, I found an empty seat in the front row. I was joined by Dr. Robert Rogers, Dean of the Liberal Arts and Sciences (LAS) College (BA, English, Michigan, PhD, Harvard; 18th century English Literature; Univ. of Illinois, rose through faculty ranks to Head of the English Department, appointed Dean in 1964). We met briefly once before and so chatted about each other's work to become better acquainted. When the proposal passed, he said, "We won a great victory today."

GSA that fall was in Washington, DC. The office staff assembled the brochure about our PhD and Master's graduate and mailed them to 350 geology departments. They then assembled individual student files with CV's, copies of transcripts, and reference letters. I was shocked at the reference letters. They barely were informative and it reminded me of my conversation with Howard Meyerhoff (Chapter 11). One letter basically said that he was the PhD supervisor, had known his student for six years, and to call his office telephone number for more information. I reviewed it with Fred, sharing my discussions with Meyerhoff. Fred was puzzled too and advised me to take CV's and transcripts only to the Washington GSA.

During GSA, I met with many of people to share dossiers. At the Illinois cocktail party, Fred took me aside and said, "George, mailing out that brochure was a terrific idea. That mailing generated more positive Illinois talk at the meeting than I ever heard before." Clearly he was satisfied we had done the right thing for our students.

At the beginning of the semester, I was visited by Linda Provo (BS Hope, MS, Illinois, PhD Cincinnati; Exxon Exploration) introducing herself as a new sedimentology student. I had not recalled her application. I reviewed her program and like many small college graduates, she was deficient in physics and chemistry. I asked her to complete the year's worth of physics and chemistry required of all Illinois undergraduates. She completed a Master's thesis and I advised her to move on to Cincinnati to try something new.

Roscoe Jackson kept visiting me during the late spring and several times during the fall. He wanted to do his PhD on Wabash River Point bars but I told him that Wayne Pryor at Cincinnati had already worked there. Each time I asked what he would do that was new, significant and different and he had no explanation.

He was taking fluid mechanics courses in engineering and in November, 1971, returned and told me, "Doc, I really want to work on the Wabash River." I said, "Now Roscoe, we've been over this before, what are you going to do that will be new, significant and different?" Roscoe replied, "That's what I want to talk about. Every sedimentological paper on river bars fails to include the engineering hydraulics. Using the lab's current meter and boat, I can take long term velocity and bed shear measurements and tie them to river stage, sediment texture, bedforms, orientation, and evolution. It means a two-year monitoring program." I told him, "Ok, go ahead. That will be new, significant and important. Incidentally, when Paul Schluger leaves, why don't you move into his carrel in the sedimentology student office?"

After Thanksgiving, I was visited by another first year student, Daniel E. Lawson (BS, Lawrence Univ., PhD Illinois, glacial sedimentology; U. S. Army Cold Regions Research Lab). He asked, "Do you think a PhD thesis on Recent glacial sedimentation would be significant?" I told him it would, but logistically, it would be expensive, difficult to arrange, and conditions would be rough, but it was doable.

A year later, Tom Ovenshine (BA, Yale, MS, VPI, PhD, UCLA, sedimentology; USGS), who was a senior at Yale my last year there, called. He was now Assistant Branch Chief of the USGS Alaska Branch and needed a summer field assistant to work in Turnagain Arm, AK, to study tidal sediments. I told him about Dan and his interests and he offered to hire him for the summer and help find a glacial system that was accessible and met his requirements. Dan worked on the Matanuska glacier and turned in a significant thesis.

During late September, Fred called a meeting of the tenure committee to review assistant professors for promotion to tenure. We agreed to recommend Dave Anderson and Dan Blake for promotion and by spring, they were promoted.

Fred then asked the associate professors to leave so the full professors could review us for promotion to a full professorship. I went home, barely sat down and Fred called. He said in an enthusiastic voice, "George the full professors voted to promote you to a full professorship. In fact, they only took five minutes." I said, "Wow, and if it happens in the spring, it will be before I turn 40. Is there anything you need from me to build your case?" Fred replied, "That's why I called. Can you meet me and Dorothy Smith tomorrow morning at 10:00 am? Bring an up-dated CV, an up-dated publication list, a list of senior sedimentologists who can write supporting letters, reprints for the last four years, and a list of your current MS and PhD students." I assured him I would be there. In fact, I left the house early, ate breakfast at a local Pancake House, and arrived at my office at 7:00 am to assemble it. I delivered everything to Dorothy and Fred by 10:00 am. The promotion was approved quickly.

Paul Schluger defended his thesis and did very well. He accepted a postdoc at SUNY-Binghampton. From there he went to AGAT Research in Calgary, moved to the Mobil Research lab in Calgary, was transferred to Mobil's office in Denver, and passed away in 1981 after editing an SEPM Special Publication.

Early in February, 1972, Dag Nummedal visited me about transferring into the geology PhD program. Although I welcomed the idea, I explained the old guard professors would require him to take undergraduate courses, retake his PhD Prelim (which he passed in October), and then finish his thesis. I explained it would take two extra years to complete. I suggested that while finishing his thesis in the geography department, he enroll or audit as many geology courses for which he could find time and try and do a post-doc in geology to make the transition. Two weeks later, Miles Hayes visited from the University of South Carolina and Dag arranged to meet him. Miles offered him a Post-Doc and Dag's geological career took off.

In mid-February, Fred called me in and asked if I would serve on the departmental executive committee representing the entire soft-rock faculty. It meant meeting weekly with Fred and the other committee members. He decided to revamp it because he felt he needed better feedback and advice. The new committee included Chapman, Graf, Johnson and me. Chapman was a New Englander and I found him very reserved, but as I got to know him, I realized he was a wise man up to a point. He was not as good as Howard

Meyerhoff, but worth listening to. By now, I knew Don Graf quite well. I found he spent too much time carping about people everywhere, and especially in the department. Johnson was knowledgeable about the university and the constant changing administrative rules.

We started to meet and I noticed a troubling pattern. Fred kept meetings focused on personnel issues. He was troubled about Ralph Langenheim, alleging Ralph was trying to undermine him with outsiders and on campus. He also criticized Ralph and Bill Hay for never following policy with their students when petitions for extensions or waiver of rules were made.

In Hay's case, the problem was compounded by a change in his faculty appointment in 1969. Bill was at Illinois nine years and developed an international research reputation in ocean micropaleontology. The Rosensteil Institute of Oceanography at the University of Miami (FL) made him a generous offer. Fred did not want to lose Bill and in a creative administrative way, arranged for Bill to spend half a year in Urbana, and half a year in Miami as a joint-institutional faculty member. Bill accepted this arrangement but it created difficulties for his Illinois students during his absence and his inability to supervise them. The problems were compounded when Bill went on ocean-going research cruises during his 'Miami semester" and literally was unreachable.

After our second executive committee meeting, Fred called a 7:00 PM faculty meeting. He told the faculty he wouldn't tolerate any more efforts to undermine his goals for the department or its programs. He disclosed that he and Mavis received many unsolicited orders at their home that one could place from unsigned tear-out post cards inserted into magazines. They recovered some of the order forms. Fred alleged that he recognized the font from a faculty member's typewriter.

He then placed a paper bag on the conference table, emptied it, and poured out a dozen 10' carpenters nails. He claimed they were placed under his car tires in his special 24-hour parking spot next to the building. He suggested a well known faculty prankster (implying Langenheim) likely did it.

That meeting was troubling to me because I hadn't noticed anything major that was amiss. Clearly there were some unhappy faculty, and although I heard things, I ignored them. I also began to wonder about Fred. Four days later, I passed his parking spot which he had vacated for lunch. A university pickup with workmen was temporarily parked there eating lunch and one could conclude that the carpenters nails we were shown could have fallen off a similar truck.

As I worked with Fred and the others on the executive committee, I also noticed Graf continually disparaged Langenheim and Hay suggesting they be fired (both had tenure). Fred and I often talked privately and it disturbed me he was talking about my colleagues with me. I felt he should find a discrete department head to discuss such issues rather than a faculty colleague. I reminded Fred I had to work with and maintain good relations with those people too. I suggested in a most understated way that Fred should seek a discrete outsider on campus to review these things. The message was not understood.

By now, I was supervising nearly ten graduate students, a heavy load. I then reviewed the transcripts and academic record of every student I supervised and classified how well they did. I then compared their performance with their schools of origin. The findings surprised me. The students who did very well under my guidance earned degrees from flagship state universities like Illinois and Kansas. The second group that required some assistance to get started came from what could be described as emerging state universities such as Temple and Lawrence University. The people who were most problematic, had the greatest academic deficiencies and needed too much help came from small colleges such as Franklin and Marshall and Hope. I decided to preferentially recruit students who graduated from large flag-ship state universities. I also realized that as a Wesleyan graduate, I could never be admitted to my own program.

I also recalled witnessing a conversation Joe Peoples had in 1958 with Marlon Billings, the famous structural geologist at Harvard and whose textbook was used by three generations of American geologists. Marlon explained to Joe that when he started at Harvard, he admitted students with straight 'A' records. He discovered once they were on their own doing research, they had trouble finishing. He also admitted some with a less promising undergraduate record who did very well. As undergraduates they earned straight 'A's' their freshman year, their grades took a nose-dive during either their sophomore or junior years, but the following year until graduation, their earned straight "A's". Billings discovered that the nosedive involved either a death in the family, parents divorcing, or a serious break-up with a girlfriend. He realized these people were more self-reliant in graduate school because they overcame adversity. Once realizing this, Billings only recruited students with what he called a 'V-profile." My best students up to this point all had 'V' profiles.

I decided also that I needed to screen students working with me after they arrived. I met with each individually and explained that my current role was provisional pending their meeting certain requirements. First, they had to take a graduate course with me, earn an "A' and be in the upper third of the class. If they earned a "B", they had to rank in the top quarter of the class.

Second, by mid-November of their first semester, I expected them to see me about a possible MS or PhD thesis project. I then helped them refine it if they chose to continue to work with me. If they couldn't meet these two requirements, I suggested they find another advisor.

Near the end of the semester, I asked what they planned to do between Christmas and the beginning of the next semester. Usually replies ranged from going to the beach in Florida, skiing in Colorado, or visiting with friends at home. I told them to report to me on the day after the New Year and explained I was going to help them prepare proposals for their thesis funding to be submitted to GSA, Sigma Xi, and AAPG, and other relevant agencies.

There was grudging assent, but they did. I also explained that the department would fund their thesis work at a higher amount if they received such grants. Being young and inexperienced, they applied, not understanding fully the significance of their application. All my students received such external funding and when they did and looked around the department, they discovered few others had. They realized they were now on their way to becoming a practicing professional. I always made a supervisory trip during their field work not only to double check their field skills and provide guidance, but also to bond with them, and continue to guide their professional development.

When they returned to campus, I told them that now that their field work was done, or would be finished next year, they were more on their own. I left it to them to keep me posted on progress and to take the initiative to see me if they had problems or needed help. I had ways to casually check on their progress, but except in one case, this system worked well for my students and made them sought after by oil companies and academic institutions. I also reviewed Prelim proposals thoroughly with my PhD students and only cleared them to take the exam when they were ready. In the process, they came across as more advanced and professional to members of their PhD committees. Years later, Phil Sandberg described my supervisory style as 'tough love" and it probably was.

One thing I stressed to my graduate students was the need to finish their degrees in a timely manner. Usually, a Master's degree required two years, but three years was appropriate if they spent their first summer as an intern with an oil company. For PhD students, I expected them to finish within five years after earning a BS degree, and three years after earning a Master's degree. I did so because I could not predict where they would find their first job after graduation. If they chose to work for an oil company, I discovered that 'slow finishers' were preferentially passed over by recruiters because likely, they would work slower in a company setting, generated poorer work, and were the first to be terminated. Those who finished in a timely manner usually did the best in a working situation according to industry contacts.

Margaret Leinen, who scored the top grade in Geology 437 during the fall term, 1971, had a different undergraduate profile that didn't fit either model. I discussed her future plans because she requested I write reference letters for admission to a marine geology PhD program. She explained her parents didn't want to her to go to college, refused to provide funds, so she applied for a scholarship and received a coveted James Scholarship to attend Illinois. To earn enough money to stay in college, she worked from 20 to 30 hours per week part-time, but after one year, lost the scholarship. She had limited time for studying focusing on courses she liked. She received little guidance and mentoring.

Her Junior year she got pregnant, married the man who was responsible and had a son in 1968. She got divorced in 1971 and relinquished custody of her son. She supported herself financially as a drafts-lady at the Illinois Geological Survey which enabled her to tap into a unique benefit, tuition and fee waivers to take courses on campus. She was enrolled as a "non-degree" student taking graduate courses including Geology 437, and with the security of a permanent job, it did wonders for her GPA.

Margaret complained that her graduating classmates were ahead of her and thought she was missing opportunities. When we talked, I had to finish a project on the Bay of Fundy and needed someone to do some math to complete it. I offered her project work and she accepted. I told her, "We'll publish, your name will be on the paper, and you'll be ahead of all your classmates." She agreed, did well and before the end of 1972, the paper was published in the *GSA Bulletin*.

Because she clearly was a bright lady and, in my opinion, had tremendous future potential, I used the opportunity to share with her some of the things I learned from Joe Peoples, Ray Moore, Bob Shrock, and Harry Hess, amongst others, about professionalism. She also knew Bill Hay and he also guided her. Later, she received excellent additional mentoring at Oregon State (ORST) and the University of Rhode Island. She earned both her Master's and PhD with distinction because for the first time in her life she had sufficient financial support to concentrate on her studies and research.

Her history was a classic variance from the "V" profile, and she is not alone. At Illinois, I met countless students who overcame similar adversity. Margaret went on to a distinguished career as professor of marine geology at the University of Rhode Island, Dean of its Oceanography School, Provost for Oceanography and Environmental Science and then as Associate Director, Geoscience Directorate, National Science Foundation. Her son earned a Computer Science degree from Illinois, went to work and designed, patented and sold the computerized airline reservation system in use today, and now is president of a new venture, Climos, Inc, a climate change company. When she left NSF, he hired Margaret to be Climos' science advisor.

During March, 1972, I took my graduate depositional environments course (Geology 477) on a field trip to the Ouachitas and Arkoma basin of Arkansas and Oklahoma. While at Sinclair, I examined these outcrops and decided they would make a great graduate field trip. Fred enthusiastically supported the trip when I asked for departmental support.

Paul Schluger advised during his exit interview that I form a food committee. Margaret Leinen headed it. I also established an environmental committee to arrange clean-up at camp sites. When we picked up a carryall to supplement the private cars we were using, Roscoe Jackson said, "I'll drive. You ride and save your energy for field instruction." It worked and I learned a valuable lesson. I never drove since on field trips, except on parts of the return travel to Urbana after the last stop. We had many memorable experiences.

While working as a Research Geologist for Sinclair Research in 1961, I visited an outcrop in the Atoka Formation in a roadcut north of a bridge on Oklahoma Highway #82 crossing Lake Tenkiller, South of Tahlequah, OK. I noticed the sandstone was organized as a fining-upward sequence and interpreted it as a meandering fluvial channel fill. A month later, I brought Glenn Visher, and a Sinclair palynologist, Bill Meyers, to the roadcut. Glenn concurred with my interpretation. Neither of us had time to examine outcrops behind the road cut.

I took the 1972 class to the same outcrop immediately before SEPM awarded me a JSP Best Paper Award for research on tidalites in the Bay of Fundy. I showed slides of that outcrop to illustrate fining-upward sequences characterizing ancient meandering channel fills in all my classes and the graduate students recognized the outcrop instantaneously. Immediately, they all took pictures with people in groups, or individually because now my slides became physical reality and they wanted their own.

While pictures were being taken, two paleontology graduate students, Dennis Kolata (BS, MS., Northern Illinois, PhD, Illinois, Texaco, Inc, IGS) and Frank Ettensohn (BS, MS, Univ. of Cincinnati, PhD. Illinois; University of Kentucky), started hammering at the lag conglomerate at the base of the fining-upward sequence. Although most of the lags were shale chips, they suddenly spotted something else and Dennis yelled out "Hey Doc, there are marine fossils in here." I checked. They were right. The shells also were concave up and it turned out to be a tidal channel (and, yes, I had missed the fossils).

It was a good lesson for the students because they discovered that even a person considered to be expert can make mistakes or miss something, professors don't know everything, and that every time one returns to an outcrop, one will see something new they didn't observe beforehand.

I went back two weeks later, and hacked my way through the underbrush behind the road cut. There the sandstones all showed herringbone cross-bedding with reactivation surfaces, current ripple superimposed on cross-bedding at right angles to dip, lenticular bedding, flaser bedding, wavy bedding, bioturbation, and a host of other features confirming the tidal origin of this fining-upward channel sequence.

On the Sunday of the 1972 Ouachita-Arkoma basin field trip, I invited Glenn Visher, now at the University of Tulsa, to show us three outcrops I had not seen. One was a spillway with the best sedimentary structures in a tidalite I recall seeing, and the others were in deltaic facies. During discussions, Glenn was his usual acerbic self and came across as dogmatic, dismissing student questions. I interjected and pointed to some of the problems I had with his interpretations. With me, he was more careful, but still rigid. The students read his paper on the Vertical Sequence concept and were disappointed with his rigid responses. During dinner that evening, we reviewed it. I also suggested we spend more time at the spillway outcrop. They agreed and we spent the next morning there.

In April, I attended the annual meeting of the AAPG-SEPM and received the 1970 JSP Best Paper Award in their opening session. Wayne Pryor was my citationist. The room seated about 1,000 people and the spotlight literally glared at me. When climbing the stairs to the stage, I noticed Ralph Langenheim, Bill Hay and Al Carozzi near the front. I felt it was a good day not just for me, but also for Illinois and its large geo-alumni network, many who sat in the audience.

During late April, my promotion to a full professorship was approved by the University administration. In retrospect, it had been quite a year.

That summer, I taught a half-session of field camp in Sheridan, WY, at Fred's request. He wanted me to evaluate the program placing me in an awkward position with Ralph Langenheim, the field camp director. Ralph referred to me all summer as "Fred's spy." I said nothing, doing everything Ralph requested from all instructors. Before leaving, Fred flew out and brought Dave Anderson because Dave was replacing me. They arrived in mid-afternoon and I was leaving the next day.

After dinner, Fred suggested we leave the junior college used as a base and talk. I only knew of one bar in Sheridan and it was a bit of wild place, mostly a cowboy and cowgirl bar and the patrons knew me. I picked a corner table at the back of the bar and gave Fred my assessment of the camp. I told him the offerings were good, but could be shortened from eight to six weeks and explained how. We were periodically interrupted by people I knew. They also knew I was leaving the next day and some of the cowgirls and waitresses came by to give me a big hug and a kiss. Fred told me on the way back to the junior college, "You really made some friends, but I didn't know you could be that wild."

I then taught a three-week short course at the PETROBRAS training center in Salvador, Bahia Province, reconnecting with people I met earlier during my training consultancy on the Candeias Formation. At the end of the course, the training center hosted a Churascaria barbecue in my honor. The legendary Brazilian hospitality and partying skills were evident.

I made a quick trip to Zion, IL, to check Julie Badal's field work. During the spring, I completed a scuba course and made some dives in her area to check on bedform development. Afterwards, I concluded Julie was not clearly defining her problem, neither coming to grips with her research, nor seeing the limitations of the area. I wondered what had to be done to salvage her thesis and to possibly suggest she go back to her research in Connecticut.

I made final preparations to present a paper at the International Geological Congress (IGC) in Montreal, Quebec. I drove there, met the Schlugers on the way, visited Linda Provo in the field because she was doing a thesis in the Enfield Formation at Watkins Glen, NY, and proceeded to Montreal. At the IGC, I reconnected with people I knew plus many more. I presented my paper on the Tidalite Concept and it generated a lot of interest.

This IGC had an unusual format. Technical sessions started at 8:30 am and continued until 1:30 PM. The afternoon was off for sightseeing, and the evenings were for social events connected to the meeting. However, in my registration packet, I found a notice about an impromptu session between 2:00 and 5:00 PM on the second afternoon summarizing the accomplishments of the Deep Sea Drilling Project (DSDP). I attended.

At the special DSDP session, I listened to talks by Bruce Heezen, W. B. F. Ryan, Bill Hay, and others. I asked pointed sedimentological questions because sedimentology was being ignored. I recall after Ryan (BA, Physics, Williams College, PhD Columbia; Columbia University/Lamont-Doherty Earth Observatory Research Associate) finished, he sat in front of me with Bruce Heezen and one other geologist from Lamont. I commented, "I knew these guys at Lamont didn't know much sedimentology." Everyone within earshot turned and laughed.

The three Lamont people, Mel Peterson (BS, Minnesota, PhD. Harvard, Scripps Institution of Oceanography (SIO), Director, DSDP), and Bill Hay converged on me when the session ended. Mel asked why I never signed up for DSDP cruises. He added DSDP seldom got people with my experience to come. I explained that I needed to discuss this with Fred Donath because their cruise schedules did not coincide with the academic year schedule. The issue of financial support also had to be reviewed. I suggested Mel send me an application form to see what could be done.

On my return, I discussed this conversation with Fred. I explained I would need a sabbatical, but as we talked, another possibility emerged. Fred would excuse me from teaching during the spring semester because it was important for the department to expand its ties to DSDP. I wrote Mel Petersen indicating availability from February 1 through August 15, 1973.

Don Graf called at my office on the Sunday afternoon after my return asking if he could meet me. He was still at home in Philo, IL, about 10 miles south of Urbana. I asked if it could wait and he told me it couldn't.

When he arrived he got to the point, "George, I have to let you know that Julie Badal and I are getting married in October. Fred knows but I would appreciate it if you said nothing to anyone else. I'm resigning from her thesis committee. That's why I had to see you today."

His disclosure took me totally by surprise. Julie was half his age and I couldn't see them together, but then I wasn't sure I could judge. I finally said, "Well, Don, congratulations, I hope you two will enjoy a happy life together. And, yes, I'll keep my mouth shut until after the wedding." I then continued, "By the way, how long have you two been dating and how did you able to keep it away from the department?"

Don explained that when they dated, she drove to his house and they visited small communities where they weren't known, attended their summer and other festivals, county fairs and so on. They hid their relationship well.

But I was troubled about a graduate student of mine marrying a senior colleague, particularly when that graduate student was experiencing difficulty with her thesis research.

I spoke with Fred Donath a week later. He asked if he should assume my role as PhD advisor and retain me on the committee because of her marriage. I assured him that I thought we could go forward as before and he accepted that. I did not realize that I had just made the biggest mistake of my entire career right there and then.

The fall started normally with my usual class load, Geology 309 and 437. George White retired in 1971 and to honor his contributions, the department decided to name the branch library as the 'George W. White Geology Library.' The dedication ceremony was attended by Dean Robert Rogers of LAS, Vice-Chancellor for Academic Affairs, Morton Weir (BA, Psychology, Knox College, MS, PhD, child psychology; Univ. of Illinois; Univ. of Illinois, Head, Psychology Dept, Vice-Chancellor of Academic Affairs, Boys Town, Univ. of Illinois, Provost) and Chancellor Jack Peltason (BA, Univ. of Missouri, PhD, Princeton, Political Science; Illinois, Head, Political Science, Dean, LAS, Chancellor; President, American Council on Higher Education).

As the event approached its end, I started to leave, but Fred called me over and told the three administrators, "I'd like to you meet George Klein.

He came here from Oxford nearly three years ago." I knew Rogers, had met Weir once, but not Peltason. Peltason asked about the administrative practices at Oxford and how they compared to the University of Illinois. A comment was made and I said, "Gentlemen, I like it here and it is a great research environment, but there are days I think this university is the ultimate-bureaucratically structured university in the western world." Peltason and Rogers laughed, but Weir took the bait and he and I went back and forth and agreed to disagree. I looked at my watch and left, not knowing they followed me about 30 seconds later. As they approached the west stairwell, I heard Peltason say, "Mort, that Klein guy has it right about this university." I did not know if Peltason said that intending me to hear it, but Fred told me later the three of them told him that they were glad to meet me and I was a strong hire for the department.

That October, I flew to Corvallis, OR at the invitation of Jerry Van Andel (BS, PhD, Groningen; sedimentology and marine geology; Shell Oil, Venezuela; SIO, ORST, Stanford, Cambridge) to present a colloquium on comparative sedimentology of modern and ancient tidalites. Jerry said that my Holocene-ancient pairs of slides were the best he had seen and suggested writing a book on it. I also decided to return on a special research leave to work on some of their submarine fan cores during the spring semester of 1974. I then continued field work in Nevada for a week and visited UCB to present a colloquium.

Since 1970, Ian Carmichael became department chairman. I noticed the department's atmosphere was more positive than in 1970. I asked what changed. Ian told me the first thing he did was form a "1980 committee" of two faculty who were to poll their colleagues in and out of the department and in and out of the university to decide where geology would be in ten years and how they could pre-empt the field. They decided that field geology was coming back, surface process studies were the key to the future, and that experimental petrology had enjoyed its best days. He concurred and they prepared a strategic plan which the university adopted and funded.

Ian also was concerned that the clerical staff was ineffective and morale was low. One afternoon, he met them and asked them to list their concerns. After two hours of complaints he asked, "How can we improve it and how should the office be run?" They made recommendations and at the end of the meeting he told them, "Alright. Let's try it for a month and see how it works." It worked so well that improvement was immediate and they stayed with the new plan.

Carmichael also implemented aggressive recruiting of graduate students. After reviewing applications, he personally visited the top four applicants at their home university, took them to dinner and talked up UCB. All four accepted offers from UCB.

What struck me was that Ian developed common sense plans and solutions and improved the place by a major order of magnitude. On my return, I mentioned all this to Fred, particularly the idea of a strategic planning committee. Fred listened, but did not seem to be too enthusiastic.

At the end of October, Mel Petersen invited me to participate on DSDP Leg 30 leaving Wellington, New Zealand in April and ending in Guam in June, 1973. Fred approved the absence from campus at full pay. I also reviewed my trip to Oregon State and Fred agreed to support my application for an Associateship in the University of Illinois's Center for Advanced Study for the spring term, 1974. That would give me a research leave for a semester at full pay, without losing time credited toward a sabbatical if it was approved.

After Julie married Don Graf at the end of October, I continued meeting with her to review thesis results every other week. I used the opportunity to outline things she should do during the time, but it was slow to register. She also complained about Charlie Collinson and how there was a high degree of equipment failure that slowed her down.

Slowly, I saw a glimmer of progress. I also asked if she should drop the effort and go back to Connecticut and she absolutely refused to consider it. She would sink or swim with this project. I was troubled about her lack of enthusiasm and her inability to see things through. I mentioned something to Hilt Johnson who reported that the IGS people were not pleased with her progress either.

John Barnes decided to do his thesis on the Zabriskie Quartzite (Cambrian) of eastern California and Nevada. I arranged for him to do some winter and spring field work, but most of it he would complete during the summer.

The semester slowly wound down and all seemed in order. That, however, changed on the first day of final exams in 1973.

LESSON LEARNED:

1) Every request to do extra duties opens an opportunity to renegotiate one's responsibilities in an academic environment.

2) When the possibility of a difficult situation arises (such as Julie Badal's marriage to Don Graf) and someone offers to assume some of the responsibilities, accept the offer, but do so carefully with conditions spelled out.

3) In academe, if one enjoys many successes like awards, well-respected publications, and grant funds, be aware that jealousies are easily aroused, even if one is friendly with everyone. Watching one's back is in order. Some colleagues keep score and can blind-side you.

4) When things are going well, be aware that the situation can change very quickly. In academe, it is very easy to go from hero to goat, and back again.

5) When I received my SEPM Award (a plaque), Gerry Freidman advised me to hang it on my office wall for two weeks only and then take it home. He explained "After two weeks, you are rubbing it in." It was good advice. I did so with all subsequent awards as well. Henry Kissinger once said *"The Reason there is so much politics in academe is that there is so little at stake."* Regrettably, it's true.

POSTSCRIPT #1. The **Tidalite concept** for sediments deposited by tidal currents (compare Turbidites – sediments deposited by turbidity currents) was developed by me in 1970 after arriving at the University of Illinois and was published in the *GSA Bulletin* in 1971. It is based on work I completed during the 1960's on tidal sediment dynamics in the Bay of Fundy which earned me an SEPM Outstanding Paper Award.

While on sabbatical leave at Oxford University in 1969, I researched a variety of examples of TIDALITES ranging in age from Precambrian to Cretaceous. I published on the Precambrian and Cambrian examples later. On my return to the USA, I spent the 1970 spring quarter at UC Berkeley as a Visiting Professor. That provided me with the opportunity to go to the Death Valley region and research late Precambrian, Cambrian and Ordovician Tidalites. This work was also published later.

On returning to Urbana, I integrated both the modern sediment work I had done with outcrop examples I had seen and concluded that these sediments of tidal origin represented a major process facies. It was a global process facies because it was astronomically forced.

At that time, the prevailing paradigm was that most major clastic systems either were turbidite, fluvial, deltas or beach-barrier systems. Tidal sediments were considered more of a curiosity. I therefore wrote a summary paper to propose the Tidalite concept. The original version was sent to "*Science*" and rejected by USGS reviewers who emphasized that tidal sediments were minor in distribution. GSA accepted and published the revised version. A paper was presented at the GSA in Milwaukee (1970).

Since publication in 1971, the term and concept was widely accepted and others found countless additional examples. The concept as developed in 1970-71 set the research agenda in tidal sedimentology for a decade at least. By then I moved to researching the sedimentology of DSDP cores and back-arc basins. Others organized TIDALITE CONFERENCES every four years. In 1996, a TIDALITE CONFERENCE in Savannah, GA, was organized and because it was the 25th anniversary of the formal publication my paper they invited me to give the keynote talk.

CHAPTER 16
The First Donath War
(Late January 1973-April 1973)

"Those who cannot remember the past are condemned to repeat it"
- George Santayana, *The Life of Reason*, Volume 1, 1905

The Monday after final exams began, Ralph Langenheim visited me around 9:00 am. He gave me a copy of a letter to Dean Rogers with the required transmittal cover letter to Fred Donath. I asked Ralph to sit down while I read it. In it, Ralph, Albert Carozzi, Charlie Collinson, and Bill Hay proposed a departmental reorganization whereby all the soft rockers (the four letter signers, Mann, Sandberg, Blake), as well as Hilt Johnson and me would form a new Division within the Department. It would be the Division of Paleoenvironment Sciences.

After reading the letter, I suggested Ralph and Bill Hay meet me for lunch to review it. He replied, "George, I leave for Nevada this afternoon and Bill went to Miami last Saturday." I suggested we meet when he returned.

The letter outlined concerns that prompted the proposal. These included:

- Fred Donath failed to consult adequately with colleagues and review their needs.
- The graduate study committee was too stringent when applying graduate program rules and appeared to single out soft-rock students.
- Dilution of placement activities with oil company recruiters by including fields of geology other than soft rock.
- Financial resources were shifted from the soft rock program to other programs.
- The graduate program was modified to focus on individually-tailored programs, and failed to adequately train geologists
- The division would be led by an elected chairman.

I realized that the group was unhappy with two areas where I had been involved, Graduate Study Committee and Placement Coordinator. As placement coordinator, I functioned alone and cleared things with Fred without a supporting committee structure. Oil companies expressed appreciation for the broad range of students they interviewed. They hired many of Dennis Wood's structural geology students. Exxon Production Research hired two of Dave Anderson's PhDs for their lab. Students in Quaternary geology (Johnson), paleontology, stratigraphy, carbonate petrology, and sedimentology were receiving offers regularly just as prior to my assuming these duties.

The authors of the letter failed to recognize that the graduate student committee makes recommendations on petitions and other issues only to the department head. It was the head who implemented these recommendations at their discretion and some recommendations were turned down.

Nevertheless, I realized I might be vulnerable in my working relationship with Fred. I had to let him know that the first I knew about a new paleoenvironmental division was that morning and not before.

I also opposed the proposal because intellectually it made no sense. The geological sciences were changing with more and more groups of people representing different areas of geological research working together addressing major problems. My upcoming DSDP cruise was a typical example. Their proposal went in the opposite direction.

After lunch, I went straight to Donath's office and asked to see him. He was in fact, almost hostile and cold when he saw me. I discussed the letter and made it clear that I only heard about the proposed division that morning and that the soft-rockers never included me in their meetings nor discussed it with me at any time. He asked me about six different ways to verify this and I was able to do so.

Fred finally smiled and told me, "George, Hilt told me the same thing so I suspected this was the case with you. Don Graf and I met this morning and he assured me you would never support this proposal because of your views on geological research and you valued my leadership." I left and clearly, my relationship with Fred was unchanged.

Dennis Wood approached me in the hall and waved a copy of the letter to the Dean. He asked, "Are you in on this?" I replied, "Absolutely not." He then said, "George, we're holding a meeting tomorrow at 10 in the morning in the conference room to write a rebuttal letter. Will you join us?" I replied, "Dennis, count me in."

I then asked, "By the way Dennis, how did you get a copy of the letter? Your name wasn't on the distribution list." Dennis replied, "A copy was left in my departmental mail box in an envelope and the typewriter font looks like Ralph Langenheim's typewriter." I replied, "You can't be serious. That's crazy and off the wall." Dennis showed me the envelope and I recognized the font.

We met the next morning in the conference room. In attendance were Domenico, Wood, Graf, Tom Anderson, Dave Anderson, Chapman, Hagner, Johnson, me, Palciauskas, Holder, Patton and Deere. Dennis opened the meeting suggesting we needed a chairman and recommended Pat Domenico. We agreed unanimously. Bill Goodman, the departmental business manager appointed in 1971, sat in as an observer. He replaced Bill Latham.

Dennis distributed a draft letter he and Don Graf prepared. We read it. Both Chapman and Hagner said they would not sign because it consisted mostly of personally-motivated attacks at Bill Hay and Ralph Langenheim. The next 30 minutes were spent toning these comments down. I chose to say little, and Chapman and Hagner still weren't totally satisfied.

I then said, "As you know, I left Penn during a big dust-up and I learned some things from that experience. First the administration doesn't care if we like each other or not and aren't interested in these kinds of personal statements. They expect us to work as professionally civilized colleagues whether we like each other or not.

The second thing I learned is that whenever issues such as this one come up, the administration wants to know will the program remain intact, will it drastically change, and if so, will it diminish the program's standing in higher education and other circles. If the administration decides to poll us on the proposal, that's what they'll look for. Consequently, we must come up with a unified message that everyone in this room can buy into to get them on our side."

Chapman interjected, "That's why we need to rewrite the letter and develop a coherent plan. George, do you have any suggestions based on your experience at Penn?"

I replied, "Carleton, at Penn we had a small department of seven people. About ten days before I interviewed here, the department faculty went to a meeting with the provost, the college dean and the graduate dean. They only had one question and it was 'What will it take to make the University of Pennsylvania's geology program one of the top tier geology programs in the USA?' We gave the administration our views."

Don Henderson (a Harvard colleague of Boucot) chimed in, "Well I heard it differently. I understood most of the time Boucot and Faul were arguing." I replied. "Don, I guess you talked with Art Boucot. That did happen, but the Provost started the meeting with that question, and then repeated it again threatening to terminate the meeting if we didn't provide answers. We got our answers in before Art and Henry went back at it and the provost ended the meeting."

Graf then said, "George, I heard you wanted to nearly triple the size of the Penn department in three or four years. Is that true?"

I replied, "Don, I don't know where you got your information. Yes, I proposed such a plan and all my colleagues except Faul supported it. I think the Penn administration found it too expensive, and a year later, when Penn Central went bankrupt, they lost 20 percent of their endowment. Just looking at all the new lab space Fred arranged to be renovated since he arrived, I know these things aren't cheap. But as a group, we answered the provost's question in terms of the national situation."

Domenico then said, "I think I see where we need to do something like George is suggesting and must lead off with it in any revised letter we send. We also need to show how the department is progressing along these lines and demonstrate that Fred is fostering that development. Any ideas?"

Dave Anderson mentioned Fred's interaction stick diagrams but as we discussed it, we got lost in details. I finally spoke up and said, "Let me make a suggestion. We should tell the administration that geology is changing and that the traditional boundaries of its subfields are breaking down. Major progress is being made when people representing different subfields of geology work together and when this happens, major new breakthroughs occur. We've seen this with many of you in this room. We can save details for discussion in any meetings we may have with the administration. We need to show how Fred fostered these interactions, whether it is dedicated space, equipment purchases, starter funds, or whatever. Moreover, we need to include a statement if grant funding to the department increased because we moved into these collaborative efforts. I don't have that information, but because Bill Goodman is here, he might know."

Bill said, "Yes, grant funding to this department has been on a steady rise since Fred came here. I can provide a graph if you need it."

Graf then asked me to repeat what I said so Pat could write it down.

Chapman spoke again, "We should use that approach in our letter but add a statement that some good work is still being done on an individual basis. We can also expect a review by the administration of Fred's leadership and we must deliver a consistent message."

By this time, it was close to lunch. We adjourned and met again at 2:00 PM. Dennis and Pat were going to revise the letter and Don Graf offered to help.

As we left the room, Chapman called me aside and said, "You really learned from the Penn experience. You know when we removed George White, we stressed that White was a decent man, but the department needed to move more into geochemistry and geophysics. We never attacked White personally. That's how we won the day." Chapman was acting department head between White's removal from the headship and Fred's appointment.

I said, "Carleton, that's good to know. But how did you gain access to the Dean?" He said, "Don Henderson knew him through Boy Scouts, and I knew him through the local Harvard Alumni Club. We contacted him at home one evening asking how to proceed without destroying the department, and he was most helpful. Bob is very genial and when a review comes, if you say what you told us today, he'll really like you." I replied, "Thanks for the tip. That's good to know."

We met again and the letter was essentially rewritten as discussed. However, it included a sentence about Hay and Langenheim that was at best benign. We agreed to sign it. Pat arranged to have it typed, including a cover letter we also were going to sign, visited each one of us individually and afterwards gave it to Fred to send to the administration.

The letter to the Dean was approximately like this:

Dr Robert W, Rogers, Dean
College of Liberal Arts and Sciences
Lincoln Hall

Dear Dean Rogers:

We write as a collective group of geology professors to express deep concern with respect to a letter dated January 17, 1973 (via Fred A. Donath) from Professors Carozzi, Collinson, Hay, and Langenheim proposing a reorganization of the geology department by forming an internal Division of Paleoenvironmental Sciences. We recommend in no uncertain terms that this proposal be declined. Moreover, their views are in no way representative of the majority of the faculty in this department. Their letter is highly critical of the administration of this department. We wish to disassociate ourselves from their views.

The field of geology is changing. It is becoming more intradisciplinary as traditional boundaries between various subdisciplines within it are breaking down. Major advances are being made when individuals representing different subsets work jointly to address major problems. We can illustrate this trend both in this department and elsewhere, including some involving colleagues in other universities or government agencies, both nationally and internationally. This trend comprises some of the new programs instituted by Dr. Fred A. Donath, department head, while previous areas of strength were improved. Thus the department is able to capitalize on current changes in the field as proven by increased external funding received during the past four years (attached graph).

The proposal to form a Division of Paleoenvironmental Sciences would reverse this trend. If implemented, it would slow the department's progress made during the past five years. In "soft-rock" geology, this proposal runs the risk of losing the department's international leadership. We therefore again recommend you decline this proposal.

In closing, we fail to understand why this proposal was made. Although we respect their accomplishments, both Professors Hay and Langenheim have created internal problems that we prefer to leave unsaid.

We look forward to a favorable response to our recommendation.

Sincerely,
(Signed by the 14 people who attended the meeting to draft the letter)

The letter was sent around January 22. We then waited. And waited.

I continued my bi-weekly meetings with Julie Graf and sensed any momentum I helped facilitate slipped away. She lacked enthusiasm for her work and seemed unresponsive to my suggestions. I called Charlie Collinson and suggested that one way to help her progress was to let her apply for GSA and Sigma Xi student grants so that she could function a little more freely than before. Charlie was very supportive and offered to write letters supporting her efforts. During our conversation, we bantered as before as if nothing had changed, despite his signature on the letter to the dean about the proposed division.

When Julie and I met again, I suggested she apply for grants and work independently from the Survey. We had all the equipment she needed. She could hire her own field assistants. The response was at best underwhelming but she went ahead and applied for grant money and received what she needed. The department supplemented her award too.

Late in January, Fred asked me about her. He heard things from Don Graf. He finally asked, "Can you see her through?" I replied, "Fred, I would like to, but there are days I seem to want her to earn a PhD more than she does."

During late February, we received a memo from Fred telling us the Dean was conducting a review of the department and his leadership. We were to contact the Dean's office for an appointment. I arranged to meet the dean during the first week in March.

If there is one thing I learned over the years, when meeting Deans, Vice Chancellors and Provosts, it was best to be dressed in a suit, dress shirt and tie, and I did so. Because Bob Rogers and I knew each other, he greeted me in a friendly way, offered coffee and he started to ask questions. Looking back, I recall the dialog went this way:

Rogers: George, I've reviewed both the letter your four colleagues sent as well as the one you signed. Did the four ever contact you about the proposal or meet with you?

Klein: No. I first knew about the proposal when Langenheim gave me a copy of the letter in mid-January.

Rogers: Because your interests are closely aligned with the division proponents, why did you oppose it?

Klein: I did so primarily because I don't think it will work and it wasn't in the best interests of the department, much less theirs. I felt programmatically it would weaken geology at Illinois and hurt us with funding agencies and recruiting graduate students. Also, geology is changing.

Rogers: Yes, I noticed that. How have these changes impacted your work?

Klein: First, I rely on help from Vic Palciauskas in geophysics. Some of what I do involves acoustics and he helps there. I interface with people in fluid mechanics in engineering. Hilt Johnson and I compare the depositional process work I do with his analysis of glacial landform evolution. I'm about to go on a Deep Sea Drilling Project cruise where I'll be working together with geophysicists, petrologists, paleontologists, people in tectonics, and geochemistry addressing specific problems about ocean floor evolution. That gives you a sample of how my work is impacted. Fred Donath and I talked about this change for hours as I suspect he may have told you.

Rogers (smiling briefly): The division proponents were unhappy with how the graduate study committee dealt with their student petitions and the new graduate requirements. Can you tell me why?

Klein: I chaired the graduate study committee for two years. We changed the requirements in response to a charge from Fred. Our proposal went through modification and was ultimately approved by the faculty. I recall some of the proponents voted for the changes.

Also, when the committee reviewed student petitions, we made recommendations to Fred. He decided in the end whether to accept them and some he declined.

Rogers: They were also upset with changes in placement procedures. Can you tell me more about that?

Klein: In 1971, Fred asked me to coordinate placement. I met with students to listen to their concerns. Many perceived that oil company recruiters were preferentially seeing only soft-rock students. I changed that to open it up to all. When I arrived in 1970, only soft-rock faculty were invited to lunch by oil company recruiters. Again, I broadened representation so other faculty could meet them.

I also did something else. I prepared an annual brochure about our graduating MS and PhD students which was mailed all over the country. I used an approach similar to the economics department. It generated many new job prospects for students and led to better placement for our graduates.

Rogers: The other concern the proponents had was they felt soft-rock geology wasn't represented on the departmental executive committee. Fred tells me that was your role. Can you explain it?

Klein: Not entirely. My instructions were to attend meetings and keep proceedings confidential. I honored that. No one instructed me to meet with the other soft-rockers and none of them ever visited me to share their concerns for the committee to review.

Rogers: So you think I should come down squarely on Fred's side?

Klein: Yes, very definitely.

Rogers: You have no problem with his administrative practices or leadership style?

Klein: No. It's not easy to manage 20 individual contractors. However, Fred might want to do more strategic planning to capitalize on the gains he made.

Rogers: Well, George, you've been very candid and straight-forward. I want to thank you for coming by. If I need to talk with you again, I'll call.

I got up, shook hands with Bob and left.

I checked my mail box, and Fred asked me to come into his office. Bill Goodman was there too. Fred asked how my meeting with Rogers went, what I was asked, and what I told the Dean. I summarized it, shared my answers, and he was really pleased. Fred then said, "George, that experience at Penn really matured you." I replied, "Fred, in more ways than you'll ever know." Both Bill and Fred smiled and I left. Having talked with Deans and Provosts at Pitt and Penn, I had, perhaps, learned to keep things focused on content.

When I cleared my mail box, I received a letter from the University of Illinois Center for Advanced Study confirming that I was awarded an Associateship during the spring term, 1974. I was cleared to spend that time at Oregon State.

I had lunch the next day with Graf, Pat, Vic, and Hilt. We all met the Dean and they were trying to read him and predict an outcome. I said I really couldn't read the situation and we would have to wait it out. And, again, we waited.

During the last week in March, Fred called a faculty meeting. That morning, he called an executive committee meeting also. He told us the review was completed and one of the outcomes was the dissolution of the executive committee and election of an advisory committee in accordance with new university statutes. He said he would let us know the rest that afternoon. I concluded he survived the review, but some restrictions were placed on him.

Fred opened the faculty meeting announcing we would each get letters from the Dean reporting that he was retained as head. He disclosed that the executive committee was dissolved and an elected advisory committee would replace it. Elections would be held later. He said that there would be some other changes. He asked us to come together, work together, and move forward. He opened the floor for questions.

Dan Blake asked if Bill Hay had been notified. Fred replied that he would call Bill that evening.

Chapman commented that we should expect that the administration will watch the department more closely for the next three years and we needed to become more collegial in our actions. If something goes amiss, they will review the department again.

The meeting adjourned.

A week later, I packed for my trip to meet the *DV Glomar Challenger* Leg 30 in Wellington, New Zealand, with a stop-over at DSDP headquarters in La Jolla, CA.

LESSONS LEARNED:

1) When dealing with university administrator, or corporate upper management, always focus on relevant content issues. If they shift the conversation to more personal items, move with them and let them lead the conversation back to content. If they don't shift back to content, take the lead to do so.

2) When dealing with colleagues who take an opposing position as did the four proponents of the Division discussed herein, keep lines of communication open with them. Not doing so makes it more difficult to repair working relationship that will require rebuilding once decisions are made.

3) Never expect the higher administration to make a quick decision. It will take them nearly three times longer than expected because of other competing issues and the need to develop a solution that keeps academic programs intact.

4) Keep the students out of controversies such as described herein.

CHAPTER 17

Illinois: An Uneasy Truce and A Difficult Decision (April 1973-December 4, 1974)

> *"A good research director doesn't direct the research. The trick is to get the right people and then turn them loose. When you give people the freedom to work, then the burden is on them to produce."*
> - Rufus LeBlanc

> *"Getting a student is a risk"*
> - Chu-Yung Chen, 10/31/1986

My flight landed in San Diego in the early evening and I drove a rented car to the motel in La Jolla where DSDP made reservations. Next morning, I saw La Jolla in its splendor. The setting was outstanding and a major change from the dour flat lands of east central Illinois. I reported to DSDP and met with Mel Petersen. He gave me a dossier on Leg 30 and reviewed the cruise's mission. He also asked me to serve as the cruises' Chief Sedimentologist. Policy prevented them from inviting me to be a co-chief scientist because this was my first cruise. He gave me my air ticket to fly to New Zealand and return from Guam.

Bill Hay told Mel about the situation at Illinois and Mel inquired about the outcome. I said the Dean sustained Fred and I left shortly afterward so was not conversant with other changes. Mel disclosed he graduated from Minnesota a year before Fred and found him a dull individual. He was surprised Fred was first appointed to a faculty position at Columbia and then the headship at Illinois.

I wrote ahead to New Zealand contacts and arranged a lecture tour at the Universities of Auckland, Otago, Canterbury and Wellington with local field trips. During those trips, I saw spectacular features in Cenozoic turbidites and photographed them. I used these slides in my university courses and short courses until 1993.

I arrived in Wellington two days before the Glomar *Challenger* docked.

Once the ship arrived, I was required to live onboard, which I did. When Leg 29 arrived, I was invited also to a party of all their scientists at the home of Peter Barrett, a faculty member at the University of Wellington. It was the usual 1970's style party – reasonable food, excellent wine, men dressed casually, and braless women wearing blouses with plunging necklines and short skirts.

I met one of the Leg 29 scientists, Peter Webb (BSc Wellington, PhD, Utrecht, micropaleontology; NZ Geological Survey, Northern Illinois, Ohio State) who accepted the headship of the department of geology at Northern Illinois University (NIU). Peter, his wife and I were sitting on bean bags while he asked about the higher education system in Illinois.

When he was through asking questions I offered some advice. The first was, "Peter, when you get there, attend all faculty meetings and at the first one, get up and say something." He asked, "Why?" I replied, "Peter, Midwesterners are a bit self-conscious about being viewed as provincial. When they hear your commonwealth accent, they'll ask, 'What department's he in? Geology! Must be a good department to attract a person like that." The Webbs roared with laughter.

I then asked, "Peter, where did you earn your PhD?" He replied, "At Utrecht." I asked if he owned the academic gown and he said "No." I advised, "Peter, buy one and march in all the convocations and graduations." He asked, "Why?" I said, "Peter, it will add class and color and the parents and guests will ask what department is that colorful fellow in? Geology! Must be a good department to attract a colorful professor like that."

Again, both of them roared with laughter and fell off their bean bags. I told them I would see him at the annual meeting of GSA in Dallas in November. When I saw him there and asked how things were going at NIU, he said "I'm seriously thinking of buying the academic gown."

The *D.V. Glomar Challenger* was an industrial ship with few frills. The galley however was sumptuously stocked and open 24 hours. Food was excellent. Everyone onboard ship ate there so one quickly met the crew members including the drilling crew.

We all worked shifts, 12 hours on, and 12 off. Shifts were staggered to assure smooth change-over. A science lounge provided a minimal library, conference table and work space. A science office was located immediately below the bridge and had carrels for a typist (Yeolady), Louise Henry, the head technician, "Gus" Gustafson, and both chief scientists. Space was functional but tight. Aft of the science lounge were the sleeping quarters for the scientists, technicians, SIO representative, and yeolady. I was assigned a room with a bunk bed and immediately took the lower one.

Slowly, the scientists arrived. The co-chiefs were Jim Andrews (BA, Amherst, PhD, Miami, FL; marine geology, Univ. of Hawaii) and Gordon Packham (BSc, PhD Sydney, sedimentary petrology and South Pacific tectonics; Univ. of Sydney). I met Packham before at IAS meetings. Others included Gerrit Van Der Lingen (PhD Utrecht, NZ Geological Survey), Jim Eades (MS, Auckland, NZ DSIR) and Dave Jones (BS. Yale, PhD. Stanford, regional mapping; USGS) listed as sedimentologists working with me, Tsunemasa (Tsune) Saito (BSc , PhD, Tohoku, foraminifera; Lamont Doherty, Univ. of Yamagata), Brian Holdsworth (PhD Leeds, radiolaria; Univ. of Leeds) and Samir Shafik (BSc Cairo, PhD Adelaide, nannofossils; Univ. of Adelaide) as micropaleontologists, Loren Kroenke (BS Wisconsin, PhD, Hawaii, Marine tectonics and physical properties; Univ. of Hawaii) and Douglas Stoeser (PhD, Oregon, volcanology; USGS). Jim Andrews also served as the SIO representative.

We left Wellington in mid-April for a 51-day cruise. Jim Andrews gave us a tour of the labs. Immediately below the sleeping quarters was a core description and physical properties lab that led out to a gangway where cores were split and cut into 30 foot sections. Beyond the gangway was the drill floor where only the co-chief scientists were permitted. The paleontology and thin-section labs were on the deck below the core description and physical properties lab.

We took five days to reach the first site. Each sedimentologist chose a site where they served as lead and I chose the last site, Site 288, on the Ontong Java Plateau. Work space was crowded. Gerrit had prior DSDP cruise experience, as had Saito, Andrews and Packham. The rest of us did not. My room-mate was Shafik who had the irritating habit of taking showers once per week and changing clothes once per week on a different day. We finally convinced him to shower and change clothes more often.

The one person with whom I developed a good friendship and working relationship was Tsune Saito. Tsune mentioned he was preparing a proposal to do a paleontological and sedimentological study on the west side of Hokkaido Island during the summer of 1975 and needed a sedimentologist. Was I interested? I indicated interest and he explained that under the NSF international program ground rules, we needed a Japanese counterpart. I nominated Hakuyu Okada (BSc, PhD, Kyushu University, sedimentology and sedimentary petrology, Kagoshima Univ., Shizuoka Univ. (Dean), Kyushu Univ.) who, after contacting him, agreed to join us.

The cruise ended in Guam. We met with the Leg 31 scientists including Arnold Bouma. The hotel had a great sushi bar which I enjoyed. I left Guam for a 22 hour flight to Champaign.

Mel Peterson arranged for me to return for the summer to DSDP headquarters to work on cruise items and any other projects I wanted to develop. After two weeks in Urbana, I returned to DSDP. I rented a furnished efficiency apartment, a car, and enjoyed the amenities of southern California. I realized that because of living costs, I discovered the area far too late. Mel Petersen moved there in 1960 before the real estate boom occurred.

Jim Andrews joined me two days later and we worked together on finalizing site reports and research relevant to the Leg 30 DDP volume. I also spent time in the DSDP core lab and discovered the cores were underutilized as a resource. I began compiling data on sedimentary structures and vertical sequences in DSDP cores and later developed a protocol for their description on all DSDP and ODP cruises.

Even though my efficiency apartment was spartan (but cheap), I liked the San Diego area and came to know it well. During the next 12 years, I returned often, at least once or twice a year. However, in 2006, my wife, Suyon (Chapter 28) and I visited and coming from Houston, an international city, we found San Diego provincial and the restaurants below par. As another friend once said, "All they can do out here is sell sunshine." But I really enjoyed the coastal scenery and the Asian restaurants, of which Champaign-Urbana had virtually none until 1980.

I returned to Urbana in mid-August and settled in for a busy fall semester. I reviewed all mail and memos and discovered Donath was taking a sabbatical to complete experiments, and Carleton Chapman was acting head.

I met a new graduate student, Jerome (Jerry) P. Walker (BS, Indiana, MS, Illinois, sedimentology and stratigraphy; Texaco, Placid Oil, Sterling Oil, Placid Oil, Nevada Power and Light, consultant) who wanted to complete a Master's with me. Jerry was the first person I admitted using my preference for graduates from flagship state universities. He worked out well, completing a thesis in New York on the Silurian and has had a successful career.

Once classes were underway, I continued meeting with Julie Graf. Her summer was a string of disappointments which surprised me because I had personally checked all her field equipment and had helped her develop a research plan for her field season. As our meetings continued, I sensed not only loss of momentum, but loss of interest on her part.

I particularly remembered a meeting in Mid-October when I asked if she calibrated her working hypothesis with the data she had acquired. She told me she didn't want to do so until she had obtained all the data, requiring another field season. I suggested she at least stop to see if everything was checking out, where data gaps occurred, of if new ideas needed to be developed. She did and progress improved for one meeting and then reverted to type.

I realized her progress was going nowhere and I was left with some painful decisions to make, decisions complicated by her marriage to Don Graf. Donath told me about feedback concerning Don's negativity and underhanded behavior from the IGS. Some of it I observed while on the executive committee and during the first Donath War.

My choices about Julie were (1) let it ride, (2) let it ride but refuse petitions for extensions, or (3) terminate. I agonized, knowing there was no one I could share this problem within the department. Had I done so, it would go straight to Don Graf. Donath was generally unavailable. I mentioned something to Hilt Johnson who was most discrete. He wasn't totally surprised but had no easy solutions.

I finally decided to go with option #3, terminating her, but allowing her to use the lab and equipment provided she could find another faculty adviser who would provide her with office space. I wrote a letter to Chapman with copies to Donath, Johnson, Collinson and others on her thesis committee explaining my decision and planned to distribute the letter after meeting with her. I also realized that perhaps I should have accepted Fred Donath's offer to become thesis supervisor to avoid a faculty conflict of interest. Declining to do so was the worst decision of my career.

On November 22, 1973, two days before Thanksgiving, I met with Julie for the last time. As we started I said, "Julie, I've reviewed your work for the last three years and see no progress on your thesis. I've tried to guide you in such a way that you could work through each of the problems that arose and you failed to respond. To be honest, there are days I think I want you to earn a PhD more than you seem to. Therefore, I have decided to terminate my role as your PhD supervisor and resign from your thesis committee because my effectiveness is at an end."

I then added, "Now Julie, under the rules of the graduate school, you are free to find another advisor if you so chose. Should that happen, you are free to use the sedimentology lab and its equipment, provided others have less priority on lab and equipment use. You will have to vacate your carrel once your new advisor finds you other office space."

She sat there shaking for about a minute, turning red in the face, trying hard to control herself. She finally said, "I think you've been unfair. You constantly placed the burden on me to get the work done." Then she paused and added, "You know, I've wanted out for some time, and thought sometime ago you did too," picked up her things and left.

I then distributed in person my letter to Chapman and everyone on her committee who was in NHB and sent the letters via campus mail to Collinson and an engineering professor on the committee.

Although I could not predict what would happen next, nothing came to my attention about my decision until two days after returning from Thanksgiving vacation. Dennis Wood saw me in the hallway outside my office and to say he was belligerent would be an understatement. Because he was a member of the advisory committee, he demanded to know why I didn't inform the committee of the decision. I explained the rules of the graduate school didn't require notification.

Dennis then said he would take over as her PhD supervisor. I wished him well at which point he proceeded to scream a string of cuss words at me, all which was witnessed by three graduate students, one who worked with Tom Anderson and the other two with Sandberg. When he turned to leave and saw the students, his face turned every color of the rainbow.

Fred Donath telephoned an hour later. He wanted to know why I didn't meet him first and I explained he was on sabbatical and unavailable, having only caught up with him at the annual GSA meeting in Dallas. He told me I should have made the effort and I agreed perhaps so. He explained about Wood and we agreed on her change in office, and access to the sedimentology lab and field equipment on a priority basis (Class usage and my own students and I were ahead of her).

Fred then said, "George, I've been hearing an earful about you from different colleagues. They find you too formal, too businesslike, complain about your wearing coats and ties, and your butch haircut. One complained 'every time I pick up a journal around here, there is a paper by Klein in it.' "

I responded, "Whoa Fred, I thought we were all supposed to be publishing our research. In my case, some of my papers were held up in the journal's back log which is why so many appeared at once. As for the rest of the comments, that's just a matter of style. For your own information, I don't necessarily agree with other peoples' style, including one colleague who swore at me in front of graduate students an hour ago."

Fred responded that he understood about journal back-logs. He then disclosed that Carleton Chapman gave a copy of my letter to him to Don Graf. I told Fred, "I find that an ethical breach and it really surprises me." He explained the Graf's were planning to write a rebuttal but he told them he would refuse to accept it because of the unacceptable way they obtained the letter.

Fred then let me know that he would meet with the advisory committee to discuss it and there may be a need to hold another advisory committee meeting to resolve everything. I asked to be kept in the loop. The advisory committee now consisted of Blake, Domenico, and Wood.

After lunch, I visited Chapman and asked if he released my letter to the Grafs. He replied, "Yes, George, I thought it would promote harmony around here and we sorely need some." I responded, "Carleton, I really think you committed an ethical breach and what you did will be more disruptive. Frankly, I expected more from a Harvard man and thought you knew better," spun around and left. I went to the mail room to use the copying machine and suddenly, someone tapped my arm. It was Carleton Chapman looking ashen and he said, "I apologize. Perhaps I acted in haste." I did not respond.

The advisory committee met and Blake was instructed to let me know nothing came out of the meeting and they had to hold another one. Wood wanted me fired and had a petition to circulate signed also by Don Graf to do so. Dan said, "George, sit tight. Just relax. It will blow over."

The advisory committee met again, and after the meeting ended, Domenico dropped by. He said, "The advisory committee voted to sustain your position because you followed all the rules of the university. Fred sat really hard on Wood for trying to circulate a petition to get you fired and put the clamps on both him and Don Graf. Fred also chastised Carleton for giving a copy of your letter to the Grafs. This should settle it."

I was relieved that it seemed to end, but remembered H.T.U. Smith's comment at Kansas about how difficult it was to set standards higher than those around you, and Tom Anderson's comment during a faculty meeting that "the hardest thing about earning a PhD at Illinois was getting admitted." I knew I had to be wary and that this was not going to go away quickly.

My neighbor visited me one evening and asked me if everything was alright in the geology department. He was John Haltiwanger (BS, South Carolina, MS, Civil Engineering; Univ. of Illinois, Associate Head, Civil Engineering). John explained that the night before Don Graf come to his house, introduced himself, and said, "We're having some problems with George Klein in the geology department. As his neighbor is there anything you can tell me about his moral character, behavior and standing in the neighborhood?" Haltiwanger explained he told Graf to leave and never return.

Haltiwanger ate lunch regularly in the faculty dining room with a group of campus administrators, including the Vice-Chancellor for Business Affairs. John said he inquired if they knew Graf and shared Graf's visit with them. That was the administrative clue that there were problems in the geology department and as is usually the case at the University of Illinois, word spread faster than a laser beam.

Domenico heard about Graf's visit to Haltiwanger from a friend in Engineering and told Donath. Fred, I was told later by Pat, called Graf in, chewed him out and told him to stop his bizarre actions which hurt the department's campus standing.

Gordon Fraser defended his PhD thesis successfully. That spring he joined BP, Alaska in San Francisco.

The semester ended, I packed and moved to Corvallis, OR, to start my leave as an Associate of the Center for Advanced Study. I arranged to make one return visit to supervise graduate students.

Oregon State University (ORST) was established in 1858 as a private school with college courses added in 1865. In 1868, it was designated as the official land grant university in Oregon. Later Sea Grant, Space Grant and Sun Grant programs were added.

The School of Oceanography expanded after World War II with funded research support from NSF, the U.S. Navy and NOAA. It operated a fleet of ocean-going research vessels and established a marine aquarium at Newport, home port for its oceanographic fleet.

Arriving in Corvallis, OR, I rented an apartment arranged by Jerry Van Andel's secretary, unpacked and moved in. I went to the Oceanography building the next day, met Jerry, and was given a temporary office. I met again with Vern Kulm (BS, Muskingham, PhD, Oregon State, Marine geology; Oregon State), Paul Komar (BS Michigan, PhD, SIO, Coastal Processes; NATO Post-Doctoral Fellow in UK, Oregon State), G. Ross Heath (BSc Adelaide, PhD, SIO, Oregon State, Rhode Island, Dean Oceanography, Univ. of Washington), and Ted Moore (BS, North Carolina, PhD SIO, Ocean carbonates and paleoclimate; Oregon State, Exxon Production Research, Univ. of Michigan), and two other faculty.

Jerry established a tradition, similar to the George and Harry's schedule at Yale. At 8:30 am, and again at 3:00 PM, everyone in marine geology met for coffee and discussion in the chart room. Discussion could be administrative issues Jerry had to let everyone know (he was the professor in charge of marine geology), the day's news, news about other colleagues, new scientific findings, recent papers, or whatever came to mind. No one wanted to miss it.

I saw Margaret Leinen who was thriving in her new environment. However, I knew she was aware of my dropping Julie as a PhD candidate and the uproar that followed. She said nothing about it to me.

During one afternoon gathering, Jerry said for no particular reason, "You know, the most vicious people in our profession are the geochemists." I saw red flags, thought a moment, and asked, "Jerry, why do you think that is?" Jerry replied "They stay in their labs in their own little world. They never go to sea so they never work with people in other fields. Now geophysicists, help the geologists with coring and dredge sampling, and we help them deploy seismic cable. Geochemists don't do that. Therefore, they never learn to work together." I told him, "I didn't know that."

I returned to Urbana in late February. I made plans to visit Dan Lawson in Alaska in July, and Jerry Walker in New York in August. When leaving, Fred called me in and asked I if was attending the AAPG-SEPM annual meeting in San Antonio. I told him I was and he asked if I could meet him on Monday for breakfast at the Hilton Hotel where he was staying. I agreed.

I returned to ORST and a month later went back to SIO for a post-cruise meeting. We reviewed the final site reports, and status of our work, manuscripts, and deadlines. Jim Andrews reminded us that the proprietary requirement on our work ended on July 1 after which we were free to publish wherever we wanted in addition to the Cruise Volume. My papers were done and I gave them to Jim for the final volume.

From SIO, I went to San Antonio for AAPG. On Sunday I went to committee meetings and attended the opening 'ice-breaker' where I saw nearly 90 Illinois geology alumni. Next morning, I met Fred. After inquiring about my leave, he talked about the department and on his own initiative assured me things had settled down.

He then mentioned a new program that Mort Weir, Academic Vice Chancellor, established. Mort was worried that the best people on campus might leave and proposed a retention program whereby each department was to nominate people who the "university could ill afford to lose" for a significant retention raise. Fred disclosed I was on his list. I asked if he needed anything else from me and he told me, "No. I have everything I need." I asked if he could disclose who else he nominated and he told me Dave and Tom Anderson, Dan Blake, Pat Domenico, Dennis Wood, and Vic Palciauskas.

Fred then pulled out a pyramidal diagram, and showed me how we all fit with his view of the department. I had trouble following his reasoning but concluded it would sell with the administration.

We then talked about trends in geology, I shared things I learned on the West coast, and concluded the meeting on a positive note. I felt that perhaps the problems of the fall were behind me. Fred was nominating me for a retention raise, but not Don Graf. Fred disclosed he was worried that since leaving IGS in 1966 for Minnesota, Graf had not published any research papers, despite generous support from NSF.

I attended the Illinois cocktail party that evening but the turnout was disappointing. Only about 35 alumni attended.

On returning to ORST, I finished my writing. In the end, I did not undertake the research on submarine fan levee systems. I focused entirely on all my DSDP work, including the public manuscripts for refereed international journals. I left early in June and returned to Urbana.

After arriving, I went over John Barnes' draft PhD thesis manuscript. I recommended changes. John accepted a job with Amoco and was leaving in two weeks. He said he would finish his thesis in absentia but never did. He told me later that once on the job, he concluded he didn't need a PhD.

After two weeks in Urbana, I flew to Alaska and met Dan Lawson. He received research funds from GSA, Sigma Xi and the Arctic Institute of North America. The USGS let him enter their storage warehouse in Anchorage to take whatever he needed: food, equipment, supplies. All equipment and left over supplies were to be returned at the end of the summer. That support-in-kind saved him a lot of money.

I was met at the Anchorage airport by Dan and by Tom Ovenshine. Tom explained that he and Dan decided that because I was there for a week, Tom would show me his work with other USGS geologists on the tidal sediments in Turnagain Arm and use a USGS helicopter to visit an inland glacier. Turnagain Arm was struck during the 1964 Alaska earthquake and we visited the town of Portage which was faulted down. The old houses were now surrounded by newly prograded tidalites. A new town was built. The distance of progradation since the earthquake ten years before my visit was close to 1 km. Moreover, when we looked at vertical cuts in creek beds, we observed multiple tidalite fining upward sequences. Tom suggested these represented individual earthquake events. He planned a drilling program to determine how many sequences existed and use radiocarbon dating to determine earthquake frequency.

After two days, Dan and I went to the Matanuska Glacier where he established a base camp. We spent three days there during pristine clear weather. As temperatures rose, meltwater discharge increased. Consequently, we were able to see original glacial till being resedimented by slumping, debris flows, slurries and turbulent flow with photographic evidence to prove it. I suggested ways Dan could establish criteria to separate true glacial till from resedimented till, and on his return, check Pleistocene glacial debris in the Midwest to see how much was true glacial till and how much was resedimented. Dan was on to major breakthrough research.

We returned to Anchorage on a Saturday and met George Gryc, the USGS Alaska Branch Chief, and Louis Pavlides, a USGS Engineering geologist who wrote the definitive papers on the 1964 Alaska earthquake. On the way to a restaurant for dinner, Louis led us on a tour of all earthquake slides in downtown Anchorage. I observed that the entire city was rebuilt on unstable slides including the Captain Cook Hotel, a 12-story High Rise built on the Fifth Street Slide.

I flew back the next day and as we climbed, the pilot was ecstatic announcing that the visibility was the best he had seen in 15 years flying this route. I took my camera from my travel bag and took pictures of glaciers from the entrance door windows knowing this was a once-in-a-lifetime event.

On my return, I spent a week in Urbana and drove east to meet Jerry Walker for a field supervisory visit. He was doing great. Clearly, the Indiana University field camp trained him well.

Once in Urbana, I settled in. Fred called me to his office, was most genial, and handed me my annual salary letter. I was awarded a 25 percent retention raise. I asked how the department did and he said, "We got all of them." I assumed his diagram convinced the administration, but several years later, I discovered very few department heads applied for faculty retention raises, and all nominations were not only approved, but there was money left over.

When I discovered this, I inquired why. Many department heads were afraid to make choices and hear objections from colleagues who didn't get them. Others in fields such as the humanities didn't expect to lose people because jobs in those fields dried up and Illinois' library was a retention tool in itself. In some departments, people gave up hope of improving programs and were happy to see people leave and do better elsewhere. I wondered if as a faculty member committed to and focused on research, and publishing, I was in the minority on campus. I recalled the comment passed on by Fred that 'every time I pick up a journal around here, there is a paper by Klein in it.' It raised fundamental questions whether the university was in decline.

As I reviewed my mail, I found a memo with the results of election to the advisory committee for 1974-1975. The new committee consisted of Albert Carozzi, Don Graf and Dennis Wood. Recalling Graf's service on the old executive committee, I had reasons for concern.

Three new students arrived: Bill Busch (BS Iowa State, MS. Illinois, PhD. Oregon State; Univ. of New Orleans), Jim Castle (BS, Allegheny College, MS, Wisconsin, PhD Illinois; Chevron Research and Chevron Services, Cabot Oil and Gas, Clemson University) and Linda Tills (BA, Wittenberg, MS. Illinois, Illinois State Water Survey). Linda's wanted to study clay mineralogy but the department didn't have one. I took her on only if she did the clay mineralogy of Wabash River sediments and collected them under Roscoe Jackson's supervision. I went over their programs and proceeded to teach my Geology 309 and 437 classes.

The students in Geology 309 seemed subdued but nothing usual happened. The Geology 437 class was a bit argumentative spurred on by Dave Rich (BS Notre Dame, PhD Illinois; carbonate petrology, Shell Oil Co., founder of Rockware) and Patricia (Tricia) Santogrossi (BS Illinois, MS Illinois; Shell Oil, Marathon Oil, Consultant, Chroma Geoscience, Statoil). Rich admitted during the fieldtrip that on arrival, Julie Graf told the new graduate students that "Klein is a terrible person; don't take his courses." But Dave was also street savvy and added, "And then we found out about her."

At the end of September, Dorothy Smith retired. She was replaced by Wanda Morrison, who had spent the previous 20 years raising children who now had left for college. Wanda was a polite and nice lady but she clearly lacked depth.

Because Dean Rogers suggested I do so, I reconnected more collegially with the remaining soft rock faculty, Blake, Mann, Sandberg, Langenheim and Carozzi. Sandberg was the most responsive. The others were more reserved and understandably so. They felt I was not that supportive of them during the First Donath War and beforehand. Over time my relations with them improved. I also continued to have my semi-regular lunches with Domenico, Palciauskas and Johnson and the atmosphere was as friendly as before.

That fall, I started teaching short courses to the oil industry, first under the sponsorship of AAPG, and then later the SEG, a private company, IHRDC, and a few initiated on my own. I contacted Paul Schluger who was now working for Mobil in Calgary. He advised, "Give them the depositional systems section of Geology 309 and add oil field examples for each including both well log data and seismic data." Paul was right and the course was a success.

My first offering was to the New Orleans Geological Society immediately after the GSA meeting in Miami. It was offered at night at the Tulane University auditorium. My local host was Ram S. Saxena (BSc Lucknow, PhD, LSU; stratigraphy and sedimentation; Texaco, Superior Oil - Exploration manager, East Region, private consultant). Ram asked for a course syllabus and I sent an outline. He wrote back and asked me to include text and illustrations. I prepared a preliminary version, he added his own chapter, and printed it.

I could sense tension in the air as the semester continued. Unknown to me, Graf, Henderson, and Chapman looked at the annual university budget book to determine faculty salaries. They complained to Fred about both my and Dennis Wood's retention raise. Fred met with me and reported this and I commented, "Fred, people will always complain about something regarding anyone they chose. These people will probably come in again and complain the next time I use the toilet."

Fred laughed, agreed, and then commented that he hadn't expected people to complain about other faculty salaries and showed the three of them my annual report and his nomination papers. They remained unconvinced.

We held faculty meetings and I contributed my share of comments. However, I could tell from the non-verbal behavior that about five people (including Graf, Henderson and Chapman) were unhappy I was contributing at all.

In mid-September, I talked with Pat Domenico about the need for strategic planning and we held several discussions. Pat proposed a great idea which we developed and presented to Fred. Fred reviewed it, asked perfunctory questions, and never did anything more. Later, he concluded that I had set Pat up (incorrect), and was challenging his authority to lead the department.

We held a tenure and promotion committee meeting. We reviewed the assistant professors and voted to advance Palciauskas.

When Holder was proposed, there was extensive discussion and I also expressed reservations. During a trip to the University of Michigan to present a colloquium a month before, I met a solid-state geophysicist who never heard of Holder, which was troubling. Holder was still publishing in physics journals and never attended meetings of the American Geophysical Union (AGU). Wood commented on poor enrollments in Holder's classes. He was advanced by an 8 to 6 vote. I voted no.

After the meeting, Fred took me aside and said, "George, I thought you supported Holder." I explained that when he arrived, I liked the idea of the type of research Holder said he would do, but felt he went on a different track that wasn't advancing geology or the department. Fred then said, "George, if I get you more information, would you change your vote?" I replied, "If you have new information, I'd like to see it and yes, I would consider changing my vote if the information justifies it." I thought that was fair, but Fred never provided the additional information.

A week later, the annual GSA meeting was held in Miami. I attended the Illinois cocktail party which was held in two adjoining hotel rooms with a connector door. The rooms were registered to Donath and Goodman. Fred and Goodman arranged for some of the graduate students to wear white waiter jackets and serve and mix drinks. I noticed all the liquor bottles had Illinois State liquor stamps. Apparently it was cheaper than making arrangements with the hotel (like other university departments).

One of the students told me the next day that they were offered a free carryall to attend the meeting to find jobs in exchange for which they were to help with some departmental duties (unspecified). Just as they were packing the carryall, Goodman came out with a box of liquor he had bought to load on the vehicle.

The students were trapped and drove to Miami. All knew that transporting liquor across any state line was a serious offense. As a defensive measure, they drove 5 MPH below the speed limit, and checked everything under the hood and tire pressures at every gas stop. They were fortunate nothing went amiss during the entire round trip.

I found this disclosure shocking because one does not make vulnerable those who depend on you for a recommendation or financial aid, as Donath did in this case. I filed the information away for future use. Fred and Bill clearly were being reckless with other peoples' lives and placing the University of Illinois at risk.

During the meeting, I visited with Arnold Bouma who was at Texas A&M. He mentioned that the Department of Oceanography was looking for a new department head and my name came up. I told him I was not ready for such a job and not interested.

Immediately after Thanksgiving, Dennis Wood accosted me in the Halls. He claimed, in front of students (Dennis loved audiences) that I was calling sedimentologists around the country about dropping Julie Graf as a PhD advisee. He called me a "fucking liar", a "bloody bastard" and a few more choice British swearwords. I told him it was all news to me. I suggested that maybe friends of hers were spreading the word.

As I write, I should remind the reader that I now live in Texas which is in the South. In the South, they say "There's something in the air." Clearly in November and early December, 1974, there was, indeed, something in the air in the department of geology at the University of Illinois.

Late in the afternoon on December 4, 1974, Wanda Morrison called and said, "Dr. Klein, Dr. Donath wants to meet with you tomorrow afternoon at 4:00 PM. Are you free?" I explained I was and then added, "Is there anything I should bring?" She asked me to hold while she checked with Fred. When she returned to the phone she said, "No that won't be necessary." I concluded the phone conversation and went home.

LESSONS LEARNED:

1) When working in international teams, it is essential to try to determine cross-cultural variables that will improve cooperation. DSDP Leg 30 would have been more successful if a mechanism existed to achieve this.

2) When difficult decisions about graduate students (or personnel) must be made, it is best to be decisive and make the decision earlier than later. Delay only compounds subsequent difficulties. It is well and good to give people the benefit of the doubt and encourage them to do better, but often it won't work.

3) When others anticipate problems that seem premature, do not dismiss them too quickly. Ask for time (a week) to evaluate and review before offering a response. When Donath offered to chair Julie Graf's committee because of potential conflict of interest problems, I declined far too hastily. Had I waited a week, there would have been more cover when things didn't work out.

POSTSCRIPT #1. The Society of Exploration Geophysicists (SEG) booked me at The Captain Cook Hotel in Anchorage, AK, in 1985 to teach a short course. When checking in, they offered an eighth floor room with a view. I told the clerk I wanted a room on the lowest floor where rooms were available and got one on the third floor. When asked why, I explained about the fifth street slide and the need to get out alive if another earthquake struck Anchorage.

POSTSCRIPT #2:

> "Those who cannot remember the past are condemned to repeat it."
> -George Santayana, *The Life of Reason, Volume 1, 1905*
> *(Spanish-born) philosopher (1863-1952)*

CHAPTER 18

The Second Donath War. Part I
(December 5, 1974-May 1975)

"When you have enemies at this university (i.e. U.I.U.C), they tend to be unremitting."
- Robert C. Bilger - 9/20/1987, *Champaign-Urbana News-Gazette*, p.3

"Where there is no vision, the people perish; but he who keepeth the law, happy he is."
- Proverbs, 29:18 (King James Version)

December 5, 1974, was a typical Midwestern cold winter day. It was cloudy with intermittent drizzle, rain and sleet. By 4:30 PM, the sun (if one could see it) had set.

I arrived at Fred's office at 4:00 PM wearing dress slacks, blazer, shirt and tie. Wanda Morrison ushered me in. Fred shut the door as I sat and the dialog went approximately as follows:

Donath: George, I've noticed your attitude around here has changed. You opposed the Holder tenure promotion, you set Pat Domenico up to jointly develop a strategic plan signaling the faculty to undermine my headship, and you are friendlier with your colleagues in soft-rock geology. Some think you are after my job."

Klein: Well Fred, I offered to reconsider about Holder......

Donath (interrupting): That's not an issue now. George, you are an outstanding researcher and an outstanding teacher, but your personality is not appropriate for this department. I must ask you to leave.

Now, Texas A & M University is looking for a head of their oceanography department (handing me an announcement about it) and it would be a good opportunity for you. I think you've always wanted to be a department head so I nominated you.

Klein: Well, Fred, I was approached about the Texas A&M job at the

Miami GSA and turned it down. They need a blue-water oceanographer and I only have one DSDP cruise under my belt so I wouldn't have much of a chance. This is a non-starter.

Second, given what you just said about my personality, it surprises me you would even nominate me. If you did, it raises substantive ethical questions.

(Fred reached over, grabbed the announcement sheet, crumpled it, and threw it in the waste basket.)

Donath: Well, I didn't mention your personality. I just told them about your research and your teaching.

Klein: You know Fred, perhaps I need to say a few things. First, I'm not interested in your job because it is too much time away from research. Second, Dean Rogers asked me to reconnect to my soft-rock colleagues during the review of your headship eighteen months ago, and I chose to do so.

Third, I think you need to face up to something. The Bible says "As you sow, so shall you reap." You sowed the seeds of dissension in this department and are reaping the consequences. If the administration becomes aware of our conversation, and likely they will, they will want to review you again, especially events since the last review. In my case, they will ask why you gave me a large retention raise four months ago and now want me to leave.

Fred, I've noticed some other things. I noticed your health is declining. You seem not to be on top of things to the extent when I arrived. Perhaps you've been head too long. Fred, you asked me to leave. I think I have the right to ask you to resign the headship and do it today.

Donath (irritated): I have no intention of resigning my headship.

Klein: Well Fred, I predict that my full professorship in geology at Illinois will outlast your headship.

Donath: That's probably true.

Klein: That reminds me. Isn't your field experimental structural geology and rock deformation?

Donath: Yes, why do you ask?

Klein: I looked at John Ramsey's new book on structural geology and rock deformation when it arrived in the departmental library a month ago. Not one of your papers was cited.

There was a short silence.

Klein: Anything else?

Donath: Yes, I did want to ask, is it true you retained a lawyer during the problems at Penn between Faul and Boucot?

Klein: Not during the active warfare between them. The summer before I went to Oxford, I visited an attorney about Penn's sabbatical leave policy to return for a year after the leave ended. I was advised to ask for a release which I received and then resigned from Penn to come here.

Donath: Also, I understand when Julie was doing her thesis work in Connecticut, you went over her maps and samples one evening in her room she rented. Did you have a liaison with her?

Klein: Good grief Fred. Doing something like that with Julie is like kissing your kid sister. Please credit me with better taste and better sense.

Donath: If you stay, you won't like it because academic standards which I know you value will start to deteriorate.

Klein: If that's the case, then it is likely you'll be reviewed again by the administration, and if that happens, I'll feel sorry for you.

I got up and left, politely thanking Donath for his time. I knew he couldn't fire me because I had tenure and basically had no grounds to do much else. Other than filing charges with the administration and the Board of Trustees, all that was open to him was harassment and reduced pay raises. Anything else would invite a complaint to the administration or the grievance committee of the faculty senate. Even an inadequate raise invited a senate grievance committee petition and review.

On my way home, I stopped at the Bull and the Bear at the Urbana-Lincoln Hotel for a drink and met some people I knew socially from campus functions. Most were in the Humanities and Social Sciences. It was clear from their questions they were aware through the campus grapevine of problems between me and Donath and they brought it up. I was very careful but one lady psychology professor who specialized in psycho-linguistics mentioned I might want to talk with Terence Brown, an African-American law professor to see if there was a civil rights angle.

I replayed my meeting when I went home and made some operational decisions. I learned long ago a job is an asset and one protects it as such. A faculty appointment at the University of Illinois had intrinsic value because its administration knew how to foster faculty research in a way very few places did. Such a faculty appointment is not only coveted, but it was also worth the fight to keep it.

The intrinsic value of a faculty appointment at the University of Illinois was repeatedly brought to my attention during overseas travel. Wherever I went in Asia, Africa, Australia, or Europe, I discovered doors were opened for meetings and appointments that normally were difficult to arrange. I got them simply because I was a professor at Illinois. Many positive things came from those meetings. That university's global reputation in science and engineering was that strong. A professorship at Illinois was therefore worth keeping and difficult to match elsewhere.

To me, another reason to resist Donath's request was to avoid gaining a reputation of being an individual who could get pushed out easily because of perceived personality differences.

I decided to call Brown and at least get a legal reading if there was one. In the interim, I knew that I should play a waiting game to see what actions Fred and his cohorts might take. If they took actions that I could use to cut them down, then I would make a move or two. Fred's hand appeared not to be as strong as he thought and he did not hold all the cards. Playing a waiting game might enable me to snatch a few cards away from him.

Lunch the next day was with Pat Domenico and Hilt Johnson. Everything was normal and nothing was said or volunteered. Because Johnson was ex-officio on the advisory committee, I suspected he knew.

The question remained who was in Fred's corner? It was easy to identify Wood, Graf, Henderson and Chapman. Carozzi was another possibility but I concluded he was not likely to do too much. I learned two years later he opposed Fred during the advisory committee meeting when it was discussed. The problem I foresaw was fence-straddlers and I had no way of knowing who they were or which way they were leaning. It was best to be wary, but be polite to everyone, say nothing and keep working on my research.

Three days later, I called Terence Brown and we met two days afterwards. On meeting, he said, "We haven't met, but I've heard a great deal about you. You're widely respected on campus for your research publications, your teaching, and your awards." I then reviewed with him what transpired and he asked, "Did you get a retention raise? I assume you were eligible." I told him I received one.

Brown finally said, "The only thing I can suggest is that you write Donath a letter telling him you're sorry to hear he is unhappy with your work and tell him if he wishes to bring charges, he does so in accordance with state and federal law covering tenured faculty. If that does not stop him from taking action, or if he files charges, come back and see me. "

A week later I typed a letter and sent it to Fred. The letter stated:

Dear Dr. Donath:

I understand from our conversation on December 5, 1974, that you appear to be unhappy with my work.

I write to inform you that it is my intention to stay at the University of Illinois as Professor of Geology. If there are charges to be brought, I trust they will be filed in accordance with Illinois state regulations and relevant state and federal statutes governing due process for tenured faculty.

> Very truly yours,
>
> George Devries Klein
> Professor of Geology

cc. Robert W. Rogers
    Morton M. Weir

After final exams, I reassessed my situation and decided I needed a separate income source to offset the possibility of retaliatory raises. I knew I had to make people aware of my short course without offending the AAPG which sponsored it because they opposed self-promotion and electioneering. Reports about my New Orleans course spread and Texaco Research in Houston invited me to give it there in January, 1975. They obtained the camera-ready copy from Saxena and reproduced my short course syllabus.

My short course syllabus needed upgrading. I revised, edited, and updated it, and concluded I could publish it myself. Once revisions were completed and camera-ready copy was prepared, I formed a business, Continuing Education Publishing Company (CEPCO), incorporated, opened a bank account for it, and rented a P.O. Box. I visited my local banker to arrange financing and brought along a copy of Robert L. Folk's famous 'orange bound' Sedimentary Petrology lab manual printed by Hemphill's in Austin, TX. I explained to the banker that Folk's manual was an academic success but a commercial failure. I said if I printed my own short course manual, I could require it be used in all my short courses, and market it commercially. The bank gave me the loan.

I spent winter break preparing the revision and once I received the loan, designed a cover, arranged printing and binding of 1,000 copies, and stored them in my garage. I bought shipping supplies, stamps, mailing labels, and was in business in my home. With the P. O. Box and residence as a physical address, any concerns about conflict-of-interest were solved. I paid the loan off in four months. I also paid myself royalties from book sales every month. I printed another 500 copies after six months.

Next, I designed a brochure, mailing them to university geology departments, oil companies, government agency, and oil field service companies. Soon orders came in. I spent nights and weekends handling the business end of CEPCO until it was sold in 1978 to the Burgess Publishing Company in Minneapolis, MN. AAPG was happy they didn't need to print my course manuals because they could buy them from CEPCO. I now had a separate income stream and the publication, under University of Illinois rules, qualified as a book, so I listed it in my CV.

I continued this activity 'under the radar' for only two months. When the University of Illinois branch library ordered copies for use in my classes, one of my colleagues discovered the CEPCO phone number was the same as my home phone.

My letter to Donath stating I was staying did not generate a response either from Fred, or from the Dean or Vice Chancellor of Academic Affairs. Other than a polite "hello" in the halls, I did not speak with Fred. I did the same with my probable adversaries and only Don Graf refused to say 'hello' back. Politeness was to be implemented to the best of my ability. I still had lunches with Pat, Hilt and Vic, coffee breaks with Sandberg, and periodic chats and banter with Langenheim.

Early in January, I received my course evaluations. My Geology 309 undergraduate sedimentology course was highly rated with an overall average of 4.8 out of 5.0. I thought the course moved smoothly but no better than in the past. Jerry Walker, the course TA, told me afterwards the students chose to send Fred Donath a message because they wanted me to stay.

The graduate course (Geology 437) generated a bi-polar distribution. When I was at ORST, I audited Ross Heath's marine geology course. He periodically assigned a paper to a student to be reviewed in class with an oral presentation. After the presentation and discussion, Ross also critiqued their presentation style to help improve them and for other students to learn from his critique. I implemented a similar system but at Illinois, my critique was viewed as personal criticism and even a personal attack.

Al Carozzi visited me during February, 1975. He had a long conversation with Donath because Fred complained Al was gone a lot from campus while consulting overseas. Al hired a graduate student to teach his classes and paid them $10.00 per hour of instruction during his absence.

Fred wanted Carozzi to reduce his consulting and Al demanded a salary increase. They reached a stalemate. However, Fred needed Al's expertise in carbonate rocks for some experimental deformation research he was doing. He offered Al a chance to work with him, and from his funding sources, pay a summer supplement. Al decided to try it for one year and to see if anything came of it. He seemed unenthusiastic. Al warned me that Fred wanted to do the same experiments with sandstones and may ask me to get involved.

In general, things were quiet. I taught my graduate depositional systems course but had to cancel the Ouachita field trip because of a family emergency. My father had a hip replacement completed in 1970, and my mother was diagnosed with breast cancer which had cleared, or so the doctors thought. The cancer spread to her bones. A major surgical procedure was scheduled during the time of the field trip and my father wanted me around. She survived.

We interviewed people for a vacancy in clay mineralogy and hired Dennis Eberl (BS, Dartmouth, PhD, Case Western Reserve; clay mineralogy; Buddhist Theological School; Northern Illinois, Illinois, USGS). He was trained by John Hower, so I called him for a reference. John told me, "George, Dennie Eberl is unquestionably the best student I ever had." I asked, "Better than Bruce Velde?" Hower responded, "Better than Bruce Velde." I concluded we couldn't lose and voted to offer him the job.

In late February, Roscoe Jackson defended his PhD. Roscoe was offered and accepted a faculty appointment at Northwestern, replacing Bill Krumbein who retired. The committee consisted of Ben Yen (Civil Engineering), Pat, Vic, Hilt, and me as chairman. Fred Donath decided at the beginning of the year to sit in on all PhD thesis defenses. The format was that committee members asked the candidate questions based on reading the thesis. In later years, a public presentation to the entire department was added, but not in 1975.

At the end of the defense while Roscoe waited outside, the committee discussed his thesis and performance and voted to pass him. Fred then said he would sign the thesis and turned around and ripped into me, attacking my qualifications and teaching and questioning my fitness to serve on the graduate faculty. He then left. Pat Domenico and Vic Palciauskas were totally shocked and asked "Whew, what's that all about?" Ben Yen sat and took it all in and reported it around the engineering school. Soon word spread all over campus. Hilt said nothing. I said, "Why don't we give Roscoe the good news." We closed the meeting.

During spring vacation, I taught a short course at Amerada-Hess in Tulsa. The course was attended by 20 people, including the exploration manager, a round-faced man of about 55. On the morning of the second day, I discussed both Fisk's and Saxena's models for deposition of fluvial deltas, and suggested that to recover the thickest sands, drilling should be located at the levee crest below which the thickest sands could be penetrated. The adjacent channel fill was likely filled with mud during an abandonment phase of channel migration. I could see the exploration manager turn red in the face and sit straight up in his chair.

He came to see me during the coffee break and my mental reaction was 'Oh, Oh. Now I'm going to get it. I proposed something that contradicts the company's exploration paradigm.' I was wrong. He said, "I'm so glad you showed that slide about the drilling recommendation on the levee to get the thickest reservoir sands. You see, the Shell Oil Company drilled a dry hole on a channel in North Dakota and relinquished the lease. I've been trying to persuade the higher ups to lease the acreage and drill an offset. Now I have some ammunition."

On my return, Fred approached me about working with him in his experimental rock deformation lab. I politely declined.

Tsune Saito applied to NSF for funds for the joint USA-Japan project on Hokkaido Island. To hedge my bets, I applied to MUCIA (Midwestern University Consortium for Internal Affairs) which sponsored international research programs. I met the University of Illinois MUCIA director. He explained the program was dominantly run by agriculture schools (Illinois, Michigan State, Purdue, Wisconsin, Purdue, Ohio State) but he liked my proposal.

MUCIA turned my proposal down. The reviews said I was well known for running such programs and mentioned my paper on Brazil. The Illinois MUCIA office had separate funds and offered to pay my plane ticket. I notified AAPG about an upcoming trip to Japan and it generated another short course. AAPG had a system for international courses that if one's international air fare was paid by a third party, they marketed the course to affiliates in countries to which instructors were travelling.

During the middle of April, I saw Dennis Wood in the halls and he came forward and harangued me. I can't recall his concerns, but he swore at me, said I was a disgrace to the department and should leave. Again, his behavior was witnessed by graduate students. He then said, "You know George, one day I'm going to beat the shit out of you and leave you dead." I said nothing and walked away knowing he threatened assault and bodily harm. Perhaps it was time to review it with an attorney. The lady psycho-linguist from the Bull and Bear happy hours mentioned one.

Two days later, Fred sent letters about next year's salary. The Dean proscribed a range in absolute numbers (minimum of $500.00) and a maximum of 2.5 % with a cap of $1,500.00. My raise was only $500.00. I now had a card to play and filed an appeal with the Faculty Senate Grievance Committee.

I also contacted an attorney in Urbana, Robert Isham Auler (BA, Philosophy, JD, Illinois, States Attorney's office, private practice), and made an appointment.

Bob was in his thirties and his office was in a newly-renovated house overlooking Lincoln Square. He explained he bought the building for a song because it was in bad shape, and hired architecture students to design the renovation.

Auler told me he wanted to major in geology but couldn't handle the math, physics and chemistry so switched to philosophy. He asked after Mike Wahl and I explained he left, which Bob didn't know. He took a Historical Geology course from Carozzi. Bob also admitted that as an undergraduate, he was hired by the Chancellor's office "to spy on the faculty" and provide them with student opinions about them.

I summarized my problems with Fred, Wood, and Graf, and gave Bob copies of documentation. He looked them over. Bob made it clear I needed to be conversant with the University's rules, by-laws and statutes and stick to them. He said, "George, if you run afoul of any one of them, the administration has an excuse to do nothing."

Auler continued to go through the papers I gave him. Finally, he said, "George, I first have to tell you that you are subordinate to the head and I have to acknowledge that to Donath and you must understand that. However, I see some things where he may have acted illegally, or close to it." Auler mentioned a pending case in Arizona, *Peacock vs. University of Arizona*, that was relevant to my problems. Peacock's complaint in many ways paralleled my situation.

Bob also mentioned that Graf would be difficult to smoke out. Graf seemed to hide his tracks very well. Wood definitely violated the law with his threats, according to Bob.

Bob then said, "What I'll do is write a letter to both Donath and Wood on your behalf and we see where it goes. Also, go ahead and appeal your salary raise and when that's decided, we can go from there."

I said, "You describe Graf very well. I understand about being subordinate to the head and I expect to honor it. I'll sit tight for now."

Bob then suggested, "Have you thought of transferring to the Center for Advanced Computation, teaching geology courses in the geology department, transferring your existing space to the Center, and doing your research there?" I told him I'd look into it, but had a hard time seeing a match.

Three days later, Bob sent out the letters to Fred and Dennis and I received my copies. Again, after nearly 35 years[4], I can only approximate what the letters said:

Dr. Dennis S. Wood   CONFIDENTIAL
Dept. of Geology
University of Illinois
Urbana, IL, 61801

Dear Dr. Wood:

This office represents Dr. George Devries Klein regarding your widely-known harassment and threats you directed against him. We shall represent him also regarding threats witnessed by others to inflict bodily harm and death. This office also represents Dr. Klein regarding your involvement in attempts to have him dismissed and your intrusion into his carrying out his academic duties, including terminating supervision of a graduate student whose performance was below par.

Dr. Klein's disclosures surprised me. Through my alumni ties to the University of Illinois, I heard you are an inspiring teacher and attracted graduate students to work with you. I therefore developed the impression, perhaps mistaken, that you were a credit to the University of Illinois. I regret this is not the case.

I therefore demand you cease-and-desist from harassing and haranguing my client again and to cooperate with him so that both of you can serve the mission of the University of Illinois. If you fail to do so, my client and I shall seek our remedies in another context.

       Very truly yours,

       Robert Isham Auler
       Attorney-at-Law

cc: George Devries Klein

I reread the letter and concluded it would likely catch Dennis by surprise but did not expect any improvement in his behavior.

The letter to Donath, as I recall[4] approximated the following:

Dr. Fred A. Donath, Head   CONFIDENTIAL
Dept. of Geology,
University of Illinois
Urbana, IL, 61801

Dear Dr. Donath:

This office represents Dr. George Devries Klein, Professor of Geology, regarding certain prejudicial decisions you made against him. I have instructed my client to follow the rules of the University of Illinois and its statutes and by-laws. I also reminded him that he is subordinate to your departmental leadership and must act in accordance as stipulated in the University's statutes as well as state and federal laws.

Nevertheless, I have read memos you sent him, his replies, and his memoranda to himself regarding meetings and discussions with you. It appears you exceeded your authority as department head in several areas, acted prejudicially against him contrary to state and federal law, and made requests outside your jurisdiction as head. Moreover, your salary recommendation for Dr. Klein for the next academic year is prejudicial, personally-motivated, and in violation of Illinois statutes governing performance evaluations of state employees. Expressed in another way, your actions are placing the University of Illinois at risk.

I demand that you quickly reach an accommodation with Dr. Klein, retract in public and private your request that Dr. Klein leave the university, and implement ways to work together with him and others in harmony for the good of the university. If you fail to do so, my client and I shall seek our remedies in another context.

    Very truly yours,

    Robert Isham Auler
    Attorney-at-Law

cc: George Devries Klein

I reread the letter to Donath again and concluded he was adequately warned but the outcome was unclear.

Four days later, I saw both Donath and Wood talking in the halls and appearing worried. When they saw me, they walked away. Their behavior assured me that Auler's letters was read. Later, I heard from Haltiwanger that both Donath and Wood visited the university attorney, Ed Madigan, who told Wood to get his own attorney and Donath that there wasn't much he could do.

On the last day of the semester, a departmental faculty meeting was held in the conference room. I decided to sit at the foot of the conference table opposite where Donath would sit so I could observe reactions of my colleagues if anything transpired that involved me. The meeting was routine and as we were about to adjourn, Donath said, "Before we adjourn, Dennis Wood has asked to say a few words."

Dennis stood up and handed everyone, including me, a sheet of paper, face down. It was a copy of his letter from Auler. He then said, "The department faces a serious situation. We have in our midst a colleague, Dr. Klein, who takes actions to undermine our programs and standing. He has been asked to leave but continues to stay. He has been disruptive of our activities.

As you can see, I have now received a letter from his attorney with allegations that are false. Clearly, Dr. Klein wishes to use the courts to achieve his goals because they cannot be achieved on a collegial basis.

Dr. Klein, I ask that you leave this university so the department can be at peace and that you find another department to destroy to pieces."

He then sat down. I quickly looked around the table while he spoke and noticed nearly 3/4 of my colleagues looked down at the table. Several turned to me and smiled.

When Dennis finished, I raised my hand and was recognized.

"Dennis," I asked, "Isn't that letter labeled 'confidential.'?"

Wood replied, "Yes, but that's for MY protection."

I replied, "Thank you Dennis. That's exactly my point. Mr. Chairman, I move we adjourn."

Chapman seconded and we left. Half of the faculty left the copy of Auler's letter on the table.

LESSONS LEARNED:

1) When things go wrong in academe, document everything one does, keeping detailed logs, records, notes, memos to oneself.

2) In an academic career, it is essential that one knows and understands the rules and by-laws of the university and department, follow them closely, and never deviate from them.

3) When things go wrong in academe, keep a low profile and avoid making waves.

4) If problems arise in a highly structured environmental like the University of Illinois that is administratively top-heavy, use an attorney if all else fails because it levels the playing field and forces the administration also to abide by their rules and by-laws.

CHAPTER 19
The Second Donath War. Part II
(Late May 1975-August 1976)

"When personal issues are involved, reason takes a Holiday"
- Cardinal Richelieu

"The veneer of civilization is very thin"
- Margaret Thatcher

"You have to trust your colleagues even if they are untrustworthy"
- Ralph L. Langenheim, Jr 4/9/1981

I continued working on my research after exams were over, and prepared for travel to Japan. After commencement, I received a telephone call from Nancy McGowen, the executive secretary to Dean Rogers. She explained Rogers wanted to meet me and to allow two hours. We set a 2:00 PM appointment the following Tuesday.

During that meeting, Rogers told me the university administration was extremely concerned about the two letters my attorney wrote and wanted to see if the problems I had with Donath could be resolved. I made it clear that the last thing I wanted to do was sue the university and I wanted this settled collegially, if possible.

Bob then told me the most astonishing thing I heard during both the first and second Donath wars. Bob said, "George, the administration has been worried about Fred for some time. He will spend six months singing the praises to the administration of whoever he views as his favorite professor, and then suddenly reverse himself and asks our help to remove his favorite professor from the university. We saw it with Adrian Richards and Mike Wahl. When it came up with Bill Hay, we recognized the pattern. We were expecting this to happen with you and in fact, one of your colleagues predicted it would."

I asked, "What can the administration do in a situation like this?"

Bob replied, "There is little we can do. All administrative positions from the department head to the President in this university are life-time appointments until retirement, so removing one requires extraordinary efforts. This university has had a difficult history of imperial department headships and efforts to remove them generally led to very bitter results."

Bob then asked me to tell him exactly what transpired since the last review. I brought copies of key documentation and gave them to him to keep. He was quite shocked about the Julie Graf problem seeing a major conflict of interest, and the way Chapman handled the situation. He was surprised at Fred's hostility after the Holder tenure review, and was shocked that Fred failed to follow-up on strategic planning. He said, "Our strongest campus departments, chemistry, psychology, engineering, agriculture and accounting do strategic planning all the time and that's why they stay so strong." He already heard via campus gossip about Wood's outbursts, and Graf's visit to Haltiwanger.

I brought up a transfer to the Center for Advanced Computation and Bob explained it would be difficult to transfer a faculty line and the associated space out of the college. Moreover, Donath would have to approve it, and he considered that unlikely.

Bob then disclosed that the Faculty Senate Grievance Committee sustained my salary raise complaint, recommended a higher one be given, and notified him accordingly. He added that Langenheim had also filed a complaint (unknown to me) and they sustained his position too.

By this time, it was nearly 4:00 PM and I asked, "Bob, where do we go from here."

Bob smiled and said, "From now on, I want you to send me a copy of everything Donath writes you and your replies. I want you to keep me aware of your travel itinerary and how to reach you quickly if I have to. Above all, I want you to know that you can come here any time to review problems in the department without obtaining clearance from Donath and I've told Fred that I am giving you that authorization."

I asked, "Anything else?"

Bob replied, "If anything else comes up or is needed, I'll contact you"

I left and returned to my office. I realized the administration was not only watching Donath, but they also were watching me. However, I felt the Dean was sincere in his desire to solve the situation but could neither predict the outcome, the time schedule, or what could be done.

Two days later, Rogers called. He explained he reviewed our meeting with Mort Weir (Vice Chancellor for Academic Affairs) and that Mort wanted to meet with me the following Thursday. Bob instructed me to call Mort's office for the time. It was set for 1:00 PM.

When entering Weir's office he was busy examining about 10 accounting ledgers spread over his entire desk. I brought a file of documents for Weir and needed a place to put them. I began moving the ledgers to make available space and he looked horrified and said in a stern tone, "Do you know what you just did? You moved the budget ledgers for next year for the college of engineering, the school of business, and the school of agriculture." I replied as politely, "Sir, not to worry, I'll put them back when our meeting is concluded."

Knowing I had only an hour, I summarized the departmental problems in 20 minutes. Weir listened carefully. He asked several questions which I answered. He finally said, "George, I understand the difficulties you face and I appreciate your disclaimer about not suing the university. That will help a lot as we resolve this. However, we cannot remove Fred because he doesn't get along with you or doesn't like you. We can only remove Fred Donath if he takes action that in any way undermines the administrative structure of this university. So far, we haven't found anything although some allegations have surfaced about his rock deformation lab. Do you know what they do?"

I replied, "Sir, I know they do research. When I came here, Fred told me he had a patent on the rock press he designed and was selling them to other universities and industrial research labs. He mentioned that the net income goes into a special account for the benefit of his graduate students. I hear rumors, but know nothing else that is concrete."

Weir then said, "Try as best as you can to improve your relationships with your colleagues. They are more on your side than you think, but they want to protect their flank too. I'm aware than several of them want Donath fired."

I replied, "Well sir if asked, I'd recommend that because Fred has tenure he be removed from the headship only and return to a regular professorship and be productive."

Weir said, "George, you're being very kind and I like that. However, some of your colleagues want him removed from the university."

Realizing the meeting was about to end, I put my file back into my briefcase and moved the accounting ledgers to their rightful place. Weir was again startled and asked, "How did you remember where the ledgers were supposed to go?" I replied, "Sir, I learned to work in tight places on oceanographic ships. If you don't remember where you put things on board ship, you might lose them." Mort smiled for the first time during the meeting. As I got up, he said, "By the way, if you feel the need to see me again about this, you're free to come any time, and Fred Donath has been told that."

I left after shaking his hand and thanked him for his time and interest.

Walking back to my office, I replayed quickly my meetings with administrators at Pitt, Penn and now Illinois. Of the Vice Chancellors/Provosts, Goddard was clearly the best, although Weir was tougher. Peake (at Pitt) was well behind them. Of the Deans, I had a harder time. Rogers clearly was careful, genial and concerned, more so than Otto Springer at Penn and Dave Halliday at Pitt. But I could not decide which was stronger. It seemed a draw.

I knew that now I had open access to the administration, that some clamps were placed on Fred but I had no idea what they were, and that an unspecified, group of colleagues wanted Fred out. I knew I had to be careful, documenting my activities on time sheets to protect myself in case someone tried to set me up. It meant walking on eggs without cracking them, something I was not known to do well. In short, it appeared as if Fred and I were in a sparring match on a frozen pond waiting for one of us to hit thin ice, fall into freezing waters, and die.

On the way home, I stopped for a drink at the Bull and the Bear with the usual crowd. The lady psycho-linguist knew I met with Weir and asked what I thought. She then said when Weir was head of the psychology department, he was tough but fair. She added that with Weir, everything depended on performance. If one worked hard, raised external research funding, and published, he was inclined to support those people.

I spent the weekend reviewing where I stood. If I learned one thing, it was that the administration would do everything it could to keep their administrative structure intact. The key to getting Fred out of the headship was to demonstrate anything that suggested or proved he was a threat to that structure.

Bob Auler called on Monday to inquire about my meeting with Weir. I explained my perceptions and he agreed. He advised me to see if at all possible if I could find out more about the rock deformation lab. Bob said, "Something's not right there. It doesn't pass the smell test."

By this time, Dennis Wood left for the UK to run his annual British Isles field course. I saw Fred in the halls but nothing unusual happened. Chapman went to Maine.

I left for Japan in late July. AAPG arranged for me to give my short course in Tokyo to the Japan Petroleum Development Corp which was an investment bank. The participants represented small oil companies and were seconded from the bank to explore in areas where the small companies had concessions. One of these companies, the Japan Bengal Oil Exploration Company invited me to dinner to show me their seismic mapping and interpretation. The entire company exploration staff consisted of three people, a chief geoscientist, a geologist and a geophysicist. Map interpretations from the shelf region showed a vertical succession of shifting channel systems from the Brahmaputra River delta, a pattern I had not seen before, but came across commonly as a consultant working later in Angola and the US Gulf of Mexico (Chapter 26). It illustrated the concept of fluvial axes developed during the 1990's on the US Gulf Coastal Plain.

From Tokyo, I flew to Kagoshima to meet Hakuyu 'Happy' Okada. He took me to dinner at an outstanding Japanese restaurant with an ocean view. The food was excellent and I ate everything they served. The waitress told Okada I was the first *gaijin* (foreigner) who visited the restaurant who ate everything. I explained to 'Happy' that if it was good enough for him, it was good enough for me.

Happy arranged to drive me around an active volcano, Sakurajima, in the middle of Kagoshima Bay next day. It periodically spewed clouds of ash, but little lava. We then flew to Tokyo, spent the night at a Buddhist-run hotel he knew and which offered excellent food, but different from standard Japanese food. We met the rest of the team of US and Japanese scientists at Haneda Airport and flew to Sapporo on Hokkaido Island.

Team members included Saito, Okada, H. Ujiie (PhD, Tokyo, micropaleontology; Tokyo Museum), Dennis V. Kent (BS CCNY, PhD Columbia; paleomagnetics; Lamont-Doherty, Rutgers), Peter Thompson (PhD, Columbia, Foraminifera, ARCO Research, Consultant), I. Koizumi (radiolaria), Howard Harper (PhD Harvard; dinoflagelates; ARCO Research, SEPM Executive Director) and T. Sato (Geologist, JAPEX). I learned quickly that to do field work in Japan, every university, research organization or other entity invited someone from an oil company as a co-investigator or participant. The oil company participant had one function: bring a company expense account and provide company jeeps, pay for gasoline, special meals, banquets, and liquor bills.

We travelled in two jeeps and measured a major stratigraphic section through the Lower Pliocene and Upper Miocene. It was mostly shale. Okada and I plotted a sedimentological log but were not thrilled with the results. On the third day at our first location, we passed a headland into a small bay. We noticed that above the shale section there was thick sandstone representing the top of a prograding delta. A small oil field was located updip, obviously extracting oil from this deltaic sandstone. Essentially, we examined a slope facies overlain by a delta. In current terms, it probably was a shelf-edge delta or an incised valley fill.

We continued work in other areas. Japanese custom is that in the evening, one wears a kimono provided by the hotel and head into town and party. Tsune took us to pachinko parlors and a few bars dressed this way.

Field work continued until we reached Wakanai, the northernmost city in Japan. On a clear day, one could look north 50 miles across the Sea of Japan and see Sakhalin Island, now part of the USSR. It was as close to the USSR/Russia that I have ever been and I have no desire to get closer. We visited a compelling monument on top of a hill overlooking the Sea of Japan.

It was a memorial dedicated to thousands of Japanese who fled Sakhalin Island on boats after Japan surrendered at the end of World War II. The Russian air force continued to fly sorties and attack innocent civilians, treating them like a turkey shoot. It was a sober reminder of what the USSR and its system stood for. There was a US Air Force base outside of Wakanai for a good reason.

Because of teaching commitments, I left early. The night before leaving, Sato hosted a sashimi dinner for me and the rest of our group.

On my return I checked mail and memos. The election results for the new advisory committee were announced. Wood, Graf and Carozzi were elected to serve one more year.

One memo concerned faculty financial allocations from the departmental budget for use to travel to meetings, course supplies and materials, and other operating expenses. In the past, we charged the departmental budget and Fred used the cost over-runs to justify increasing the departmental budget. However there was a major change. Each allocation was based on enrollments in our classes, and how much overhead the department recovered from our grants and contracts under a campus-wide overhead sharing arrangement. The memo gave times when different departmental groups (Hard rock/geochemistry; structure and geophysics; hydrogeology and geomorphology; and soft-rock) would meet to receive more information.

When I met with the soft-rock group, Donath and Goodman handed a sheet to each of us with our individual allocations. I checked mine and concluded that the overhead calculation was too low. Sandberg noticed the same thing on his sheet. We were told that there would be no change in our allocations.

I returned to my office and checked my grants statements and could prove I was under-allocated. I wrote Dean Rogers a memo attaching all the paperwork and requested his intercession. He called and asked me to see him. When I did, he explained he level-funded the departmental budget to send Fred a message and would do so until Fred stepped down. However, department heads could allocate their budgets for expenses any way they chose.

During the meeting, I brought up the 'Miami liquor run" involving the previous years' alumni cocktail party at GSA when students were required to transport liquor from Illinois across several state lines. Bob turned pale and said he would ask an associate Dean to investigate. I told Bob that the 1975 annual meeting was in Salt Lake City, UT. My Mormon friends told me that hotel cleaning ladies are required to check stamps on empty liquor bottles to see if any came from out-of-state. If they did, the guest was arrested and prosecuted.

The investigation confirmed my allegation to the Dean. Because rooms were rented for the 1975 Salt Lake GSA, Fred and Bill Goodman had to buy their liquor from a Utah states liquor store (after buying a permit to do so). Carozzi and Wood tended bar. I chose not to attend but Langenheim reported what transpired.

I also applied for a sabbatical leave for the fall term, 1976 and that was approved.

During GSA, I met Ken Eriksson of the University of Witwatersrand in Johannesburg, South Africa. He told me the South African Geological

Society was holding a conference with a field trip on Tidalites next August and wanted me to be their keynote speaker. He explained he talked to Wood who invited him to give a colloquium and would discuss details with me when he arrived on campus.

Dennis did everything to discourage that meeting, but Eriksson prevailed. I told Eriksson I could only make the trip if my travel, hotels and meal and expenses were covered. He suggested that if I came to South Africa and presented short courses and gave colloquia at universities, he might be able to arrange all my expenses but no honoraria. I suggested he contact me again when all was in place. He returned to South Africa and arranged sponsorship from three universities (Witwatersrand, Natal at Pietermaritzburg, Natal at Durban), Soekor (national oil company) and the South African Chamber of Mines. In addition, he arranged visits to two mining operations in Precambrian paleo-placer gold. The trip required a seven-week stay. I again notified AAPG who arranged for me to give my short course at PETROBRAS in Rio de Janeiro on the way to South Africa, and in Lagos Nigeria, to the Nigerian Geological and Mining Society on my return to the USA. Consequently, my summer was booked.

Dean Rogers met me just before Thanksgiving. He explained that he was undergoing surgery and would be unavailable until mid-January. Jack Stillinger (BA, Texas, PhD Harvard; 19th century English literature; Illinois) would be acting Dean. Rogers told me he would prefer I did not involve Jack (whom I knew socially) and to hold all memos and responses until he returned.

Rogers then said, "George, I've decided I need to appoint a special panel to investigate the administrative practices of the department. The university rules stipulate that the department head must concur. If Fred turns me down, would you, on my instruction, request such an investigation?" I assured Bob he could count on me to make such a request, particularly because the University rules stipulated I could. I was never asked, but knew Bob would use my assurance as leverage with Fred, who stalled as long as he could.

Late in November, Wanda Morrison left to accept another campus job. She was replaced by Mrs. Sue Lawyer, a mid-thirtyish woman who worked in a law office in Champaign. She was well-dressed and kept her distance from the entire faculty.

The fall semester ended without further incident. A week after the new semester started, Dennis Wood scheduled an out-of-town colloquium speaker who canceled. Wood called and asked if I could give a colloquium on short notice. I said I could prepare a semi-popular talk about DSDP rather than a scientific presentation because of time constraints. He accepted.

When I completed a draft announcement, I visited Dennis in his office. It was the first time I had done so since Thanksgiving, 1973. I gave him the announcement sheet and then said, "Dennis, you surprise me. You harangue and swear at me in the halls, and you seem to want to have me fired, so why do you want me to give a colloquium?"

Wood replied, "George, personality issues aside, you are the best speaker in the department when presenting colloquia and you always have something significant to say. I invited you out of respect for your accomplishments even if I don't want you around. I know you'll do a good job and won't let the department down." I left. When the announcement appeared, Langenheim told me, "Good move George. The new students who didn't take your courses will now know you don't have horns." The colloquium went well and the students were interested in what I said.

Donath telephoned early in February, 1976. He explained politely that Julie Graf was defending her thesis in three weeks and they invited John Southard at M.I.T. to serve as an external examiner. Fred wanted to arrange for me to meet Southard after Julie's exam but not before and I suggested lunch. I then asked Fred, "What's John doing the night before the exam?" Fred replied, "Nothing has been scheduled." I said, "Look, can you arrange for him to meet with all my graduate students. It's a great opportunity for them to discuss their research and meet him." Fred replied, "That's a great idea. Ask your students to be in the sed lab at 7:00 PM and I'll bring him there." The call was very civilized on the surface.

Julie's committee now consisted of Southard, Wood, Donath, Carozzi, Mann, and Johnson. Mann told me what transpired. Carozzi went to see Donath two weeks before the defense, and told him the thesis was unacceptable both scientifically and in form and recommended the exam be postponed. Donath said no. Tradition was followed on the morning of the exam and the committee met first without the candidate present. Both Mann and Johnson said they concluded the thesis was not ready to be defended and recommended postponement. Donath again said no. The exam started and was long and tortuous. When it ended, Julie was excused while the committee deliberated. The committee finally agreed after a long discussion to accept the thesis pending major revisions.

Donath then circulated the required signature cards from the graduate school certifying she passed. Johnson, Mann, Carozzi and Southard refused to sign until they read the revised thesis. Wood then went ballistic with his usual swear words aimed at everyone including Southard. Donath restrained Dennis and the exam ended.

I took Southard to lunch. Southard was surprised I was still at Illinois. He explained that Dennis told him I left the university and that's why they needed an external examiner. I then said, in a resigned tone, "Well, I guess she passed." John replied, "Not entirely. She must revise the thesis according to directives we gave her and only when she does will I and several others sign the required certification card for the graduate school."

I then asked John about his meeting with my graduate students. He replied, "George, they're a real good group, one of the best I've seen in the country." We returned to NHB to retrieve his luggage and Dennis took him to the airport.

Soon after, there was a knock on my office door and I opened it to see Hilt Johnson. Hilt said, "George, I just want you to know that nobody on Julie's committee is second guessing you anymore. Not even Fred. Not even Dennis." I replied, "Well, if she makes her revisions she'll still get the degree. That's unfortunate for those students who work hard and meet requirements." Hilt replied, "Yes, that's true," and left.

Although I haven't written this until now, it was very clear from the outset that Donath's attempt to get me fired was motivated by a back-door political payoff to Graf to get Julie a faculty appointment at my expense. With Julie's performance at her thesis defense and the marginal nature of her thesis, that attempt had blown up in Fred's, Dennis's and Don Graf's faces.

Ram Saxena telephoned to offer a unique field course for our students. If we travelled as a class to New Orleans, Texaco would take us on a two-day field trip in the Mississippi Delta at their expense, including meals, housing on a crew boat, and high speed boats. I accepted immediately and offered it as one of the standard field courses. I limited enrollment because of space specified by Texaco. My own students received priority which left room for 11 more. A signup-sheet was posted in the Wanless Room.

Three days later, Donath sent a memo to the entire faculty stating that stand-alone field courses could be given but that both the number of students and the time spent leading them would not be incorporated into teaching hours calculated for financial allocations. I forwarded the memo to Dean Rogers with a request he reverse Donath's ruling. Dean Rogers did. We held the course in October and it was a success. Twice as many people signed up as could be accommodated.

Sometime in March, I received an order for CEPCO at the department from another university. That university addressed it to CEPCO, c/o Dept. of Geology, University of Illinois. Because of conflict of interest issues, CEPCO could not accept mail there. I gave the unopened envelope to Mrs. Lawyer and asked her to forward it to the company address which I provided. A week later, and because I never received it, I inquired what transpired. She told me she had given it to Bill Goodman who decided it was not to be forwarded. I reminded her that I made a request that did not require Goodman's involvement and advised her to be very careful.

Without my knowledge, she talked to Goodman who encouraged her to file a grievance about my comments with the office of Personnel Services which oversees the non-academic staff. I first knew about her complaint when I received a notice instructing me to attend a hearing about her complaint. I notified Auler who agreed to come.

Three days before the hearing, Donath called a faculty meeting. He produced a document with a series of comments about me and an allegation about harassing Sue Lawyer. He talked for half-an-hour. When he was done, I stood up and said, "Mr. Chairman, this meeting violates university statures about filing charges against tenured faculty. University rules require you to file such charges with the Vice Chancellor of Academic Affairs and not the Faculty. I therefore move adjournment." Chapman seconded (much to my surprise) and we left.

We were on the third floor conference room and as I walked out with Langenheim, Ralph said, "George, if that's all he can come up with in one-and-half years, he has nothing." I walked down the stairs and noticed that Donath, Graf, Wood, Carozzi, Chapman, Goodman, and Henderson entered Dennis Wood's office for a pow-wow. At least I now had confirmation about who was in Fred's inner circle that wanted me out, although I suspected Carozzi attended because he was a member of the advisory committee.

The hearing was attended by Mrs. Lawyer, Goodman, Auler, me, the director of personnel services and the associate director of personnel services who happened to be Afro-American. The Director of Personnel Services, protective of his people, asked me questions that were designed to trap me and I quickly shifted the problem to the real facts, explaining I had made a routine request to forward a letter because of conflict of interest issues. The associate director then said he fully understood and suggested his boss try a different approach. That associate director mentioned that he and his wife owned the local book bindery (where my books were bound) and to avoid conflict of interest, she ran the business and all their business correspondence went there.

Auler made a few comments and suggested to the Director that they meet privately. The associate director and I talked about different binding options while Goodman and Mrs. Lawyer sat in silence. When Auler and the Director returned, the Director said that he saw the problem more clearly now. He thought the best solution was that Mrs. Lawyer and I signed a statement that the entire dispute arose because of poor communication (the usual liberal cop-out) and poor supervision of non-academic staff in the department of geology. He handed Mrs. Lawyer and me copies and asked us to sign.

I read it and signed my copies and asked to exchange my copy with Mrs. Lawyer to sign her copies. She read it and refused to sign. That upset the Director of Personnel Services who ordered her to sign it while giving me additional copies to sign (we ended with seven signed copies; one for Mrs. Lawyer, one for me, and the rest for the administration). The two argued for five minutes and he finally told her, "Sue, if you don't sign, I'll drop your complaint against Dr. Klein and order you to write a personal apology to him with a copy to Chancellor Peltason." Goodman sat there motionless. She finally signed.

Bob Auler told me as we left the building that we just bloodied Donath's and Goodman's noses and exposed them to the administration. They don't approve of non-academic staff being manipulated into issues where they don't belong so there would be consequences.

Three days later, Sue Lawyer resigned and returned to the law firm in Champaign where she was previously employed.

Donath sent a benign memo to the faculty saying the issues between Mrs. Lawyer and me were resolved. Langenheim, Sandberg, Mann, Domenico and Johnson asked what happened. I showed them the signed document. I knew from their non-verbal reaction that they realized Fred and Bill seriously exposed themselves to a campus sector that could hurt them.

In late May, the faculty received a memorandum stating that the Department was to be reviewed in the fall by a select committee appointed by the Dean. The chairman was Douglas Applequist (PhD. Cal Tech, Chemistry; Illinois), Robert Scott (PhD, Univ. of Washington, Director, School of Social Sciences, Illinois), and Charlie Wirt (Head, Dept. of Mining and Metallurgical Engineering). Two days later, Dean Rogers wrote saying that my unrestricted access was suspended until the committee filed its report.

The semester ended and I flew to South Africa. My first stop was Rio de Janeiro where I gave a short course to PETROBRAS which was now centralized. I reconnected with several of my friends from Salvador.

My host told me that Al Carozzi recommended PETROBRAS become a multinational company. PETROBRAS sent exploration teams and bought concessions in Ecuador, Algeria and Iraq. My host just returned from Iraq the previous week. They made a successful discovery and immediately, the Iraqi government confiscated the concession and asked PETROBRAS staff to leave in 48 hours.

I then flew to Johannesburg and met an American on the flight. He worked for a steel company and said I would enjoy South Africa. He said, "Just like America in the fifties."

On arrival in Johannesburg, I went through immigration. They asked the purpose of my visit and I explained it. They pulled me aside to an interrogation room for 30 minutes while peppering me with questions. When I finally cleared customs, Ken Eriksson and his wife met me and asked why I was detained. I told him I didn't know.

While in Johannesburg, I stayed at a Holiday Inn. I went to my room, unpacked, and then went to the hotel bar. I was totally shocked that the bar patrons were integrated and discovered later that the apartheid laws had numerous exceptions. When Holiday Inn wanted to open a facility in South Africa, they told the South African government that a lot of their clients were African Americans. Consequently, they received an exemption from the apartheid laws.

My first short course was in Johannesburg for the South African Chamber of Mines. It was well-attended, but these hard rock mining geologists weren't prepared for a depositional systems course.

My second stop was Pietermaritzburg to present my short course to the branch campus of the University of Natal. My host was Dave Hobday (BS, Witwatersrand, PhD, LSU; sedimentology; Post Doc at Oxford, Natal-Pietermaritzburg, Texas BEG, Royal Oil in Sydney). That course was followed with a weekend in the Drackensburg with the Eriksson's, Dave Hobday and his wife, and Tony Tankard (PhD, Capetown; stratigraphy; Capetown Museum of science, Univ. of Tennessee; Petrocanada, Consultant, Calgary) and his wife. Eriksson's and Hobday's wives were American; they met at Cincinnati and LSU respectively.

From there, I went to the main campus of the University of Natal in Durban two days after the riots in Soweto, a 'black township' (South African terminology). I ate dinner at the home of their economic geologist. He was a German immigrant who as a teenager fought on the Russian front in 1944 and miraculously survived without injury. At dinner, the department head and his wife joined us but because there was a big rugby game on television, I talked alone with the host's wife afterwards. She was a South African sociologist and we talked at length about that country's apartheid and racial problems, the fierce tribal divide in the 'black' population, and long-term outcomes. It was an enlightening chance discussion.

My next stop was Capetown where Tony Tankard hosted me. I gave a talk to their local geological society, toured Table Mountain and visited the oceanography department at the University of Stellenbosch. One of their faculty, Tom Supko (PhD Miami, FL; marine geology; DSDP; Univ. of Stellenbosch), hosted me. Stellenbosch was in the heart of South African wine country and Tom took me on a tour of the Neiderberg estate, including their wine-tasting room. Neiderberg was the best South African wine I tasted and it was my wine of choice.

I returned to Johannesburg and was met by Laurie Minter, the chief geologist of Anglo-American Corporation. He arranged a visit at a gold mine at Carleton. The routine in South African mines is that the geologists enter the mine with the morning shift of mine workers and map new rock faces that were exposed by blasting at the end of the day shift. I joined them and observed Precambrian channel point bars which hosted paleo-placer gold in their channel floors. At 11:30 am, we returned to the office, took a shower and changed into street clothes. During the afternoon, the geologists plotted their new data on company maps. I had lunch with the mine manager, an Afrikaner of Dutch heritage and was appalled at his attitudes towards the 'black' miners. When we drove away, I told Laurie, "That mine manager's views are to the right of Hitler." Laurie agreed.

It was instructive to see how principles of sedimentology were used in gold mining operations. After exploratory drilling and assaying was completed, a shaft would be drilled to the gold-bearing horizon, normally a channel floor. From paleocurrent mapping of cross-bedding, pay shoots were extended updip and downdip along the channel floor. Once the channel was mined out, drilling was undertaken at right angles to the channel direction (i.e. parallel to depositional strike) until another channel was encountered. A cross-cut was made and a new pay-shoot was opened parallel to the new paleochannel axis and mined.

After leaving Carleton, Laurie drove me to Welkom, a mining community. Welkom was a new town of 25,000 people built after a gold discovery. I was scheduled to present a talk that evening to the local geological society in a private dining room in the only hotel in town where I also stayed. Welkom was located in the heart of the Orange Free State, which is populated mostly by Afrikaners (of Dutch descent).

My room was on the third floor and at the scheduled time, I left, and took the elevator to the ground floor. The elevator stopped on the second floor and an Afrikaner couple joined me.

As the elevator started, the couple talked in Dutch about the American and made pointed comments about my color-coordinated shirt and tie. I said nothing.

When the elevator reached the Ground Floor, we exited, they turned to the right, and I turned to the left to the dining room. After two steps, I spun around and said, "Tot Ziens" which in Dutch means "So long."

The couple stopped, turned around and in accented English asked if I spoke Dutch. They looked a trifle shocked and embarrassed. I replied in poor Dutch that I'm an American citizen but was born in the Netherlands and left as a child.

My parting comment to them in English was, "You need to be a little careful what you say about people from other countries. Maybe, just maybe, they know enough of your language, and they may not accept such comments kindly."

Next day, I completed another mine visit. On reaching the mine shaft, I asked Laurie to stand next to the elevator cage to take his picture. I did so and also caught 'hell' from the mine superintendent. In South African mines, timing of elevator trips was everything. Miners are grouped by their tribes and each is segregated to avoid conflict. A delay in scheduling shaft trips risked the possibility of two separated tribal groups coming too close and attacking each other. My picture-taking risked such a possibility. Fortunately, it didn't happen.

After lunch, we examined the company's core collection. I spent the evening at the hotel and was taken to the airport next day to fly to Johannesburg on a company DC-3. At the airport, the airplane was surrounded by armed guards and I went through an intense security screen. I boarded with three other passengers, all company employees, sitting in a small cabin in the front. The rear 2/3 of the plane was sealed off and the entry door to the rear compartment was separated from us by an armed guard.

I asked another passenger what kind of plane it was and he said, "Oh, this is the company 'gold plane.' We use it to ship refined gold bars from the Welkom Mine smelter to company headquarters in Johannesburg." On arrival, a large coterie of armed guards met us and we were escorted to the terminal by two armed guards.

The tidalite conference began the next day at the University of Witwatersrand and I gave my keynote talk which was well received. Harry Roberts (BS, West Virginia, PhD, LSU; Coastal geology; LSU) also gave a great paper. John Rodgers was on a world tour and attended also. My keynote address was about "A tidal circulation model for cratonic shelf seas." Rodgers said afterwards it was the best statement he ever heard on cratonic sedimentation. Coming from him, I had reasons to be pleased.

The conference was followed by a field trip. I saw spectacular Precambrian tidalites and stromatolites (some taller than a house). People asked me to verify their observations. Whenever I spoke extemporaneously while answering questions, I drew a crowd. I also had to correct some interpretations.

During the evenings, conversations kept drifting to the Soweto riots and its aftermath. I listened. I found most South African geologists, regardless of heritage, were willing to give the South African 'blacks' a chance for a better life and opportunity. But they felt powerless against the harsh views of the Afrikaner majority government.

John Rodgers, Harry Roberts and I talked privately one evening. John mentioned that he saw similarities between the French Canadians and the Afrikaners, both having lost wars to the British. Each had their own cities

(Montreal, Johannesburg) and overcome their occupation through a strategy of "La Revenge de la crèche." In South Africa, World War I provided leverage to the Afrikaner community. Once the vote was restored to the Afrikaners in 1919 (after agreeing to serve in the army), they controlled parliament. South Africa led the fight for the Treaty of Westminster, passed in 1933, to allow commonwealth nations the right to leave the British Commonwealth, and South Africa did so in 1947.

After the field trip, I returned to Johannesburg and left on an overnight flight to Lagos Nigeria, via Libreville, Gabon. At that time, Nigeria refused direct flights to and from South Africa. I was booked to fly with UTA (a French Airline) from Johannesburg to Libreville, Gabon, and connect to Lagos with Air Afrique. I left Johannesburg at 8:00 PM and arrived in Libreville at midnight.

I assumed I could stay in a transit lounge until my 7:00 am flight to Lagos. When arriving in Libreville, the runway accommodated a Boeing 707, but the terminal was no bigger than the one in Champaign, IL. There was no transit lounge. I was the only passenger to deplane and was met by a UTA representative, a multi-lingual French lady. She asked if I had a visa and I told her I planned to stay in the transit lounge. She explained there was none. UTA arranged a free motel room nearby. Air Afrique arranged for a taxi to take me there for the night and back to the airport in the morning.

However, I needed a visa to enter Gabon. The UTA representative arranged an entry visa for me. While waiting, I admit I was nervous being the only American at midnight in the airport of an African tropical city with no way to contact anyone if things went wrong.

Landing in Lagos the next day, I was the first person to deplane and went to the immigration counter. The immigration officer sat on an elevated platform behind a podium. I gave him my passport and he said, "Ah, an American." I replied, "Yes," and he leafed through my passport. At about the point I expected him to see the South African Visa and entry stamps, I heard a voice from behind him saying "Dr. Klein, you are here" and a Nigerian came forward, introduced himself as the chief geologist with the Nigerian National Oil Co (NNOC) and said something in Swahili to the Immigration officer. The Immigration officer smiled, stamped my passport and said, "Dr. Klein, I hope you will enjoy Nigeria more than the last country you stayed in for so long."

The drive to the hotel took two hours, and in retrospect, I wish I took pictures from the car. We went through small to larger sized villages with open air markets, camels, and people dressed in national costumes. It was colorful, and almost reminiscent of an earlier time.

Once in the hotel, I had the weekend to myself. The chief geologist, his associate chief geologist and their wives joined me for Sunday buffet featuring western and Nigerian dishes. I tried the Nigerian dishes which were far better than the western food. The wives of my hosts were Jamaican. While studying nursing in London, they met their husbands who were earning an applied Masters in petroleum geology from Imperial College.

The course was to start on Monday at 11:00 am with an opening ceremony scheduled for 10:00 am. It featured the Minister of Petroleum, Colonel Muhammadu Buhari. I met Colonel Buhari who was about 33 years old, fit, with thunderbolt markings etched on each side of his face. He was from the Hausa tribe in the desert region of Northern Nigeria. Buhari completed military training in the UK at Aldershot. He was extremely well spoken and gave a flawless address stressing the importance of training nationals so they could move through the ranks of multinational corporations. When the ceremony ended, I said good-bye and he said, "No, I want to hear some of your course."

The chief geologist took me aside and said Buhari wanted to be sure my course was the real thing and not watered down. Buhari stayed and at lunchtime, he invited me to lunch in the dining room. He talked about the need for modern training for his people and asked for advice which I gave. To my surprise, he opened up about himself. He admitted he missed his mother's cooking, and found Lagos too hot and humid. When he left, I concluded Buhari was headed to the top and in fact, he became President of Nigeria via a coup in 1982. I heard later from a Nigerian contact that he was overthrown during a coup four years later. That coup occurred during a public holiday weekend when he chose to go home to enjoy his mother's cooking and left his entourage behind.

From Lagos, I flew to New York. My mother was dying of cancer and while in South Africa, my father wrote instructing me to call from Kennedy Airport to see if I could visit her. My mother's physician checked the WHO daily reports to see if I was exposed to infectious diseases, but he cleared me, so I visited before flying back to Urban.

George Devries Klein

LESSONS LEARNED:

1) A university administration will only remove a department head (or other administrator) if they take actions that threaten the administrative structure of the university.

2) Deans provide oversight of all departments in their college but through the shared governance arrangement in universities, are powerless to interfere directly in departmental problems. Their only recourse is budgetary. If campus guidelines or statures and by-law are violated in a department, they will step in.

POSTCRIPT #1. I was invited to serve as External PhD examiners for Ken Eriksson and Borg Flemming (PhD Stellenbosch, Coastal processes; South African Institute of Oceanography, now Director, Senckenberg Marine Lab, Germany) who I met at Stellenbosch. Their rules were stringent. I had to read the thesis and accept it as is, or it was rejected. I passed both of them. Ken came back to the USA and accepted a faculty appointment first at the University of Texas at Dallas, and then moved to VPI where he is still today. Flemming eventually replaced Hans Reineck as director the Senckenburg Marine Station in Wilhelmshaven, Germany.

POSTCRIPT #2. Tony Tankard asked me to sponsor his immigration visa to the USA. I agreed but told him that if he didn't find a job quickly, I'd form a consulting firm and he would do all the work. He accepted a job at the University of Tennessee, found he didn't like teaching, took a job in Canada with Petrocanada and became a well-respected geological consultant in Calgary.

POSTSCRIPT #3. After receiving her PhD, Julie Graf tried for two years to raise research funds and received none. She accepted a job with the local USGS office doing surface hydrology. In 1982, she transferred to Tucson, AZ, and Don Graf took early retirement to move with her. During his entire appointment at Illinois between 1969 and 1983, Don Graf published only ONE paper.

CHAPTER 20
The Second Donath War. Part III (August 1976-May 1977)

"Truth kills those who hide from it"
- Old Lebanese proverb

"For he that soweth to his flesh shall of the flesh reap corruption; but he that soweth to the spirit shall of the spirit reap life everlasting"
(Basis for "As ye sow, so shall ye reap")
- Galatians 6:8

"Beware, lest stern heaven hate you enough to hear your prayers"
Or in the vernacular: "Be careful what you wish (or pray) for"
- Anatole France

On my return to the office I went through the usual mail and memos. Advisory Committee elections were completed and for the coming year, the committee consisted of Domenico, Henderson and Sandberg. Hilt Johnson resigned as Educational Coordinator and Don Henderson replaced him.

I also received a memo about the Applequist committee. Hearings began after Labor Day. I phoned Applequist and set the Tuesday afternoon after Labor Day to meet with his panel. I explained I was leaving afterwards for a sabbatical at SIO. I called Auler who agreed to attend.

To prepare for the meeting, I wrote an 85 page statement with a cross-referenced appendix of about 200 pages of memos and other documents[4]. I reviewed the 85 pages with the committee. During the middle of the afternoon, Bob Auler told them about potential legal irregularities. When I was done, Scott asked questions about the allocation process involving individual faculty accounts. He asked me to forward more information.

Wirt finally spoke and said, "George, you've done nothing wrong, so please understand we appreciate that. There are faculty on this campus who show up drunk in class, patronize prostitutes, gamble, fudge their university expense accounts, take secretaries with them on trips and stay in the same room, for openers." I thanked him for his comments. Bob and I left and gave Applequist the entire document in a folder.

Three days later, I left for SIO. I examined all cores from Pacific back-arc basins to integrate findings into a back-arc depositional model. I stayed in the same efficiency apartment and enjoyed San Diego's many amenities.

Happy Okada joined me at the end of October. Saito arranged for Ujiie and Okada to come to Denver for the GSA meeting and to meet for a final discussion and review the manuscript about our 1975 field work. Because I was driving from San Diego to Denver, I offered Okada a ride. Happy had never seen deserts. We made stops so he could photograph dunes, arroyos, deflation armor and salt lakes. At Zion National Park, he photographed eolian cross-bedding in the Navajo Sandstone. We spent the night in St. George, UT, and continued across the Colorado Plateau making more stops, including the Book Cliffs.

Happy told me years later he showed those slides to his classes because he never saw such geology before or since. While driving, I provided him with regional geological background. We arrived in Denver the Saturday night before the GSA meeting.

Linda Tills and Bill Busch presented papers on their Master's research and acquitted themselves well. Langenheim told me the Applequist Committee had two more interviews to go. Their report was to be presented to the Dean in Mid-December. He said the department was quiet and subdued the entire fall because of the Applequist Committee's review.

After GSA, I returned to Urbana and went through my mail and memos. Nothing unusual turned up. In fact, Donath and Goodman hardly wrote any memos at all. While in San Diego, I discovered Purdue University's Geoscience Department was looking for a department head so I applied. They wrote back asking me to call when I returned to arrange an interview date. I arranged to visit the first week in February, 1977.

At the end of January, there was a knock on my office door and when I opened it, Douglas Applequist was there and returned the file[4] I gave him when testifying before his panel. I asked, "All done?" He replied, "We turned in our report and met with the Dean back in December. I think you'll be hearing something soon." I thanked him for coming by and he said, "George, before I go I want to say something. Now is a very good time to reintegrate with your faculty colleagues. All but one are on your side." I replied, "Really, How do you explain it?" He said, "You're all scientists." And he left.

In early February, I went to Purdue for my interview. I visited the department and met with Don Levandowsky, the acting head (PhD Michigan, Economic Geology; Purdue). He reviewed budgets and current issues facing the department.

The department was organized into two divisions, Geology and Geophysics, and Atmospheric Sciences, with each division led by an associate head. I met heads of Physics, Chemistry, Biological Sciences and Civil Engineering and the Dean of the Science College, a PhD Mathematician with degrees from Brown University. I found the campus to be well maintained, the architecture was awful, and the place was clearly no frills.

I also met the faculty and gave a research colloquium. The geophysicists were in a separate building which was a bad sign. The lead geophysicist, Bill Heinz (PhD Wisconsin, gravity and magnetics; Michigan State, Purdue) acted as if they were a separate entity requiring special needs.

On Friday morning, I read the daily student newspaper and found one section most telling. Near the back, they published a list of all faculty publications of which the newspaper received notices since the previous Friday. Clearly, the message to everyone on campus was that research was Purdue's primary mission.

I met the search committee for an exit meeting. I felt I could do the job, but was not convinced I won them over. Later they offered it to David Atlas, a meteorologist who declined. Then they offered it to Frank Harrison, a government geological oceanographer I knew, who first accepted and then declined. They offered it a year later to Gerry Friedman who also accepted it, and then declined. They appointed Levandowsky.

In late February, Dean Rogers' office called to let me know I was to meet in the College conference room with the departmental advisory committee to negotiate terms for a move to the Center for Advance Computation. I met with the Center's director, Dr. David Kuck, but had not reached an understanding.

I met with Domenico, Henderson and Sandberg and was surprised how unprepared they were. They were unaware of my proposal to transfer to the

Center for Advanced Computation. I mentioned continuing teaching my geology courses in the geology department, transferring my budget line to the Center as a hard-money line, and ceding my office, student office area and lab space to the center. They discovered I was serious because I did not know outcomes of the Applequist Committee review. Moreover, neither the Dean nor Fred gave them instructions as to what they could negotiate.

Finally, Pat Domenico said, "George, we can't go forward with this now because we have to review it with Fred. We'll get back to you." We left together and walked back to NHB and talked as if all was normal and collegial.

Dean Rogers called two days later and suggested we meet for lunch at the Urbana Country Club. He asked about the sabbatical and my trip to South Africa. He then said, "George, The Applequist Committee recommended Fred Donath be removed as head but he refuses to accept their recommendation. Fred met with Weir and also refused. We are looking for ways and means to remove him now. We heard disturbing things about his rock deformation lab. Can you tell me what you know?"

I replied, "Bob, all I hear are rumors. Last summer, one of the part-time students claimed that the machinists were manufacturing a large number of copper jackets needed for the experiments, but no experiments were underway, and the copper jackets seemed to disappear. I also must tell you that this particular student likes to exaggerate, so I don't know how reliable his information is. I know Fred and Fruth manufacture rock presses and sell them but I had understood the sales went into a special account. I understood the university is reimbursed for labor, materials and overhead, and the rest, I was told, goes into a special account in the department to be used by Fred for professional trips to meetings or travel by his students. But please understand this is all hearsay and rumor."

As we left, Bob said, "George, I'd like you to stay in the department. The Center for Advanced Computation is a poor arrangement for you. Just sit tight."

Carleton Chapman announced his retirement after serving as a professor at Illinois for 39 years. I did not attend his retirement luncheon.

At the end of February, my father called to let me know my mother was near death and asked me to fly out for one last visit. I made the trip and because of her weak condition, could only visit for fifteen minutes. Three days later, she passed away. I returned for the funeral and my father asked me to give the Eulogy. It was extremely difficult.

When George Mandel, my sister and two nieces visited my dad afterwards, he asked to be left alone and not to call. He would call us. However, my Dad called a week later because I was going to London during spring vacation to teach a short course. He asked me to visit on my return trip. I did and realized he was long way from overcoming his grief, as was I.

In late-March, I received an inquiry from the US-AID mission on Seoul, South Korea. That agency awarded a major grant to Seoul National University (SNU) to upgrade its science departments into internationally-recognized graduate programs. Dr. Yong Ahn Park, Professor of marine geology in the Department of Oceanography (BS, SNU, MS. Brown, PhD, Kiel; marine geology, SNU, Korea's representative to UN Law of the Sea Working Group) invited me because of my research on tidalites. US-AID wanted me to come as a visiting professor for one year to teach classes.

As my research progressed during the 1970's, I had my eye on the Yellow Sea of Korea which is characterized by a macro-tidal coast (i.e. tidal ranges in excess of 2 meters). Its sand bodies were mapped by fathometer during the Korean War and described by Ted Off in the 1962 *AAPG Bulletin*. I superimposed maps at the same scale of the areal extent of the Zabriskie Quartzite, the Eureka Quartzite and the Wood Canyon Formation, and other cratonic tidalites onto a map of the Yellow Sea. The Yellow Sea was larger than these ancient counterparts. Perhaps the Yellow Sea might be that elusive counterpart to cratonic shelf seas which geologists had not identified.

I saw the invitation as a chance to see for myself. Moreover, a well-known paper by J.J.C. Houbolt from the Shell Research Lab in Rijswijk, Netherlands, on the inferred tidal circulation pattern on North Sea subtidal sand bodies also caught my attention. I saw the Yellow Sea as a place to actually measure current velocities and direction over such sand bodies and test his model. Clearly, this was a research opportunity.

I wrote back expressing interest in coming to Korea but to upgrade their graduate programs, it could only be achieved by doing joint research, not by teaching classes. I was willing to undertake a joint research program, outlined what I wanted to do and estimated costs for US-AID support. I received the usual perfunctory factotum, bureaucratic response from the program's director that they wanted me to teach classes and I wrote back that I would come only if I undertook this research project with Park and his colleagues. I heard nothing immediately, except Park wrote saying he liked the idea and would see what he could arrange.

Early in April, I received a memo from Fred telling me that John Holder was going on sabbatical next fall and I was to teach a course he offered to undergraduates, Geology 103. It was a cultural course for non-majors. When news of this decision spread through the department, several faculty, including Langenheim, told Donath it was a big mistake because likely, I would turn students off and ruin future enrollments. Fred stonewalled them

That same week, Dan Lawson defended his PhD thesis. He interviewed for faculty appointments at Penn State (as did Dag Nummedal) and SUNY-Stony Brook, but nothing developed for him. Through Tom Ovenshine's network, Dan applied for a job at CRREL (Cold Regions Research Engineering Lab) of the US. Army Corps of Engineers, interviewed, and was hired. He made his entire career there working in Arctic regions of the world.

A week later, I received a call from Ellen Abel who replaced Sue Lawyer. Donath wanted to meet with me three days later at 11:00 am. I called Auler who agreed to come and notified Ellen.

Ellen turned out to be a good hire. She worked previously in the geography department and learned to deal with difficult people. She was told by the head of geography to be guarded because as she laughing relayed to several of us at lunch one day, "Geologists are notorious sex maniacs so you have to be careful." She added that she was pleased we were all so civilized to her.

When Auler and I met with Donath, both Bill Goodman and Phil Sandberg were present. I asked why Phil was present and Fred said, "He will be acting head during the summer." Fred then disclosed that the student office next to my office was no longer under my jurisdiction and would be converted to a microprobe laboratory. He concluded the meeting by telling me he was cutting off my telephone service.

Auler and I left and Bob told me he would call the university lawyer that afternoon. I was responsible for the 1977 annual SEPM symposium in June in Washington, DC and needed a phone. Auler called that evening telling me the university lawyer and I were to meet with Weir the following Tuesday. Auler then advised me to ask Weir to arrange an investigative audit of Donath's rock deformation lab, and to tell Weir that only he can authorize such an audit but I could not.

Auler, Weir, the university attorney, Ed Madigan, and I met. I admitted to Weir that what I had heard was rumor and hearsay but he needed to know. I said, "Sir, I request you authorize an investigative audit of the geology department. I lack authority to arrange it. I understand you do." Weir asked Madigan what he thought. Madigan looked very grave and said that my disclosure places the university at serious risk. Weir turned pale. I realized then that the audit would either result in my being fired or Donath's removal.

Weir knew that if I was proven correct, the administrative structure of the university was at risk and he, Rogers and the president would be out of a job. Weir thought a minute and said, "I'll authorize it." Turning to Madigan, Weir added, "Please prepare all relevant legal documents immediately for my signature." We concluded the meeting and as Bob and I were leaving, Weir said, "Oh, George! I'll immediately arrange for you to get your phone service back" and the next day it was restored. Ellen Abel told me later that Weir's office also charged the department's budget for the restoration costs and Goodman was incensed.

That afternoon, the auditors were in NHB interviewing Dotson, Bauerle, student help, Fruth, Fred, and Goodman. They took away the departmental files with all the accounting data and returned them a week later.

The department had a new Chief Clerk who handled accounts, Murle Edwards. She came from the Department of Astronomy, and as I discovered much later, was placed in the geology department personally by Dean Rogers. Murle cheerfully helped the auditors find all the files they needed. Murle was from Harrisburg in Southern Illinois, worked in New York City, and returned to Illinois to finish her BA degree at Illinois. She referred to herself as "Murle the Pearl," As a good observer she always knew "where the bodies were buried."

Within two weeks, the auditors gave their report to Weir. The rumors and hearsay were true. Fred was removed from the headship and Sandberg became acting head. The university, however, let Fred stay as a professor, and for reasons I never understood, allowed Goodman to keep his job.

The Second Donath War was over. I had the unique distinction of being one of the very, very, very few professors in the history of the University of Illinois to get a department head removed from their administrative duties. News of my perceived success spread all over campus and until I left in 1993 People still mentioned it.

LESSONS LEARNED:

1) A job is an asset worth fighting for.

2) By retaining an attorney, the playing field at Illinois was leveled for me. If I tried to remove Donath alone, it would have taken much longer to succeed. As it was, it took more than two-and-a-half years.

3) As Postscript #4 (below) shows, the administration realized its review procedures were outdated and new ones had to be established to avoid the difficulties of terminating a headship of someone who needed to be removed.

4) A University administration will only remove another administrator if actions they take will undermine the administrative structure of the university.

5) Shakespeare said it best: "To thy own self be true, and it must follow, as the night the day, thou canst be false to any man" *(Hamlet)*

POSTSCRIPT #1. Fred Donath moved his commercial operation off campus and rented space, formed a corporation, CGS, Inc, and hired Les Fruth. While a professor at Columbia, he met Frank Press (BS, CCNY, PhD. Columbia, seismology; Cal Tech, M.I.T, Science advisor to President Carter, NAS) at an alumni function. When Press became Carter's science advisor, Donath re-established contact and Frank networked Donath to the Nuclear Regulatory Commission. They gave CGS a big contract. He took a year leave-of-absence at no pay and worked full time on the project in Los Alamos, NM. The next two years, Fred taught a beginning course which met at 4:00 PM on Tuesdays and at 8:00 am on Thursdays. He flew a rented plane to Los Alamos, NM, worked there through Monday and returned Tuesday in time to teach his class. In 1979, Fred applied for another leave-of-absence, was turned down and resigned in 1980. He hired Bill Goodman in 1980 as CFO.

When Ronald Reagan became President of the USA, he terminated all outside contracts. Fred faced an uncertain future and was headed for bankruptcy. He sold CGS to Earthtec in California, became their Vice President for Research, and moved his family. As part of his compensation package, Fred was partially paid with inside stock. In March, 1988, Earthtec went public, its stock rose from $1.00 to $12.00 per share in one day and Fred was a millionaire. He established the Donath Young Scientist Medal at the Geological Society of America and a fellowship fund at Minnesota.

I saw Fred and Mavis at the 1988 GSA meeting and told him, "Fred, establishing that young scientist award was an extremely generous thing to do." He smiled and we chatted and any remaining hostility he and I may have had melted away, at least for awhile.

POSTCRIPT #2. Bill Goodman stayed in the department of geology until 1980, although many of us wanted him removed. In 1980, he became CFO of CGS (See Postscript #1) but when that company was sold he was out of a job. He worked for a family firm for a while. Bill grew up in Champaign, IL, and never left. In 1985, he was appointed business manager in one of the departments in the School of Agriculture. I never understood how he came out unscathed from the financial issues with the rock deformation lab. I surmise he probably followed the standard cop-out dictum of just following orders.

POSTSCRIPT#3. Why did Fred Donath lose the Second Donath War? Looking back as I wrote this, I realized something I should have spotted earlier. During the First Donath War, Fred received support from 2/3 of the faculty and although the administration may have had reservations, they were not going to remove a head receiving such faculty support. During the second Donath War, Fred's uncertain support dwindled to Chapman, Henderson, Wood, and Graf. Everyone else sat on the side-lines and during the administrative review by the Applequist Panel, opened up to show that Fred's support was less than a quarter of the faculty. That lack of confidence enabled the administration to encourage Fred to step down but he fought them until the investigative audit disclosed irregularities that threatened the administrative structure of the University of Illinois.

POSTSCRIPT #4. During 1976, Mort Weir asked the Faculty Senate to develop a new review procedure for administrators. They recommended that all campus administrators be automatically reviewed after five years of service. They had to step down after ten years of service. The recommendation passed and was ultimately approved by the higher administration and the University Board of Trustees in 1977. It ended the tradition of 'Imperial Headships." Under the grandfather clause, it exempted those who were still serving as department heads and had been appointed to indefinite terms.

CHAPTER 21
Illinois: Transition (1977-78)

"Time will tell"
- Well-known Southern USA proverb

Sandberg immediately became acting head. An announcement was sent to the faculty and graduate students. Final exams started and the semester came to an end.

The DSDP Chief Scientist, Dave Moore (PhD. SIO; Marine geology; Naval Electronics Lab, DSDP), telephoned. He asked if I could join DSDP Leg 58 as a co-chief scientist between mid December and early February in the northwest Pacific (Shikoku basin and Daito-Ridge-and Basin). I told him I'd check and let him know. I would need approvals and asked Dave to give me two weeks.

I discussed Moore's invitation with Sandberg who said the Dean would have to approve, but he was sympathetic to my going on Leg 58. I also asked if I could be excused from teaching Geology 103. He explained that could not be done this late in the year although he also shared Langenheim's concern. Phil asked me to make the best of it for the sake of the department.

I was about to bring up some other things and he told me that that until I met with the Dean, he couldn't negotiate some of the outstanding issues between me and the department. He added, "After seeing the Dean, we'll need to have a very long talk."

Phil would do well. His father was professor of geology at LSU and served as Department Head. Phil learned much from him while growing up. When the allocation process started and the Dean froze the department's budget, Phil told me that they did the same thing at LSU for three years to create a paper surplus to use as an incentive for his dad to become head.

I visited Dean Rogers. He immediately approved my going on Leg 58 and congratulated me on the opportunity. I brought up Geology 103 and he said that was up to Sandberg.

I asked if he had any advice about how I could leave in early December to meet the *D.V. Glomar Challenger* with existing class obligations. He suggested that time spent on field trips could be credited as instructional hours. He advised I run some Geology 103 Saturday field trips to offset the time away in December. He suggested also that by Thanksgiving, I probably had 85 percent of the data needed to calculate final grades. Rogers suggested writing each student a letter with the proposed grade for work completed by then. If they accepted by countersigning the letter, they were excused from class for the rest of the semester and that was their final grade. If they wanted to try improving their grade by taking the final exam, they could but must accept the final grade including the final exam. He advised I ask a colleague to sign the grade sheet and complete it in my absence.

Rogers then brought up my salary grievance. He told me he could not give me any back pay but he could give me a raise starting in the fall. He wanted to know what I thought was fair. I thought a while and asked what the salary range was for a full professor in the LAS College. He gave me the data. I asked for a raise to the top 20th percentile. He agreed.

When we were done, he said, "George, I appreciate everything you've done for this college. I hope now you can return to the department and work as you always have and enjoy good relations with your colleagues. I'll look forward to seeing you in the future, but you must understand this will be the last time you can see me without going through Phil or his successor." I told him I appreciated everything he had done for me.

I called Dave Moore to accept the Leg 58 co-chief scientist assignment. Dave asked me to fly out to DSDP in early August to meet with my co-chief scientist, Kazuo Kobayashi. Kazuo returned to Japan from Pitt and now was a professor of geophysics at the University of Tokyo.

I spent the summer working on research papers and preparing for Leg 58. I visited DSDP, had a good visit with Kazuo, and met Stan White (PhD. Univ. of Washington, field geology, Cal State System, DSDP, Pennzoil), the Leg 58 SIO rep. I also discovered that three more Japanese scientists would participate on Leg 58.

My return trip was routed through Tulsa, OK at the request of IHRDC in Boston who contracted me to teach short courses and wanted to videotape mine. I went to Phillips Petroleum in Bartlesville to give the course in their video studio. I had given IHRDC my slides ahead of time and many were redrafted and duplicated for the offering. The video-taping went well.

Once back at Illinois, I checked to see if there were any Japanese language courses offered. There was a beginning course and the instructor approved my auditing it. I learn enough to understand their grammatical structure which proved critical when editing manuscripts for the DSDP volume.

Four teaching assistants (TA's) were assigned to Geology 103 and two of them served in the course the previous year. I set some ground rules. After exams were graded, they were to be sorted by lab section and handed back by the students' TA. Labs were to remain unchanged. The course outline was too traditional. Instead, I lectured on coastal processes the first week. By the end of that week, the class had to submit a newspaper clipping about geology. I tallied them and lectured on the topic with the greatest number of articles. When that topic was finished, I repeated the process. Each submitted article was graded and counted for five points towards the final grade.

We discussed field trips and decided on three one-day trips on different Saturdays. I then explained I anticipated a great range of student talent, behavior and ethics. If students visited me at the podium after class and I sensed a problem, I would scratch my head. The TA's were to come to the podium to assist if needed and witness what transpired. That procedure reduced many problems during the semester.

I also taught Geology 309 and 437 and both generated strong enrollments. Between these two courses, Geology 103 and auditing Japanese, it was a busy semester. Phil Sandberg facilitated me in every way to get it done.

During the first class meeting of Geology 103, I provided explanations about the three required field trips and my excused absence policy. Absence due to illness must be certified by a family doctor or the university health center. If there were other issues, such as a death in the family, I required verification.

The first field trip generated the usual number of absences. Half could not be verified so those students got zero. My toughness was noticed.

During the second trip, we had fewer absences. One female student, a senior, saw me after class and asked to meet in my office. I told her to wait. After talking with the rest of the absentee students, I told her, "Well, tell me. What's your excuse?" She said, "On Saturday morning, I had very serious menstrual cramps and heavy bleeding. I couldn't get out of bed." I scratched my head and the TA's converged behind her and she did not notice them. I asked just as the TA's were in earshot, "Did you go to McKinley (Health Center)?" She replied, "No, I just couldn't move." I replied, "Well, I may know more about this than you realize. If it was as bad as you said it was, you should have gone to McKinley. I'm not accepting your excuse."

She shrieked, turned around, saw the TA's behind her, and ran from the room. I then said to the TA's, "I bet I'm the first male professor on this campus since she arrived three years ago who didn't accept that excuse." She never appealed or filed a grievance.

After the first exam, I summarized how students did by college of origin. I then said, "Folks, I had to start a new grade book and noticed your school and college of origin. The engineering students and LAS students did well. The School of education ranked the worst. That's why Johnny can't read and you can't write."

The class roared with laughter except the education students. One reported the comment to a professor in the School of Education who phoned Henderson. Henderson asked Dave Biehler who confirmed my comments and Henderson said something negative.

Biehler told me about Henderson's comment so I went to see him. I said, "Don, I understand you were unhappy with my teaching in Geology 103." Don replied, "No. That's not correct. Actually, I think you're very creative with the way you're handling the course." I replied, "Well, I heard you got a complaint from the School of Education." Henderson replied, "Oh. That. I agree with you, but you really shouldn't say that to the class." I replied, "Don, I hear you, but how else can we give those people the message?" I left his office with both of us in good spirits.

During October, a search committee for a new head was appointed by Dean Rogers. It was chaired by Henry Wirt, head of mining engineering, and the committee members were Sandberg, Dave Anderson, Tom Anderson, and Domenico. Advertisements were sent out with a February 1, 1978 application deadline.

Carleton Chapman's position was filled by Jim Kirkpatrick (BS, Cornell, PhD, Illinois; experimental petrology; Exxon Production Research, Post-doc at Harvard, DSDP, Illinois, Michigan State – Dean of Natural Science). He arrived in January, 1978, because he was the SIO rep on DSD Leg 55, which sampled the Hawaii-Emperor Sea Mount Chain, a very significant cruise.

With the change in leadership, I reconnected with colleagues from whom I had become estranged. Carozzi and I talked, but he was a bit formal. Graf never spoke to me again going back to Thanksgiving 1973.

With Dennis Wood, it was easier than expected. When the semester began, Reg Shagam called. He was spending the year as a visiting professor at Franklin and Marshall. I inquired if he was attending GSA in Seattle and he said he would like to but needed money for the plane ticket. I said, "Reg, let me try something. I'll call people I know in Big 10 geology departments and see if we can schedule you for colloquium visits and share the air fare." He agreed.

I spoke with Dennis Wood who was colloquium chairman and when I described Shagam's research and how well he interacted with people, he agreed. I called people at Northwestern, Wisconsin and Indiana and all agreed. I explained that if each department could contribute $200 as an honorarium, Reg could make the trip. All accepted this arrangement.

When Reg arrived, he first visited with Sandberg, then Dave Anderson, and finally Dennis. Reg met with Dennis's graduate students and all went well, entertaining them with stories about my time at Penn. Dennis told me later he really appreciated Reg's visit. Some of Dennis's students thanked me also.

The third Geology 103 field trip was held near the end of October. I deliberately scheduled field trips on Saturdays on away football games. One female student explained she couldn't attend the third field trip. I asked, "Why?" She responded, "Well, I'm going with the band to the University of Wisconsin for the football game. I'm an 'Illinette." I asked, "What's that?" She explained she was part of the dance team that performed during half-time. I asked, "Is that an extra-curricular activity?" She replied it was. I told her if that's the case I could not give her an excused absence.

She went to Madison, WI, and on her return, I confirmed I would not give her an excuse. Three days later, the band director called on her behalf. I asked if the activity was extra-curricular and he confirmed it was. I told him I wouldn't give her an excuse.

I then received a call on her behalf from an associate dean in the LAS College, but he too confirmed that her participation with the 'Illinettes' was extracurricular. He disclosed that her father was the president of a bank in Rantoul, IL (15 miles north of Urbana), a loyal alumnus, and a contributor to the university foundation. I said that to the best of my knowledge, there was nothing in the university rules, by-laws and statutes that stipulated special privileges for alumni children or "Illinettes." I told him I wouldn't give her an excuse.

Two weeks later I received a call from the campus ombudsman. He explained that actually, she was enrolled in 'Band 103' which carried academic credit so there was a conflict. I asked why neither the Band Director nor the LAS Associate Dean knew this fact. He had no explanation. I finally said in exasperation, "As far as I'm concerned, Band 103 is a course with no standing and isn't equivalent to a science or other main-stream academic courses. I won't grant an excuse" and hung up.

One thing that always irritated me about mid-level and lower-level university administrators was that they never supported or backed-up faculty on issues of this kind. I recall reporting a cheating incident at Pitt, the lower-level administrator made it appear as if it was entirely my fault. I developed my own way of handling cheaters and my methods were never appealed or challenged. I did so without benefit of a university administrator (See Postscript #1).

The lower-level administration in this 'Illinette' incident denigrated all serious academic programs which enhanced the university's reputation, particularly in science and engineering, at the expense of someone, as I often put it, who "wanted to wiggle her fanny at half time and got academic credit for it." It did not win me friends at the lower and middle level of the university administration, but because of my publication and research track record and high scores on teaching evaluations, they left me alone. They knew I was not taking their nonsense or tolerate the inevitable immaturities of the academic environment. I learned later they respected my position.

I also received a call from the Alumni Association director and explained the whole thing was so poorly handled that the case was closed. He told me that her father would file a complaint with the administration.

Early in November, I gave a paper at the GSA meeting in Seattle, flew to New York to see my Dad and arranged for him to pay my bills while at sea.

On the weekend before Thanksgiving, I calculated final grades for Geology 103 and on Wednesday before Thanksgiving each student received two copies of a letter which I signed with their calculated final grade, based on 85% of their work (the class was warned there would be a critical announcement about final grades and the final exam). They were required to return a countersigned copy if they accepted the grade. All but three students signed to accept their recommended grade.

The TA's and I went through the countersigned copies and indeed, the "Illinette' signed to accept her grade. I told the TA's, "That's it. A letter's a contract. By signing and accepting the grade, she signed away all her rights to appeal."

When returning from my DSDP cruise in February, the Alumni Association Director called again and told me that the 'Illinette's' dad was furious with his daughter for signing the acceptance letter for her final grade and felt I 'snookered" his daughter and the rest of the class. After assuring me he calmed her daddy down and that should end it, he asked, "Where did you get the idea to do that?" I replied, "Dean Rogers advised me to do so because I was leaving campus early." The word "Dean" ended the discussion.

I went to Japan to join DSDP Leg 58. DSDP arranged hotel accommodations in Yokohama where I was to board the *D.V. Glomar Challenger*. Leg 57 returned the next day and I moved on board ship. Dr. Noriyku Nasu (BSc, Tokyo, PhD, SIO; marine geology; Univ. of Tokyo, Director - Ocean Research Institute), hosted dinner at a Chinese restaurant for the Leg 57 team and I was invited also. It was the best Chinese meal I ever ate because the Japanese cook with very little or no oil so one could taste the flavor of the food.

Once on board ship, I met Roland Von Huene (PhD, UCLA, marine tectonics; USGS), the co-chief scientist on Leg 57, to review their accomplishments. Such meetings between departing and arriving co-chief scientists assured scientific and logistic continuity.

As the week continued, the scientific staff arrived. The SIO rep was Stan White. Dave Fountain (PhD, Univ. of Washington; geophysics and physical properties; Univ. of Wyoming), Doug Waples (PhD, Stanford, Organic geochemistry; Post-doc with Lopatin at Moscow State University; Colorado School of Mines, consultant), Nick Marsh (PhD candidate, petrology; University of Leeds), Doris M. Curtis (BS, CCNY, PhD. Columbia, stratigraphy; Shell Oil, Univ. of Oklahoma, Shell Oil, first woman president of SEPM and GSA), Dorothy Echols, (ABT, Columbia, micropaleontology; Washington Univ., St. Louis) and Herve Chamley (PhD Paris, clay mineralogist; Univ. of Lille) all arrived the next day.

Louise Henry was our yeolady and Gus Gustafson was the lead tech. I knew we were off to a good start because I worked with them on Leg 30.

Henry J. B. Dick (BA, Penn, PhD, Yale, Igneous Petrology; Woods Hole Oceanographic Institute) arrived a day later. He completed my undergraduate Physical Stratigraphy Course at Penn and had been problematic (See Chapter 12). Dave Moore asked if Henry would be a problem because Dave was aware that Henry and I had previous difficulties. I told Dave, "Look, I haven't seen Henry in nine years. I assume he grew up and is more responsible. Let him come."

Henry, during his senior year, applied to several graduate schools, including Yale. I told him I would support his applications elsewhere, but advised against going to Yale because I thought he would have difficulty given his style. He was awarded an NDEA four-year graduate fellowship at Brown and I advised he accept it. He was admitted to Yale and although he told me on board ship it was tough, he made it.

Kazuo Kobayashi arrived on Saturday as did the other Japanese scientists. They were Histake Okada (PhD Tohoku; nannoplankton; Post-doc at WHOI and LDGO; Yamagata University), Atsu Mizuno (PhD, Tokyo: sedimentology; Japanese Geological Survey) and Hajima Kinoshita (PhD Tokyo, paleomagnetics; Chiba University).

The Russian petrologist, Gennady Nisterenko (PhD, Moscow; igneous petrology; USSR Academy of Sciences) still had not arrived. Dave Moore explained that Russians always arrived barely on time, and the minute the ship landed, they took the first flight back to Moscow. We were scheduled to leave on Sunday at high tide at 2:00 PM. We had no information about Nistorenko's arrival. Kazuo immediately contacted an igneous petrology graduate student at Tokyo to be at dockside with his gear to take Nisterenko's place on a standby basis.

Noriyuku Nasu, who had connections to the foreign ministry, found out when Nisterenko was arriving, met him, and got him dockside by 1:30 PM. We sailed at 2:00 PM with a four day trip to the first site in the Shikoku basin. In retrospect, the Japanese graduate student would have worked out better.

Stan White, Kazuo and I held a meeting while waiting Sunday morning for Nisterenko. One issue was determining who would be the 'day co-chief' and the 'night co-chief." The plan was we would be on duty either from 12 noon to midnight (day), or midnight to 12 noon (night). White took a Japanese coin and asked Kazuo to identify which was heads and which was tails. Once identified, White flipped the coin and Kazuo, who called "heads" said the coin showed 'heads' and picked the day assignment.

I agreed provided we could make a time change, the day shift from 8:00 am to 8:00 PM, and the night shift from 8:00 PM to 8:00 am to overlap work by all scientists. We would alternate day and night shifts on each site.

Once underway, I sat one evening at dinner with the four Japanese Scientists. I tried to converse with them in Japanese and they stonewalled me. I tried again later and got nowhere. I concluded that the Japanese operated on two planes when dealing with westerners. When they spoke English, they were genial and cordial. I called this the "Outer Japan." The second Japan I called the "Inner Japan" where language and centuries of customs, culture and paradigms operated and no foreigner was permitted to enter. When trying to converse in Japanese, they saw it as a threat to "Inner Japan."

Drilling commenced at the first site and soon I discovered a benefit from working the night shift. By 10:00 PM, most of the scientists, Lou Henry, and Gus were asleep. Between 10:00 PM and 4:30 a.m., I had the science office to myself.

The co-chief scientists were responsible for coordinating all site reports which must be finished before the ship reached port. As co-chief, I had to be on deck when a new core came out of the drillhole to make preliminary observations and help the paleontologists get samples from the core catcher. Core description forms were stored in the science office when completed. With the office to myself, I began writing the Site report as soon as data was available. The sedimentologists and paleontologists would review, correct and add material once they were done, and the igneous petrologists, Waples, Kinoshita and Fountain added their sections within a day after we reached the bottom of the hole.

Because of the short time (usually a day-and-a-half to two days) between the start of pulling out of the drill hole, sailing to the next site, and the start of drilling on the new site, site reports had to be completed and reviewed as a group before the first core from the new hole was on deck. We had little time to complete this before work started on a new site. I proposed to Kazuo and Stan that I be the 'night co-chief' for the remainder of the cruise. Both accepted enthusiastically.

The igneous petrologists were the last to file their part of the site reports. They needed to complete XRD and thin-section analyses. Their rocks were the last we recovered. With the short time available, it was difficult, and they kept asking for delays for our site summary meetings before the next site was being drilled. Henry Dick was the petrologist on the first site, and Nisterenko was chosen to be the lead petrologist on the second site.

Kazuo and I decided to meet with Nisterenko to convince him to get his report to us faster. For reasons I did not know until the cruise was over, Nisterenko avoided me during the entire cruise.

The three of us met in the science office. Lou Henry was typing our final report for the first site and Gus was completing his paperwork so both listened to the conversation. Kazuo asked me to do the talking. The dialog, as I recall, went as follows (phoneticized spelling mine):

Klein: Gennady, we have a problem. We had a cascading delay on the last site report because the igneous petrology part was delayed. We must keep on schedule.

Nisterenko: Chorch, Zat is eempossible. Ve need za time to look at sin-sections and get zee XRD. Zen I need time to write it up.

Klein: Gennady, I can't allow that. Now to help you, I've asked Doug Waples to translate your report.

Nisterenko: Eempossible. I nefer had to turn in a report on a deadline in Russia.

Klein. I noticed you spent ten years working on the extensive plateau basalts of Siberia. Weren't you required to file annual progress reports to assure support for field work the next year, such as the Geological Survey of Canada requires?

Nisterenko: Nefer.

Klein: We must have your report, and we need it a week from tonight at 8:00 PM.

Nisterenko (sneering): Eempossible. Eet can't be done!

Klein: Gennady, you will have that report on my desk a week from today at 8:00 PM.

I then slammed my fist on the desk in total frustration.

The effect of that gesture was amazing. Gennady completely changed expression, got red in the face, and stammered."Yes, I'll haf it done" and left.

Lou Henry and Gus witnessed this and spread the incident to the entire Global Marine contingent who operated the ship.

Nisterenko gave me the report as promised. I reviewed it and asked questions. He was missing key literature although I hadn't taken an igneous petrology course in 20 years. I asked about differential crystal settling, resorbtion, inclusions, and other things and he didn't seem to know the answers. I mentioned papers by Walker and Sparks and he was unaware of them.

I finally asked, "Gennady, how did you manage to work in Siberia for ten years and not know these papers?" He replied, "Chorch, vould you like to see za report?" I replied, "Sure." Gennady went to his cabin and came back with his published monograph. I went through it and although the text was in Russian, the illustrations and captions were bilingual: English and Russian. I asked questions, and checked the references (also bilingual). The references were pre-1950.

Gennady then said, "Chorch, I deedn't know you ver eenterested in zese things. Vy?" I explained that I had to map basalts in my thesis area and gave a journal club presentation at Yale on Walker's paper on the Palisades sill.

It troubled me that Nistorenko was so far behind in his understanding of his field. I discussed it with Henry Dick who said Gennady was very good at description but his knowledge of petrology was no better than what he learned as an undergraduate from Butler.

We continued as scheduled. However, we never reached our target depths. Because it was winter, we would get cold-air out breaks off the Siberian platform about every five or six days. For safety reasons, the drilling supervisor and chief safety officer, Cotton Guess, an experienced driller from Mississippi, ordered the drill crew to pull out of the hole. The ship's meteorologist was very accurate in his forecasts.

On the last site, Site 446, we were an estimated 50 feet from the basalt basement and received a weather advisory that another storm was on its way. Cotton ordered drilling to be terminated. We were all very disappointed. The storm, however, took a turn away from us. A day later, I went to the drilling platform to thank the drill crew for everything they did to assure a successful cruise and Cotton said, "George, maybe I acted too quickly on the last site." I knew he felt bad and said, "Cotton, I'll never second guess you. The safety of the ship, the people, and the equipment were at stake. You did the right thing."

Captain Dill told me later that Cotton was surprised by my reaction. In the past, distinguished international members marine geologists swore at him and called him nasty names. Dill explained that Cotton had a ninth grade education and learned everything on the job. The ship's officers read the dossiers and CV's of all the scientists and Dill, who also came from Mississippi, said, "George, I appreciate your recognition of the expertise on this ship. With your education and record and where you work, it means a lot to us."

We had five days to finish our work before arrival. During the cruise, Gennady held a few parties for the scientists, serving vodka and canned Russian snacks. I was never invited. Before reaching Naha where the cruise would end, he had a gift for everyone. All the scientists received a set of slides of Moscow street scenes taken for the 1980 Olympics.

My set of slides was different. They were slides of 13th and 14th century Moslem architecture from Samarkand and Tamerlane. I said, "Gennady, thank you. How did you know?" (See Chapter 11) Gennady replied, "Oh Chorch, I don't speak English zat vell." I responded, "After 51 days with us a sea, you're doing very well."

I reported this incident to Captain Dill who was on the bridge with his senior officers. I was following instructions given by Dave Moore. Dill told me that he receives a CIA briefing on every Russian scientist prior to each cruise but to assure scientific collegiality, he was not allowed to share anything with the US co-chief unless something happened. He explained Gennady was a trusted communist party member and a high colonel in the non-cloak-and-dagger side of the KGB as part of his regular job at the USSR Academy of Sciences. He then added, "George, We're real proud of the way you stood up to that Rooskie back at the second site." That was my first clue that he knew.

That evening, I received a telegram from Ambassador Mike Mansfield in Tokyo, inviting all Leg 58 scientists to a reception at the American consulate in Naha. I polled everyone but couldn't find Gennady. I commented, "Oh well, he'll be taking the first plane to Moscow." Waples said, "No, he has a three day visa to look at Japanese volcanoes."

I finally found Gennady and he told me he could attend. I asked him what he would be doing in Japan. He told me he had a three day visa. He was visiting three Japanese volcanoes, all overlooking major U.S. military installations including the Yokosuka naval base. I reported this to Captain Dill.

The ship arrived in Naha and to my surprise was met by Ujiie who now was a professor of marine geology at the University of Naha. He and his colleagues invited us to dinner that evening at a local ethnic Okinawan restaurant and gave us instructions to give to a taxi driver to find it.

After checking into the hotel, taking a decent shower, and a short nap, we all headed to the restaurant. The University of Naha reserved the second floor reached by a stairway.

We started with cocktails, enjoyed a superb dinner, and replenished our drinks. I noticed that like a stereotypical Russian, Gennady got drunk before dinner.

As the dinner party ended, I realized as the 'foreign' co-chief scientist, I had to undertake a task so the evening could close. I asked Hisatake Okada who was talking with Gennady what the norms were. Hisatake explained them and my comment was "Yeah, just like they do it in Holland where I was born."

Suddenly, in a drunken rage, Gennady exploded and shouted "You veren't born in Holland, you ver born in Nazi Germany!"

I replied "Gennady, I think I know where I was born. Besides, let me show you my US Passport which shows I was born in the Netherlands"

Gennady shouted: "Ve know zat ze American government forges passports for people like you!"

I told him I would talk to him later, gave a little speech of thanks to the hosts, and we left. I also realized that this misinformation was the reason he avoided me during the entire cruise.

When walking down the stairwell, I noticed Gennady walked next to me. The stairwell had a curved wooden ceiling, so I tapped Gennady on his shoulder and said, "Gennady, when you get back to Moscow, please be sure to correct my file."

My comment reverberated up and down the stairwell, and everyone laughed. However, I met with the deputy American Consul in Naha who took notes, and said he would arrange extra surveillance of Gennady while he toured Japan.

I flew from Naha to Tokyo, spent a day with Nasu, and flew home. After going through my usual routine, I visited Sandberg. The department had settled down and to Phil's credit, he solved about 85 percent of all the problems left behind by Fred Donath. The mood was upbeat and the best I recalled since my arrival in 1970.

When I returned and saw Dennis Wood, he asked if I could give a colloquium on the results of DSDP Leg 58. I agreed to do so a month later. Nearer the time, I gave him a title and an abstract.

Several days later, I went to the mail room while Fred Donath was copying class materials. He smiled and said, "I read your colloquium abstract. You must have had a great cruise." I told him, "Sure did." We chatted amiably and in a professionally civilized collegial manner almost like before December 5, 1974.

Henderson was also in the room. He looked shocked and his jaw dropped, and left. I turned to Fred and said, "Guess the whole department will know about this conversation." Fred laughed.

We soon started interviewing the four finalists for the vacant headship. The search committee requested me to assist. Because I was single, they asked me to host all dinners at the end of the first day. I made the arrangements. Given the few choices of restaurants, let alone decent ones, in Champaign-Urbana, I picked a steak house on top of a new ten–story high rise office building in Champaign which also had a great bar.

The first candidate was John Hower (BS, Physics, Syracuse, PhD Washington University, St. Louis with Frederickson, Clay Mineralogy; Pan Am Research, Montana, M.I.T, Montana, Case Western Reserve, NSF Program Director for geochemistry, Illinois). I previously talked with Hower at GSA meetings. Hower evaded questions about future directions of geology, growth possibilities for the department, and how he viewed the future of the department. I recommended that if he got the job, he should fire Goodman (something I mentioned to all candidates).

Hower gave a colloquium his first day. The talk wasn't great, but Hower had a unique way of establishing instant rapport with the audience and won over many people. Afterwards, we went to the Levis Faculty Center for a reception which the department hosted. I noticed John had two drinks whereas the rest of us had one. I then drove John to the restaurant and assumed we would just order dinner, but politely offered drinks. Hower ordered another drink but because no one else did, I ordered one to keep him company but did not finish it. When the meal was served, Hower only ate half of it. The rest of us cleaned our plates. I couldn't put my finger on this, but sensed it might indicate a hidden problem (See Chapter 22 and 23).

The second candidate who I knew well was Robley Matthews (BS, Texas, PhD, Rice; Carbonate geology; Brown University). He became my first choice. He had clear ideas about the future of geology, how Illinois could capitalize, how to solve the remaining problems in the department, and was very upbeat. Matthews was chairman at Brown University and obviously was up to the job.

However, one of the graduate students, Sharon Mosher (BS, Illinois, MS. Brown, PhD, Illinois, structural geology; Univ. of Texas, President of GSA; Dean, Jackson School of Geoscience, University of Texas @ Austin), earned her Master's at Brown and expressed reservations about his approach with students. In the end, Matthews split the department. His talk was very good. Dinner went well.

Bill Phinney, who I met in Nova Scotia and who tried to interest me in coming to Minnesota, was the next candidate. He was now a section leader at the Lunar Research Center at NASA. Bill wanted the department to become more like NASA and had some good ideas. I met with him before his colloquium and took him to an auditorium in Noyes Lab (Chemistry building) for his talk. We waited for a large class to leave. The student projectionist had trouble setting up the system and Bill became more nervous about the delay. I tried to calm him down. He gave his talk but it had three parts and without transitions from one part to the other. Moreover, the talk was longer than an hour. He was more relaxed at the reception and dinner.

Phinney made a fatal error. All candidates met Dean Robert Rogers. On returning to NASA, Bill wrote him a five-page letter explaining what he would do to turn the department around. Rogers did not approve of receiving an unsolicited letter and eliminated Phinney from consideration. Ultimately, Rogers made the final decision but generally, would concur with the departmental recommendation. He forwarded Phinney's letter to the search committee, as Phil told me later. Moreover, Phinney split the department, but along different fault lines than Matthews.

The final candidate was someone whose name I can't remember. He was a mineralogist who earned a PhD at Chicago while Don Graf was a post-doc there. He was Provost at the New Mexico Institute of Mining and Technology, in Socorro, NM. He had done no research for ten years. During my interview, he had no ideas, and seemed bored.

I asked how big the faculty was at NM Tech. He told me 100 faculty members. Therefore, the size of the faculty of that university was less than the size of the faculty of the Illinois English, Psychology, and Chemistry departments and most of the departments in the school of Engineering. I quietly mentioned this to others at the reception. At dinner, Graf came along, didn't speak with me, and neither did the candidate.

At the end of all the interviews, Matthews was my first choice, and Hower a possible second choice. I eliminated the other two for recommendation to the committee.

Each faculty member then met with the search committee for individual assessments. When my turn came I was asked who my first choice was and I told them Matthews and explained why. They asked who my second choice was and I told them, "John Hower, but it would be a weak second choice. There's something about his visit I couldn't put my finger on. I think it could mean difficulties (See Chapters 22 and 23). He also had no vision for geology or the department." Wirt asked, "George could you live with Hower as head?" I told him I probably could.

I then turned to Sandberg and said, "Phil, quite frankly, you are as good, if not better than the candidates we interviewed. Why don't you just continue as a permanent head?" Phil made it clear he had no interest in doing so.

I then told them that Phinney was too tense during his visit and not up to the job. As for the Provost from NM Tech, I said, "He was a washout, and a total waste of the money to bring him here."

Wirt checked his watch and then said, "George we have a little time. What would you suggest we advise the leading candidate to request from the administration?" I replied, "First, he should try to get every single faculty member a large salary increase between 10 and 15 percent." All smiled and nodded in agreement. I continued, "Second, we need a new building. NHB is barely functional and to renovate it would be counterproductive. Third, the new head should fire Bill Goodman. Last, the new head needs to form a strategic planning committee to determine where we go from here." Wirt and the committee thanked me as I finished and left.

Shortly afterwards, I received a letter from Dr. Yong Ahn Park at SNU in Korea. He was attending the 1978 IAS meeting in Jerusalem and wanted to meet me to discuss their AID visiting professorship. I replied stating I'd be happy to see him and suggested lunch.

The search committee met with the dean and the decision was made to offer the headship to John Hower. To me, John was either everyone's first or second choice. When he returned to meet with the dean, he asked to meet me. He apologized for not sharing his vision of geology explaining that as head of a section at NSF, he had to be careful with his opinions lest they be interpreted as NSF gospel. I told him I could understand that and to tell me now. He refused. Hower accepted the headship.

During the summer of 1978, I worked on Leg 58 research and then flew to the IAS meeting in Israel. Again, I notified AAPG because Illinois paid my way, and they booked short courses in London with Shell before the meeting and in Cairo, Egypt with the Gulf Arabian Oil Company (a Conoco subsidiary) after the meeting. This was two months after Sadat and Begin signed the Camp David accord, but no air service existed yet from Tel Aviv to Cairo.

I called my Dad and asked about the most secure airports in Europe through which to connect to Cairo. He advised London or Zurich. The London part of the trip went well. I then flew to Israel. It was hot and I stayed in the shade as much as I could. I visited the holy sites, was shocked I had to pay a Greek Orthodox priest to see Christ's tomb (so I didn't see it), and concluded that the local markets looked no different than during biblical times except for the transistor radios.

Yong Ahn Park and I met and held a productive meeting. We planned a research mission and he assured me that I would not teach for a year, but just come for a short period. He requested I offer my short course and I gave him instructions on ordering the short course book. We had an agreement and all looked good. We agreed I should come in the spring or summer, 1980.

Glenn and Betty Visher also attended and explained they were celebrating their 25th anniversary with a round-the-world trip. Their first stop was Rome where, as they explained, they had an unfortunate experience. They booked their flights with the TWA-Hertz "Fly and Drive" plan and picked up a rented car at the Rome airport. They received directions to their hotel. On the way, a motor cyclist with a passenger started to overtake them. The passenger pulled an ice pick from his jacket and pushed it into their left rear tire. Glenn drove into an abandoned gas station to change it. While getting the equipment out of the trunk to change the tire, the motor cyclists returned and grabbed Betty's handbag and tore off her necklace and sped off.

After changing the tire, they discovered they received wrong directions to the hotel and eventually found it. They first reported the incident to the Italian police who said, "Mrs. Visher, it's all your fault for wearing too much jewelry." They visited the American Embassy and after explaining their problem to the receptionist, were ushered to the office of an embassy official. That official told them "You're the seventh couple to report such an incident today. We think the people at the airport Hertz Counter are in on it." Betty received a new passport and arranged to get credit cards terminated and new ones issued.

As for my trip to Israel, I admit I was less than impressed. I read the *Jerusalem Post* with all its articles about the petty arguments raging through the country and not much else. The service at the hotel (a Ramada Inn) and the staff attitudes were poor. After reading how the Israelis made the desert bloom and were unified against hostile enemies, I admit being underwhelmed. The country seemed like an Eastern European socialist republic.

On the morning I left Jerusalem, I checked out and the rude person at the front desk told me to come back in an hour to settle my bill. I arranged for a taxi to drive me to Tel Aviv so I asked to speak to the manager. When he came and I explained the problem, he opened the door behind him to a conference room and the cashier was sitting there having coffee with her boyfriend. He asked her to settle the bill.

I then talked with him and he told me, "I can't fire anyone." I then said, "Sir, I must say, if there was one country in the world where I thought I could immediately settle a bill, I thought it would be this one." He turned red in the face.

Tel Aviv was more cosmopolitan with better restaurants. I left for Zurich the next day, spent the night there and flew to Cairo the following day, a Sunday. I was met by a driver from the Gulf Arabian Oil Company (course sponsor) on deplaning before I reached immigration. He asked me to go to the duty free shop and buy cigarettes for which he would pay me. Sensing danger, I refused.

I was taken to the Sheraton Hotel and when checking in, the front desk clerk asked me to leave my passport overnight and was told me that the police need to check it. I said, "When the police come, give me a phone call and I'll meet them. I can show them the passport myself." They never called.

The hotel restaurant was actually quite good. I only ordered Middle Eastern dishes and the waitress realizing I was American was impressed. I got great service and great meals.

Cairo was a dirty city. I went to the famous archeological museum and the exhibits were dusty and poorly curated. Clearly the country was poor. On every street corner, there was a huge photograph of President Sadat. Clearly, the lid was on. I was happy to leave and fly back to the USA.

Once back in Urbana, I knew a new era was about to begin at Illinois.

LESSONS LEARNED:

1) On board a ship, the captain will know everything that happens and who's responsible.

2) During the Cold War, Russian scientists were closely watched by the US Government, and for this I am grateful.

3) When negotiating with a Dean after a controversy like the second Donath War, be reasonable but go high on a salary settlement.

4) Never be afraid to take a stand for academic integrity with lower and middle level administrators.

5) When working on a drill ship, pay close attention to all safety requirements and policies, even if it results in missing a target objective.

POSTSCRIPT #1. To deal with students who cheated, I used a simple approach. I always gave essay exams and when I read two that were verbatim, I automatically gave both students a zero. At the end of class when exams were returned, both descended on me to protest. I asked a few questions, watched their non-verbal behavior and soon identified who had been wronged and who the cheater was. I then offered to regrade both exams, gave the honest student the proper grade, and left the alleged cheater's exam at zero.

The alleged cheater could file a grade grievance or go to an associate dean to go to bat for him. None ever did. Moreover, because the exam counted anywhere from ¼ to 1/3 of the final grade, the best course grade the student could earn was a 'D.' Every undergraduate geology student knew that with a 'D' in a geology course, they had no chance for admission into a graduate school or finding a geological job. Within two weeks, the cheaters dropped my course and changed majors. I therefore solved the problem without benefit of the administration.

POSTSCRIPT #2. Years later, I shared my experience with Gennady at the Okinawa restaurant with Nahum Schneidermann (BS, Hebrew University; PhD Illinois; micropaleontology; Univ. of Puerto Rico, Gulf Research, Chevron; International Negotiator for Chevron). Nahum was born in Poland but after the German invasion, the family fled east and was housed in Kazakhstan where he learned Russian and the local language. After World War II, the family settled in Israel. Nahum made his career in the USA. He was, in fact, part of the Chevron negotiating team that met with the Kazakhstan government that gave them a major concession in the Tenghiz oil field.

Nahum explained that during World War II, the Russians were paranoid that their southern Muslim republics would be infiltrated by Nazi operatives and stage an uprising. My question in 1963 (See Chapter 11) triggered that paranoia and led to the mistaken entry in my KGB file which Gennady exposed in a drunken moment.

POSTSCRIPT #3. Henry Dick's assessment of Nisterenko's petrologic skills reminded me of a conversation I had with Tor Nilsen (BS, CCNY, PhD, Wisconsin, sedimentology; Shell Research, USGS, private consultant) who participated on DSDP Leg 38 in 1974 in the North Atlantic. Tor worked with four Russian sedimentologists. None knew what a graded bed was, much less recognized them, and none knew about the turbidite depositional model.

CHAPTER 22
Illinois: The Hower Years, Chicago, Korea (1978-1983)

'Truth kills those who hide from it"
- Old Lebanese proverb

"Where there is no vision, the people perish; but he who keepeth the law, happy he is."
- Proverbs, 29:18

"Academia has its drawbacks as well. Job security sometimes comes at a price of stagnation and limited vision."
- Lee C. Gerhard

It is my personal view that the decline that began after the first Donath War continued during the Hower years in the geology department at Illinois. The choice of John Hower meant the department and the Dean picked a man of narrow and limited vision, a man who was most uncomfortable with field geology, field work and field trips, and a man who above all ended up antagonizing the very people who wanted to help him the most, while toadying up to those who would eagerly undermine him at a moment's notice.

The Hower headship represented a period of stagnation. The geology department at Illinois never recovered when I left in 1993. Although I did not maintain much contact after leaving, I am left with the impression that positive improvement began only with the headship of R. Stephen Marshak (BS, Cornell; PhD, Columbia, structural geology; Illinois) starting in 1999.

During Hower's headship, the department lost both Pat Domenico and John Holder to Texas A&M, Graf to retirement, Palciauskas to Chevron Research, Wood to Robertson's Research, Eberl to the USGS, and Donath by his own choice. I considered moving and was invited for interviews. However, it was difficult to replicate the facilities I had at Illinois, so I stayed.

Replacing the people who left occupied much of the department's time even after Hower stepped down as head in 1983. When Palciauskas left, he said, "George, the geology department at Illinois will always be conducting searches." He was right. In time I referred to the turnover as "The curse of Vic Palciauskas." It started with Hower.

The mood was optimistic when Hower arrived in August, 1978. Troubling to some was his reliance on Bill Goodman and Don Henderson who continued as Education Coordinator.

Henderson had an unfortunate career. Shortly after arriving in 1952, he contracted polio and never fully recovered. George White promoted him to Associate rank and Donath promoted him to a full professorship. Henderson taught a required mineralogy course which received mixed reviews, and optical mineralogy. With limited health, there was only so much he could do. Later, he served as educational coordinator because he could remain useful and no one else wanted the job. But "Hendy' was also a negative influence because he always opposed change and wanted things the way they were when he first arrived. His views were always disproven by a proactive approach.

Hower convinced the administration to give him a new position and we agreed that it should be in the field of Geodynamics. We interviewed a very promising person doing a post-doc at Brown who used the offer to convert his position there into a faculty appointment. We also interviewed Albert Hsui (BS, Lowell Tech, PhD Cornell; basin modeling and geodynamics; NASA, post-doc at M.I.T, Illinois) and offered him the job. Albert accepted and arrived in January 1979. He developed a program that attracted students and recognition. He and I cooperated on basin research projects also.

When Holder and Palciauskas left, Albert skillfully guided the search for replacements. One, a seismologist, Wang-Ping Chen (BS, Taiwan; PhD, M.I.T., seismo-tectonics; Illinois) was an outstanding hire but his true development and career trajectory advanced well after the Hower years. Another, a mineral physicist, Jay Bass (BS, Brooklyn, PhD SUNY-Stony Brook, Mineral Physics; post-doc, Cal Tech; Illinois) also was an outstanding hire but the pay-off came much, much later. He arrived after Hower stepped down as head. Hsui's appointment ultimately led to many good things for Illinois, but Hower was not the beneficiary.

Hower chose not to appoint me to any committee assignments his first year. Because of some personal financial issues, I asked if I could teach at field camp during the summer of 1978. He eventually agreed but could only pay $5,000 for the six-week course. I proposed taking a week off while the field camp went on a national parks tour so I could write papers and visit friends in Denver. He agreed. However, I felt he shorted me.

After three months, the mood in the department started to sour. The job also soured Hower. During a February, 1978, faculty meeting, he disclosed that when he took on the headship, his friends said it was a bad decision because he would deal with so many difficult faculty members. He admitted that was not the case and that the all faculty were helpful. His biggest headache was dealing with the non-academic staff.

Jim Castle defended his PhD thesis early that fall. I invited Hower to attend to get the feel of departmental defenses because he was new. He was surprised I asked, but came.

The fall term started normally and nothing usual happened. I attended GSA in Toronto and saw Dick Hay from UCB. Dick just received the Kirk Bryan Award for his research at Olduvai Gorge. I suggested dinner on the Tuesday of the meeting. While waiting in the hotel lobby, I noticed excruciating pain around my left hip and upper left leg. I sat down and it went away. The pain reappeared two months later so I had an orthopedist check it. After looking at X-rays, he told me I had arthritis in my left hip and the cartilage in the hip socket separating the hip from the femur was slowly dissolving. A hip replacement was in my future.

In mid December, I received a phone call from IHRDC in Boston asking if I could teach a one-week short course in Libya starting around January 12, 1979. I agreed and sent my passport to a Washington, DC passport clearance agency so it could be translated into Arabic to get a Libyan visa. Two days before leaving, I called the passport agency and they still hadn't received my visa from the Libyan embassy. I therefore routed my trip through Washington to get it in person with the understanding that IHRDC would reimburse my plane ticket if all fell through. I looked at the weather forecast and saw a very large storm was headed to Champaign on my departure day. I departed 24 hours earlier spending the night in Washington. When arriving at the passport agency, they handed me my passport and said, "Dr. Klein, I don't know who you are or who you know, but this was the fastest passport processing by the Libyans in our experience." I thanked them for their help and headed to Dulles Airport.

A month later, I read a newspaper article that President Carter's brother, Billie, received funding from Libyan interests and featured a picture of Billie urinating near a Plains, GA, airplane hangar. His Libyan friends stood in the background. I was the beneficiary of a brief thaw in US-Libyan relations without knowing it at the time.

During the flight, I read a textbook by Gerry Freidman and John Sanders to review for the *Journal of Sedimentary Petrology*. Many illustrations were photos of geological features in Israel. Gerry, a devout Jew, spent considerable time doing research in Israel and training their geologists. I also knew it was unwise to take the book to Libya. I checked it in with the concierge at the Sheraton Hotel in London because I was staying there also on my return.

The course was in Benghazi and after clearing customs in Tripoli, I connected on a local flight. The course went very well and nothing unpleasant happened. I was cut off from the world, but two of the out-of-town course participants stayed in the same hotel and told me that the Shah of Iran fell and fled the country.

While waiting in Tripoli to board my return flight, a group of British expats were loudly complaining about the quality of their life during their assignment in Libya. On boarding the flight to London, I noticed the rear three rows on the starboard side were converted to a temporary bed with a plywood board and a mattress to transport an older person, a local tribal chief. An hour before the flight was due in London we started to circle.

Suddenly, the tribal chief let out a loud groan and the flight attendants raced to the flight deck. After two minutes, the captain spoke in English for the first time and explained we were making a medical emergency landing at Charles de Gaulle Airport in Paris because London would not clear him to land immediately. On landing, I inquired if I could deplane and try to get to the US from Paris. I was told that Arab Libyan Airways was unwelcome in France so only the patient could leave.

We took off and landed at Heathrow after a one-inch snow fall. The runways were cleared by people using coal shovels. No snow removal equipment existed at the busiest airport in the world!

On landing, immigration and customs was a major problem. First, airport personnel went on strike, so only supervisors were on duty. Second, a steady stream of planes arrived from Iran after their revolution, which slowed things down. I finally retrieved my luggage, headed for customs and spotted the expat Brits who complained loudly in Tripoli. In frustration with the delays, I went up to one of them and asked, "Happy to be home?" He replied, "O yes, jolly good." I then said, "You know, this country surprises me. The airport staff can't remove snow, the immigration and customs people can't handle the passenger load, I'm nearly two hours behind schedule. How did you run an empire?" I then left.

As I sensed malaise in the department, I applied for a CIC (Council of Interinstitutional Cooperation – a consortium of Big 10 Universities and the University of Chicago) visiting exchange professorship at the University of Chicago in the department of geophysical sciences to work with Fred Ziegler (BA, Bates, PhD, Oxford, paleogeography and paleoecology; Post-doc Cal Tech; Univ. of Chicago) during the 1979-90 academic year. That application was approved. The time off campus was credited to future sabbaticals.

During the spring of 1979, the Leg 58 scientists met in La Jolla for a post-cruise meeting to finalize the site reports and set target deadlines for the final manuscripts. It was a productive meeting. We had already published a short paper in *Nature* concerning off-ridge volcanism in the back arc basins we drilled.

The spring semester ended and I drove to field camp, arriving on June 1, 1979. Snow was visible on the mountain tops and temperatures were near freezing. Everyone arrived and instruction began. I recommended to John Hower he visit field camp and become familiar with its operations. Fred Donath made an annual visit particularly because many of the participants came from other universities and he, like George White before him, saw it as a graduate school recruiting tool.

When John arrived he was totally detached. He gave a boring, perfunctory talk to the students, and did not go with them on their daily mapping tasks. Instead, John borrowed a vehicle to collect some shale at a nearby outcrop. He never paid the camp any more visits while head. Graduate recruitment suffered.

During field camp, I received good news on the financial front. When my mother was a university student in Berlin in 1926, one of her aunts died and left a bequest of about $300.00. That aunt was my mother's least favorite person. She went to a modern art gallery and bought a water color by the expat Russian artist, Wassily Kandinsky. By 1979, Kandinsky was a recognized name artist even after his death. When my mother passed away, she bequeathed the water color to me.

My father advised selling it because the water colors were fading. He learned how the art world conducts business when his sister, my Aunt Olga, remarried an art dealer, and arranged to auction the dealer's collection when her second husband passed away. We tried to auction the Kandinsky during the summer of 1978 and failed to reach minimum bid. In 1979, it sold at auction and the net proceeds after commission was close to $40,000.00.

I moved to Chicago, renting a one-bedroom furnished apartment in a high-rise in the Hyde Park section. It had a fantastic view to the north of downtown, Grant Park and the north shore. The University of Chicago ran a free neighborhood bus service and it stopped half a block from the apartment building.

The University of Chicago was founded in 1890 and opened its doors in 1892. The first head of the geology department was T. C. Chamberlin, an eminent geologist who resigned from the presidency of the University of Wisconsin to accept his new assignment. The department became prominent quickly with the addition of Stuart Weller (Stratigraphy; see Chapters 6, 7), Sam Williston (Chapter 6, 7), and Edward Bastin, an economic geologist. Later Francis J. Pettijohn (See Chapter 5) was added as was William C. Krumbein (See chapter 4) who left for Northwestern. In 1961, the departments of geology and atmospheric sciences merged to become a department of geophysical science.

The Hinds Laboratory where the Geophysical Sciences Department was housed was relatively new (10 years) and well maintained. The department was home to four groups: geology, geochemistry, geophysics, and atmospheric sciences.

During the fall quarter, I offered no courses, gave a colloquium, and two guest lectures in an undergraduate class. During the winter quarter, 1980, I taught a sedimentology course and led a spring break field trip to the Arkoma basin and Ouachita Mountains. During the spring quarter, I went to Korea.

I enjoyed being in an urban environment, using public transportation having sold my car. I had a new graduate student at Illinois, John L. Shepard (BS, Univ. of South Florida, MS, Illinois; Shell Oil; regional VP, Eastern Gulf of Mexico), and periodically visited him to provide thesis supervision. He actually didn't need supervision, but it was a pleasure to see him forge ahead on his thesis. He finished in 1980. John was my most outstanding Master's student who made his career at the Shell Oil Company.

Yong Ahn Park wrote in September explaining the research project I proposed was approved, funded, and all arrangements to which we agreed were in place. I wrote back telling him to expect me in April, 1980. During October, however, I saw both a TV and newspaper report that the President of Korea, Chung Hee Park, was assassinated by the director of the Korean CIA and all universities were closed. I wondered if my research program was still going forward. In November, Park wrote again and assured me everything was in order.

I attended a DSDP Panel meeting in October just before the annual GSA meeting in San Diego. I met there with Hower and he told me he was terminating Dennis Eberl. I reminded Hower that he considered Dennis his best student and was told, "Dennis has changed. He makes wild statements in his classes." He also disclosed that Dennis would appeal.

I found out later from one student, Kathie Marsaglia (BS, MS, Illinois, PhD, UCLA, Sedimentary Petrology; Univ. of Texas @ El Paso, Oil service company in Houston, CSU-Northridge), that Hower kept undermining Dennis in class. Eberl won his appeal and showed me his complaint. He included a list of citations to published work by the faculty to prove he did as well as his peers. I discovered mine was the highest publication citation rate of the entire faculty, and John Hower had the second highest but only half as much.

My left hip bothered me again and I arranged to see an orthopedist in Chicago. He recommended using a cane.

The field trip to the Ouachitas and the Arkoma basin went very well and I decided to revive the trip at Illinois the following fall semester as part of my depositional systems course.

During the first week of April, I returned to Urbana, picked up the sedimentology lab's current meter and prepared to go to Korea. I was looking forward to the trip but recalled my previous experiences with my Japanese colleagues on the *Glomar Challenger* and in Hokkaido and had reservations. Those reservations were quickly shattered within two days.

On landing at Kimpo Airport in Seoul, Yong Ahn Park met me and drove me to the Koreana Hotel, my home for at least five weeks. He had to leave after I checked in to get home before the 10:00 PM curfew.

The Koreana was originally built to accommodate visitors to the American Embassy which was three blocks away. However, as more hotels were built, it became a destination for Japanese tourists and locally was referred to as the 'Geisha House." When I checked in, the only room available was on a floor of Japanese tourists. The people at the front desk were most apologetic, offering to move me the next day. I told them I would take it and stay there until my first trip out-of-town. I found out in time those Japanese tourists really get rowdy, running up and down the hall, even half naked. Whenever I stepped onto the hallway, they suddenly reverted to 'outer Japanese."

My first day was mostly to myself except for an afternoon meeting with the director of KORDI (Korean Ocean Research and Development Institute) a government agency. They provided partial funding for our joint research. I had dinner that evening at a European style restaurant and was the only customer. The food was terrible. The next day, I went to the Lotte Hotel which had an underground shopping area with several restaurants. I went to the Korean one which had a bilingual menu and the food was excellent. I ate there many evenings.

I finally reached campus the second day, met Dr. Sung Kwon Chough (BS, SNU; PhD, McGill; marine sedimentology; SNU) and the Dean of the Science College, Dr. Koh (BS, SNU, PhD. Nebraska, physics; SNU). I expected this to be a courtesy call but he told me he wanted me to prepare a written evaluation of both the departments of geology and oceanography before I left. I also met a professor of mechanical engineering from the University of Washington who served as the director of the SNU-AID program. He reimbursed me for my air fare, per diem, and had made direct billing arrangements with the hotel. However, he was very cool to me and Park explained later he disapproved the research project variance I requested.

I then met the geology and oceanography faculty and toured science department facilities. The physics department was in bad shape with a high-end instrument in a dust-filled room lacking air conditioning. The chemistry department was headed by a Korean who earned his PhD from Harvard. He used the AID money well. He built three new clean-room laboratories, and hired two new PhD's who also were trained at Harvard. Of all the science programs supported by the SNU-AID program, he utilized the money the best for the long-term.

I found it easy to communicate with my Korean colleagues. They were more direct and open than the Japanese. As I came to know them better, I became more pleased that I had made the trip. Moreover, due to the economic development plan of former president Chung Hee Park, the Koreans were optimistic about the future and their ability to live a better life.

Yong Ahn Park arranged side trips to the Folk Museum at Suwon, the Admiral Yi Memorial commemorating the Korean victory over the Japanese during the 16th century using the 'Geobacksen' (metal-plated Turtle Ships), a weekend trip to the famous Bul Gak Sa Temple near Pusan, and several other places. I also visited famous monuments and palaces in Seoul that were open to the public. I also bought an English language history book about Korea.

Overall, I was content and realized after two weeks, I hadn't been this content in ten years. I liked the country, and much later, it became my second homeland (See Chapter 28 on Marriage to Suyon Cheong) although the language was difficult and almost beyond me.

After two weeks, Park, two students and I boarded a Greyhound-type bus for Mokpo to meet our rented fishing boat for our research field program. The trip provided me with an opportunity to see more of Korea, and to experience something only a pragmatic society would consider. We stopped at a rest stop and I entered the men's room. In front of all the urinals were metal milk cans about 1 meter tall. All male restroom users were required to urinate into one of the milk cans. Park explained it was a national effort to expand the urikase industry and the cheapest option for feedstock was to collect human urine.

Our field research program was suddenly curtailed by Park and the funding agencies from one week to four days because boat charter fees were high. On arrival, we were met by J. H. Chang, a marine geologist at KIGAM (Korea Institute of Geology and Mining), the national geological survey. KIGAM also provided some financial support.

The first day we left Mokpo Harbor. I looked at the highlands around the city and saw a massive installation of antenna. These were part of a U.S Military facility serving as a listening post for broadcasts and messages from China. We continued into an estuary and suddenly, entered a narrow canal and stopped at an unpretentious house. It was our lunch stop, a *Maccali* House that served local rice wine (*Maccali*) and rice and veggie snacks.

We then headed for Do Cho Island where we spent the night at a local hotel. We did some preliminary sonar mapping while underway.

The next two days we measured current velocities and directions at two locations on Odanam Sato, a tidal sand body. I wanted to determine whether the tidal currents were zoned by time-velocity asymmetry like in the Bay of Fundy (Chapters 11, 12 and 15). They were. However, I noticed at both high and low tide that current velocity was maintained as the tidal currents rotated from flood stage to ebb stage and ebb to flood stage. On the last day, we completed a sonar survey and observed that tidal bedform orientation near the crest was consistent with the rotary tidal current change.

On returning to Seoul, I asked to see a 1964 Korean Navy bathymetric chart. We plotted our locations and mapped the limits of Odanam Sato on the chart using our sonar data. We discovered that Odanam Sato was also migrating westward like an extremely oversized dune. I knew then we had enough for a paper and it was published later in *Marine Geology*.

I taught my short course and it was well attended. I lectured for an hour and Park translated my remarks. I met one of the graduate students in the course, Yong Il Lee, (BS, MS, SNU; PhD, Illinois; Post-doc at RPI; SNU, later Dept. Chairman) who was admitted to the University of Illinois starting the following fall. I agreed to supervise him as a PhD candidate.

Towards the end of my stay there was a major change in Seoul. Students were rioting both downtown (I could see them from my hotel) and at the entrance gate to SNU. Riot police with tear gas were in full force. I ate dinner one evening with the Director of KORDI and he drove me back in his official car with an agency flag. Shattered office window glass was on the streets. We took a detour over Nam San Mountain and people were streaming out of Seoul to get home, not unlike newsreels of refugees leaving urban areas during World War II. All public transportation was cancelled.

We were stopped by an MP when reaching the Seoul City Hall. I was ordered to walk the remaining block to the hotel because the car was ordered out of the area. The MP summoned a colleague to escort me. This happened on the Thursday before my departure.

On Sunday morning, I received the daily English language newspaper. The headlines screamed: MARTIAL LAW DECLARED. The article reported that under Martial Law all public institutions, including universities, were closed and curfew hours were extended. Curfew was changed from 10:00 PM to 5:00 am to 9:00 PM to 6:00 am.

As I continued to read the article about Martial Law, I noticed a box summarizing its stipulations. One caught my eye immediately. It read "Article 8. Everything must be done to facilitate foreigners in the conduct of their business." That was critical because I had a 10:00 am appointment with Dean Koh the next morning to review my final report, and was leaving the following day.

On that last Monday, I took a cab to the gates of SNU where a long line of faculty and graduate students waited to be allowed on campus. I tried something different and went to the head of the line. A Korean marine looked at me and I handed him my passport. He waved me through, so Article 8 was understood. I walked across campus and saw armed soldiers all over the place, double-timing. They were extremely disciplined and I concluded they could take any Latin American country in three days if they chose.

I met Dean Koh and he asked how I got on campus and I explained it. Only Deans, Provosts, and the President were allowed on campus. He went over the report and liked it. As I was leaving to clear my desk, the SNU-AID coordinator came by. He was admitted because he came in a US Embassy car with an American Flag. He offered me a ride back to my hotel. I previously reviewed my draft report with him and he was enthusiastic, telling me he had been saying the same things to Dean Koh for years.

Yong Ahn Park took me to the airport. As we took off, one of the American passengers mentioned the government had fallen and the new president was General Chung, Doo Wan.

I returned to Chicago in late May and continued writing papers, made a trip to Urbana to sign John Shepard's Master's thesis, and saw Hower. He asked if I was attending AAPG in Denver and when I said yes, he said he wanted to talk to me there.

Hower and I met for lunch on the Sunday of the AAPG meeting. After ordering, I asked why he wanted to see me, fully expecting him to ask why I was interviewing at other universities. I had my answers ready. But he surprised me and asked, "George, have you considered changing your style?"

I asked him what he meant and he talked about how I worked with people, my structured approach with people and students, and my formality. I told him that we all have different styles and we have to accommodate ourselves to them. I explained I wasn't happy with the style of some of my colleagues, but wasn't asking them to change it.

I thought his question was out of bounds and said, "John, let me be frank. I'm paying for that style" waving my cane "because the kind of arthritis I have can be correlated to interpersonal stress.

Furthermore, I have some things to say. First, you never appointed me to any departmental committees your first year. If you value my role in the department, it's time you did. Second, I think you made a major blunder with your attempt to terminate Dennis Eberl. In the process, you alienated many faculty who on a good faith basis wanted to extend their good will and help you succeed. Now some question your ethics thinking that because Dennis rebuilt the lab with new equipment, you wanted to take it from him. That may not be a style any of us want to emulate, if true. Perhaps fundamental questions could be asked.

Also, I have a new student from overseas and need dedicated office space for all my students. It's time I got such space again. You talk about style, but John, there are more serious concerns."

When the waiter brought the check, I paid it. I stood up and said, "John, this is on me." Hower stood up meekly realizing he had just been chastised, made a serious error, and meekly offered to pay the tip, which I accepted.

During the meeting, I attended the Kansas alumni party but because of poor attendance, I left for the Illinois alumni party. About a half-hour later, Charlie Kahle (PhD, Kansas, carbonate petrology; Bowling Green State University) arrived and said, "You have to go back to the KU party." I asked "Why?" and Charlie said, "I'll tell you on the way over."

We left and Charlie explained I was the 1980 recipient of the Erasmus Hayworth Alumni Achievement Award and they needed me there to present it. I was a bit stunned, although Dellwig requested my CV in March. While walking back, I quickly organized my thoughts to give an acceptance speech. I used the opportunity to emphasize how Moore, Dellwig and Hambleton had taught me much, summarizing the things I learned from them (Chapter 6).

The next day, I saw Hower and he asked why Kahle asked me to leave. I smiled and said, "John, the Kansas Geology Department just gave me the 1980 Erasmus Hayworth Alumni Achievement Award. It's an annual award for alumni who they think accomplished something. It's a humbling experience to be so honored by them." He stood there with a stunned look on his face and failed to congratulate me.

Bob Rogers retired from the Deanship on July 1. He was replaced by Bill

Prokasy (PhD, Wisconsin, social psychology; Univ. of Utah; Illinois (Dean), Univ. of Georgia (Academic Vice President)). Prokasy always wore expensive business suites and as Jim Kirkpatrick said, "he looked like the exploration manager at Shell Oil."

During July, I attended the International Geological Congress for one day in Paris, France. I was invited to give a paper about Leg 58 in a special session. Hower arranged travel money paying my round trip ticket from departmental funds. I was now teaching short courses for the Society of Exploration Geophysicists (SEG) and like AAPG, they arranged overseas courses if my airfare was paid. They arranged a presentation to the geophysics group at Shell Research in Rijswijk, Netherlands, before the Paris meeting and in London at BP immediately afterwards. The course at BP went smoothly.

I discovered problems existed at Shell/Rijswijk Lab. The geophysicists previously asked repeatedly that their geological group provide training and the geologists refused. They decided to invite me instead. I personally knew some of those geologists and realized this could create concern.

During the last afternoon, I barely started my afternoon module on deep water gravity sedimentation, turbidites and fans when suddenly the door opened loudly and a large person came in. I recognized him as Epo Oomkens, a Shell sedimentologist who published excellent work on the Niger Delta. The geophysicists snickered while he sat down and loudly slammed his notebook on the desk. I continued to lecture. Finally, he asked, "Aren't these sediments called 'fluxoturbidites'?" I replied, "At one time they were. But now, we understand these processes better and refer to them either as debris flows, slides, or slurry flows." Oomkens got red in the face, stood up, walked out, and slammed the door behind him while the geophysicists roared with laughter.

Moving back to Urbana in August, I prepared for the fall semester. I offered my Graduate Depositional Systems course, including the Ouachita-Arkoma basin field trip. Two weeks after the semester started, Hower told me the department could not subsidize the trip and when I did a calculation, I realized each student would have to pay $250.00 to go.

I called Jerry Walker who was now with Placid Oil in Houston. I decided to sell ring-side seats to oil companies charging $500.00 per person. Jerry networked me to the exploration manager, Ed Hawkinson, who asked "How many people can you take?" I replied, "How many do you want to send?" He said, "Five." I immediately accepted because if I had said a smaller number, we would have no participants.

Murle Edwards established a separate account for the sedimentology lab's exclusive use and any surplus funds were used for the lab for professional items. I told the class each had to contribute $25.00 per person towards field trip expenses because they should not depend completely on the largess of others. I told Hower and instead of expressing gratitude for raising money for students, he sat in stony silence, confirming my earlier assessment that he was hostile to field geology.

The trip went extremely well. I asked Kathie Marsaglia to head the food committee and we ate well. Morale on the trip was high. Everyone came back in good spirits.

Yong Il Lee arrived with his wife of two months to start his PhD program. He made good early progress. However, he was slowed by a graduate school requirement to take four ESL (English as a Second Language) courses.

Near the end of the semester, I was talking with Dennis Wood in his office. For ten years, Dennis ran a field course each summer in the British Isles. It was a popular course and attracted students from all over North America. It was also a great graduate student recruiting tool and enhanced the department's visibility.

During the previous nine years, he submitted an itemized expense statement without documentation and Donath, Sandberg, and even during his first year, Hower, approved reimbursement. This time, Hower demanded receipts and Dennis didn't have any because in the UK they weren't required. He paid various hotels for overnight stays, a bus company for a charter bus, all meals and so forth out of personal funds. His account was partly replenished by the University from student fees for room, board and transportation. By now, his checking account in the UK was overdrawn and the Vice President for campus administration was also on his case. He asked if I had any advice. I recall the dialog went approximately as follows:

Klein: Could you write the bus company, the hotels and all other providers asking them to write you a letter verifying the dates you stayed there or used their services and how much you paid?

Wood: That's a lot of letters. Where am I going to find time to write them?

Klein: We have five secretaries in this department. Ask one to do it for you. You need to do something else. You must send a copy of every letter you write to the banker in the UK, to the Vice President here, and to Hower with a cover letter explaining you are trying to resolve the problem by obtaining letters of certification proving the money was spent.

Wood: Why?

Klein: Dennis, by doing so, you are showing good faith to the banker and to the University Vice President dealing with these issues. That way, they will stay off your back knowing you are taking the initiative to solve it yourself.

Wood: I'll try and let you know. Thanks George. I hope this helps.

While we were talking, Don Graf walked by, saw us talking and went to Hower and told him about what he overheard. When I left Dennis's office, I walked by the department office and because I used a cane, everyone knew I was coming. Hower stepped out of his office and asked to speak with me.

Hower said he understood I was talking with Dennis and wanted to know what it was about. I explained Dennis shared with me his problems about reimbursements for the summer course in the UK. Hower told me Dennis had no receipts and without receipts, he would not approve the university reimbursing Dennis. I told Hower I suggested a way Dennis might solve the problem and he should at least let Dennis try to do so. Hower then told me that even if Dennis did so, he would not let Dennis teach that course ever again.

Then Hower said to me, "You know, George, you're gone a lot from campus teaching short courses." I replied, "Oh?" Hower said, "How do you meet your teaching obligations?" I replied, "John, glad you asked. If you excuse me a moment, let me go to my office and get my logs of class meetings the last three years."

I left to get them. I was advised by a professor emeritus, Harold W. Scott (BS Illinois, PhD, Chicago, micropaleontology; Montana School of Mines, Illinois, Michigan State), to keep a log of class meetings. I showed them to Hower and he looked at them and turned red in the face. He said, "George, you claim 40 hours in Geology 309 when it is only requires 30 hours of instruction, and 55 hours in Geology 437, and Geology 477, when they only require 45 hours of instruction. How do you explain it?" I replied, "John, I'm allowed to claim instructional hours during field trips as a matter of campus policy. Moreover, every time I come back from a short course, I bring back a list of job leads for our students." The conversation ended.

Dennis received his certification letters within a month, submitted them, and received his reimbursement check. Obviously, Hower was trying to squeeze Dennis financially in a way that did not set well with any of us, and revealed not only the narrowness of his science, but also his bias towards field work and an underhanded meanness.

Hower appointed me to the graduate admissions committee and after early February, we had our work cut out for us. Overall, I was disappointed by the applicants and those that had "V' profiles, although few, were admitted after I explained the concept. They turned out well.

I spent Christmas with my father. He sold the house in Scarsdale in 1978 and bought a one bedroom condo in a planned unit development in Connecticut that catered strictly to the affluent elderly. He mixed well with the people there.

Late in February, 1981, I received a call from one of his neighbors. He was taken that afternoon to a hospital in Waterbury, CT, with chest pains and had a heart attack (his second). I called my sister because I just returned from teaching a west coast short course and came down with a bad cold. She immediately went to see him, called and asked me to come as soon as I could. I went on Thursday and spent Friday and most of Saturday with him. By then, his case was terminal and under the terms of his living will, he was sedated. He died Sunday morning. We had the memorial service on Tuesday and again, I had to deliver the Eulogy. It was a very difficult time.

On my return, I received a phone call from the Virginia Institute of Marine Sciences (VIMS). It was part of the College of William and Mary but located in Yorktown, VA, about 10 miles from Williamsburg. They advertised earlier for a director and I applied. They wanted me to come for an interview and I suggested the Monday and Tuesday of spring break. From there, I would go to my Dad's condo, rent a truck and take my share of his personal property to Urbana.

On the flight to Norfolk, VA, I read the entire dossier the VIMS search committee sent and realized it was a mistake to interview. VIMS was a successful and widely respected coastal oceanographic institute. The previous director expanded into deeper water oceanography and bought a ship for $7 million with funds borrowed from the state of Virginia. The ship was junk and sold for scrap. The new director had to revive the institution and pay off that debt.

I was asked to give a talk on my perceptions about the future of marine science to both a faculty and senior staff meeting and immediately afterwards to a separate group of graduate students. I discovered on arrival that the first candidate who came from SIO just said, "It's all up to President Reagan."

My talk stressed how the institute could capitalize on its past work and go global by comparing the variability of different coastal settings. I showed how Jim Coleman at LSU had started working in the Mississippi Delta and went global and discovered the significance of the variability of deltas with differences in river discharge, wave energy flux and tidal energy flux. I suggested VIMS develop a similar program in all its activities (marine biology, chemical oceanography, coastal sediment processes). It caught their attention.

I said at the end of the talk, "Now I'm sure you have questions for me and I'll try to answer them. However, I have a question for you. Can someone please explain how a free-wheeling, soft-money research institute like VIMS can function within the framework of a traditional liberal arts college?" People laughed and applauded. I fully expected the question would force them to reject my candidacy. It did not.

I met various staff, and the new financial officer, who was personally selected by the College president to manage funds. We reviewed the problems he discovered and discussed strategies to solve them. I was told that he would work with the new director and had primary fiscal control until the situation stabilized. He was extremely knowledgeable and I concluded I could work with him.

I also met the Dean of the Law School, former U.S. Senator Spong, and asked how the law school functioned within William and Mary's structure. His answer was ambiguous.

I had lunch with the search committee on the second day and it included a representative from the governor's office, a distinguished Virginian from the McDonnell family. We talked but the only members who said anything were VIMS and William and Mary faculty. I said at the end, "Don't you really think you would be better off hiring a Southern good old boy than an immigrant now living in the Middle West?" The faculty all laughed but McDonnell said, "Dr. Klein, please understand that Virginia has changed. I'll ask the governor to have one of the heads of a scientific agency who was recruited from a Northeastern state call and assure you that the change is real." That call came a week later.

I then met the university president and the provost for an hour discussion. The president asked me if I had questions and I asked, "How does a free-wheeling, soft-money research institute function within the framework of a traditional liberal arts college?" The president began to answer and he and I talked in an animated discussion. The University of Illinois is an excellent training ground for understanding structure and form in academe, so I was prepared. What I did not know until leaving that evening was that the college president earned a PhD in Business Administration from Harvard specializing on organizational structure.

At 1:50 PM, I checked my watch and said, "Well, I understand we only have ten minutes left. What would you like to ask me?" He replied, "No, let's talk more about this," and we ended the meeting at 2:30 PM. On the way out, he introduced me to the person who was kept waiting. It was the Vice President for Development, the campus fund-raising division. I realized then that it was highly likely I would be offered the job.

Three weeks later, the president of William and Mary called to offer the VIMS directorship. He went over the terms and I was shocked that for a 12-month salary, they paid so poorly. I asked if it could be increased. He told me not now. I said I'd let him know. I did the math and concluded that in comparison to my salary at Illinois, it was same as my UIUC salary plus the 2/9 summer supplement I received from grants and contracts.

I discussed it with John Hower and explaining I wanted to stay at Illinois but had certain conditions. It included an appointment during the 1982-83 academic year at the Center for Advance Studies to be combined with a sabbatical. Hower suggested I take the VIMS job, but offered to discuss it with the advisory committee which now consisted of Domenico, Blake, and Hsui. They told Hower to give me what I wanted and he did. I turned VIMS down.

A week later, I received a call from the "Virginia Pilot", the Norfolk, VA, newspaper inquiring why I rejected the offer. They planned a story about the search for the new VIMS director because everyone who was offered the job after me also turned it down. I explained that "Illinois matched the offer." They kept asking more questions and I basically answered them. Finally I said, "VIMS is a great opportunity, but Illinois matched the offer." They sent me the article later and it reflected my statements accurately. They parsed my quote to: "VIMS is a great opportunity for someone else."

Dennis Wood returned to the UK accepting a job at Robertson's Research in Wales. He was disenchanted with Hower, seeing no future in staying. Robertson Research agreed to sponsor his British Isles Field course to North Americans. It was a serious loss for Illinois, but Hower basically pushed him out.

Shortly afterwards, Dennis Eberl resigned to work for the USGS.

John Holder also resigned to go to Texas A&M. Within a period of four months, the department lost three faculty members.

However, the year ended on a positive note. NSF funded my research to integrate all my back arc basin work. I was able to fund a Research Assistantship for Yong Il Lee to complete his PhD on sandstone diagenesis of back arc basin sands.

Meanwhile, Ellen Abel quit and was replaced by Pat Lane, who in my view was a yahoo. She was fascinated by mice and had many mice dolls in her office. Whenever she left for vacation, she put a four-foot mouse doll in her chair. She also asked Hower to hire one of her friends, Caroline Roberts, a totally incompetent typist. My general impression was that these two ladies realized long ago that their high school senior prom was the high point of their life.

One day while Pat Lane was on vacation, I brought my camera to photograph her large mouse doll. Caroline Roberts saw the camera flash, and ran out from her adjacent office. She asked, "What are you doing?" I responded, "Mrs. Roberts, you just wouldn't understand." When Pat Lane returned and heard about my picture taking, she took the large mouse doll home and left it there.

A strange incident occurred that summer. One evening, I read the local newspaper ('News-Gazette'). An article reported that earlier that afternoon John Hower was driving his Audi at a speed faster than 65 MPH on an exit ramp from one of the freeways passing through Champaign-Urbana. He lost control, the car flew though the air for 100 feet, rolled, and John walked away unscathed. He spent the night at the hospital for observation.

I assumed his accident was as reported. However, Murle Edwards had a more prophetic view. She asked if I thought it was a suicide attempt and I told her, "No." She was convinced it was.

Late in July, I ate lunch with Pat Domenico and Vic Palciauskas. Both were upset about Hower, but they never fully disclosed their concerns. Both thought Hower was unethical, couldn't be trusted and seemed to want to drive out the productive people in the department. Pat was the author of an extremely successful geohydrology text book. Pat funded his research through contract and consulting work and this did not satisfy Hower who wanted everyone to apply for NSF funding only.

That fall, I applied for funding for my sabbatical the following year (1982-83) and appointment in the Center for Advanced Study. It was approved. I also was awarded a Japan Society for the Promotion of Science Fellowship to spend three months at the University of Tokyo Ocean Research Center. Nori Nasu invited me and arranged the fellowship. Basically, my funding was a semester at full pay, a Center for Advanced Study Associateship at full pay, and a Japan Society for the Promotion of Science Fellowship to cover expenses in Japan.

The academic year moved forward without incident. Dave Anderson taught structural geology while a search for Wood's replacement was underway. Hower now had clay mineralogy all to himself so he temporarily froze Eberl's budget line to keep the other two. Hower also agreed with the administration to close the next three faculty budget lines that were vacated due to retirement, death, or departure in exchange for three new lines immediately. He believed he could ultimately get the university to reverse the agreement when financial prospects improved. It was a gamble that cost the department long-term.

During the GSA meeting that fall, I spoke with Dick Mitterer (BS, F&M; PhD, Florida State; Organic Geochemistry; Post-doc, Geophysical Lab, Univ. of Texas at Dallas) the head of the department of geology at the University of Texas at Dallas (UTD). Ken Eriksson went there from South Africa and left the previous spring for VPI. They were looking to replace him in another year. I asked if they would like me to come during winter break to teach my sandstone depositional systems short course. Within two weeks, it was approved.

In early January, 1992, I drove to Dallas. I went to the department and Dick introduced me around. The faculty had about ten people. All were cordial and I was impressed with the place. Because so many of their students worked in the oil industry, classes were taught at night. That meant during the day, research went on uninterruptedly. I liked the arrangement.

While in Dallas, I visited GSI (Geophysical Services, Inc) an arm of Texas Instruments. I met Alistair Brown and invited him to give a one-hour lecture as part of my short course dealing with 3-D Seismic images of the Gulf of Thailand and elsewhere. His talk, along with my course was very well received.

I also visited the Mobil Research Lab to meet Dick Moiola and Shan Shanmugam with whom I developed a long-term friendship.

UTD was attractive because it only offered upper division undergraduate courses and graduate degrees. They were an upper division university into which the community colleges in the Dallas area sent their graduates. Consequently, the campus was more serious and did not suffer from the immature undergraduate folderol of most campuses.

I asked Mitterer if they could possible consider replacing Ken Eriksson with me. The faculty liked the idea. Dick raised it with their Dean of Science and the three of us met for lunch. It went well. Could they fund the appointment? In the end, they couldn't. It was a great disappointment because I thought UTD and I were a good match and so did they.

During the spring of 1982, Pat Domenico was offered an endowed chair at Texas A&M University and debated whether to stay or go. He told me during lunch that he discussed it with Hower who gave him no encouragement to stay. He was reluctant to move to Texas, having made many friends in Champaign-Urbana through his golfing connections. As he stalled Texas A&M, they interpreted it to mean he wanted more money and finally, they gave him an offer that was too good to turn down.

Pat told me he was going to fix Hower once and for all with the administration. Academic tradition, sanctioned by the American Association of University Professors, required a tenured professor to resign before May 15 for a job starting the following fall semester. Pat sent a letter of resignation to the Vice Chancellor of Academic Affairs on May 12, 1982, and wrote a sharp critique of Hower, explaining it was a major factor in his decision. In the process, Pat exposed Hower to the administration.

The administration forwarded the letter to Hower for his comments and Hower accepted the resignation, but knew Pat undermined him. Pat's departure was a serious loss for Illinois because his geohydrology program was one of the signature programs in the USA, if not the world.

During the summer of 1982, I returned to Korea to negotiate additional research funding to work in the Yellow Sea. The country was clearly changing. During 1980, I thought that Korea was more like the post–war Korea, slowly recovering economically from the Korean War. In 1982, it appeared like a new Korea that was being transformed through the economic policies of former president Chung Hee Park. In just two years, the transformation was stunning and people clearly looked better off.

On my way to Korea, I visited San Francisco. I contacted Dick Hay and had dinner with him. I was aware that Hay was on a list of possible candidates for a name chair donated by Ralph E. Grim, the pioneering clay mineralogist. Grim remarried a wealthy widow and it was her money that funded the Grim Chair. However, I did not know who the search committee was interviewing and during dinner with Dick and Lynn Hay, I said nothing. However, Dick was contacted to meet in San Francisco with Hower and Dean Prokasy the following week. I was completely unaware of that meeting.

The Hays flew to Urbana while I was in Korea after meeting Prokasy and Hower. When I returned, I called Dick at his home and Lynn answered and proceeded to give me an earful for failing to disclose why Hower and the Dean wanted to meet Dick. I made it very clear I was unaware of the trip but knew Dick was under consideration. I finally spoke to Dick and convinced him.

Dick disclosed that the Dean and Hower did not present themselves well and he had reservations about Illinois. I therefore called Ralph Grim and asked what he thought about Hay being the first recipient of the Grim Chair. Ralph told me he wanted Hay to come. I said, "Look Ralph, if you really want Dick here, you need to keep after Hower and the Dean, and you personally need to call Dick and let him know how much you would really like to see him accept the Grim Chair."

During August, I attended the IAS in Hamilton, Ontario, and took Yong Il Lee with me. Kathie Marsaglia, who had completed a project on storm deposition with me, was also there and both gave papers.

The most memorable event was the banquet. I sat at a table with Peter Timofev, a Russian member of a DSDP Panel I chaired, a colleague of his named Krylov, the political officer, an Egyptian geologist who knew them, and some other people. Krylov looked like a former defensive end with the

Green Bay Packers.

When dinner ended, I asked where Peter and Krylov stayed and it was the same motel where I stayed. I offered them a ride in my car, a 1980 Buick Skylark, 4 door sedan, with cloth seats and a sun roof. As I drove away, Timofev said, "This is a very nice car. George, you must be an important member of the Republican party." I replied, "Peter, this is an ordinary American car. Anyone can buy one." Then he asked, "George, do you know President Reagan personally?" I told him I didn't but I knew Peter wouldn't believe me. I realized that by implication, cars for those who could afford them in the USSR would be of much less quality. I'm sure this information was added to my KGB file (See Chapter 21).

In September, Dick and Lynn Hay made another visit to Urbana. Dick and I reviewed my teaching offerings so that we would avoid duplication. Dick was more of a sedimentary mineralogist, whereas I focused more on physical sedimentology and environments. Before he left, he accepted the Grim Chair starting in the Fall of 1983.

Hower announced at the first fall geology faculty meeting he was concluding his headship on July 21, 1983. By then, his five year term would end. John decided to return to teaching and research. After the Second Donath War, the University put in place a review system for all administrators (Chapter 20, Postscript #4). If an existing department head, dean, vice chancellor, provost or president, wanted to be renewed after serving a five year term, they had to undergo a faculty and administrative review. Hower chose to forego the review knowing he would not survive.

I spent the fall running my Ouachita-Arkoma basin field trip for Indiana University, going to La Jolla to examine new cores to complete my integrative paper on back arc basins, and spent the rest of the time writing up my results. I also considered the search for the headship and decided to apply.

The search committee consisted of Douglas Applequist of Chemistry as chairman, Phil Sandberg, Albert Hsui and Tom Anderson. In January, 1983, I met them. They asked what I proposed to do if appointed. I stressed the need for strategic planning and the need for a new building. I explained how Percy Allen raised money to get one at the University of Reading. I said, "I propose we raise the money ourselves. I believe I can raise money from oil companies and once such funding is committed, the geology alumni will come on board." Applequist said, "That's never been done at this university." I replied, "That doesn't mean the administration won't consider it, and it would have to be a condition of appointment."

The only other candidate I knew who had applied was Kirkpatrick who was just promoted to a full professorship.

A surprise candidate was appointed. It was Dave Anderson. By now, it was early spring, 1983, so I talked with Dave informally about his goals. He stressed increasing enrollments in freshman geology and not much else.

Vic Palciauskas resigned at the end of December to go to Chevron Research in La Habra, CA.

In mid March, Bill Kanes (BS, CCNY, PhD. West Virginia; Exxon Production Research, Exxon Exploration in Libya; Univ. of West Virginia, Univ. of South Carolina, Univ. of Utah) contacted me. He was director of an International Geological Research group at the University of South Carolina. He conducted projects in North Africa and expanded globally.

Bill invited me to present a three-hour seminar on tidalites to his group. He asked if I would join them and head up a program in Korea. He explained that US Ambassador to Korea, Walker, previously headed an International Relations Institute at South Carolina. Bill arranged for me to meet him when visiting Korea after my sabbatical in Japan.

Perhaps mistakenly, I spoke to Hower and asked that if it were to happen, would Illinois grant a release from my obligation to return for a year after completing a sabbatical. He said he'd let me know.

Two days later, he went to Washington for a national panel meeting and stayed at the Cosmos Club. Hower wrote me a handwritten seven page letter on Cosmos Club stationary with a copy to Dean Prokasy. In his diatribe, he personally attacked my character, my selfishness, my social life, and other things long forgotten. He also claimed to have supported my efforts while head and stated that my request about a release was an act of ingratitude.

I was totally stunned when I read the letter. However, Dean Prokasy received his copy at the same time and immediately called me. He asked that I ignore that letter and not discuss it with John. Prokasy emphasized that John's letter did not represent his or the administration's view. I thanked him and promised to keep my mouth shut. I also realized that the University of Illinois' administration had a long memory and did not want a replay of the Second Donath War in a new context.

We finally filled the structural geology position offering it to Stephen Marshak. On the morning of the day I left for Japan, I interviewed a candidate for the geophysics slot vacated by Holder. He was Jay Bass (BS Brooklyn College, PhD, SUNY-Stony Brook; mineral physics; Post-Doc, Cal Tech; Illinois). Jay was extremely bright and I thought would be a great addition to the department. The faculty voted to hire him.

Nori Nasu met me at Narita airport on a Friday evening. Nori's institute driver took me to an apartment in a building for international sabbatical itinerants at the University of Tokyo. It was in a neighborhood of foreign embassies.

Over the years, Nori housed many ORI affiliates there and acquired a complete set of dishes, cooking equipment, coffee maker, bed sheets, towels, and other household items. When Nori took me to the apartment, his secretary had already unpacked everything the day before. He took me to a nearby bank the next day to help open an account from which to draw my fellowship funds and gave me samples of forms filled out for deposits and withdrawals. Nori knew how to facilitate visitors.

I inquired about Nori's history. He grew up in Kyushu where his father was professor of aeronautical engineering at Kyushu University. During the mid-1930's Nori's family moved to Tokyo because his dad was appointed minister of aviation overseeing everything from design, construction and testing aircraft. Nori, as the son of an important government official, went to the best prep schools and met the children of his father's counterparts. Over time and after the war, they entered various sectors of the Japanese government and industry so he had a network at the highest level in every ministry in the government of Japan. The only place he didn't seem to have contacts was the Imperial Palace.

Monday, I met Nori and he gave me an office and a tour of the building. During the next four days, we lunched together in a different neighborhood restaurant, a tempura place, a soba noodle place, a curry place, and a tonkatsu (breaded deep fried pork chop) place. He explained that from then on, I was on my own for lunch. Food near ORI was inexpensive because it was located in the wholesale food district. It was also an hour away from the main campus of the University of Tokyo.

I discovered quickly that in Japan, except for hotels, the all purpose Japanese restaurant that exists in the US is nonexistent. Restaurants offer a limited cuisine selection only: tempura, sushi, soba (noodles), tonkatsu, beef dishes and so forth. When going out, you needed to know the meal choice beforehand before choosing a restaurant.

I started my research quickly, scouring the ORI library for reference information. Art Bloom from Yale days networked me to Dr. Yonekura in the department of geography at Tokyo University. Yonekura made age determinations of all the terraces around Japan and constructed a map of timing and rate of tectonic uplift of the Japanese Islands. He gave me all his reprints, and a key report published by the Disaster Research Institute of Japan which supported his research. It had extensive maps and data. It was an extremely unique data set.

One of the main controversies at the time was the role of sea level change and the accumulation of submarine fans. If sea level was low, submarine fan evolution was favored.

Stratigraphic models were constructed based on this paradigm by the Exxon Research Group of Peter Vail, Bob Mitchum and John Sangree. Using ocean seismic data, they were able to establish a global sea level chart. Two friends, Dick Moiola and Shan Shanmugam plotted all well-dated major deep water turbidite/fans on the Exxon sea level curve and fan deposition coincided with low stands of sea level. Their data was good, but nearly all their examples came from passive margins. The question that needed testing was when do submarine fans accumulate in active margins and at what accumulation rate? In particular, when did they accumulate adjacent to areas of rapid uplift like the Japanese Island chain?

I knew the answer existed in two DSDP sites on the Toyama Submarine Fan drilled during Leg 31 in the Sea of Japan. The ORI library had the Leg 31 DSDP volume which included a complete set of core photographs with complete core descriptions. I determined the number and thickness of individual turbidites and establish biostratigraphic baselines tied to the numerical geological timescale. I discovered a direct correlation in both the number of turbidites and shorter recurrence intervals of turbidite deposition with increasing uplift rates. These were Pleistocene turbidites, a time of high frequency, glacially-driven, sea level change. The sediment flux clearly was controlled by uplift rate and masking glacially-driven sea level change. I wrote two papers which appeared later in *Geology* and *Tectonophysics*.

I visited Kyoto, Shikoku Island, Hiroshima, Nagasaki, Nara and Yamagata during my visit. Tsune Saito was now the department head of the geology department at Yamagata. Hisatake Okada was still there, so we had a good visit.

After three months, I went for a two week trip to Korea and then returned to Urbana in the middle of August. When I entered the NHB to clear my mail box, I passed the door of the department head's office and the name on the door read 'David E. Anderson, Head.'

The John Hower era officially ended.

LESSONS LEARNED:

1) When screening people who present themselves in such a way that one senses deep problems that may by psychological, always disclose one's concerns and the basis for them to a search committee and decision makers.

2) When interviewing people for administrative positions, never hire or recommend people who lack vision or are unwilling to disclose their vision.

3) When interviewing, be careful about asking questions that are intended to be humorous and designed to eliminate one from consideration. It may work in reverse (as it did at VIMS).

CHAPTER 23
Illinois: The David Anderson Era (1983-1988)

"Never get complacent - that's when things you do go wrong"
- Ed Scharlau

After spending two days reviewing incoming mail and memos, I visited Dave Anderson and brought my CV and publications list with me (I had done the same with John Hower). I told Dave that although we were colleagues for 13 years, there was probably a lot he didn't know about me so I thought he should see my CV. He accepted it graciously.

We talked about the coming year and Dave was upbeat. Dick Hay arrived as had Steve Marshak. Jay Bass was coming the following year after another year on his post-doc at Cal Tech. Dave said he was working with the dean to replace both Domenico and Palciauskas. He reminded me Graf was never replaced and he needed approval to hire one.

Dave spent the summer taking the Trans-Siberian Railway across Russia and connecting to the PRC. He talked extensively about the trip and later presented a slide show for the department. The contrast between dour, poor Russia and bustling, well-fed China was more striking than I expected.

I taught my usual fall offerings, Geology 309 and my Depositional System Course (Geology 477). I arranged field trip support again by selling ringside seats to Getty Oil Company who sent two people.

When I gave my mid-term exam in Geology 309, I was appalled at the decline in student writing skills. I discussed it with Dave Anderson. He explained that most of the same students enrolled during the previous spring in mineralogy. Half of the students took their first essay exam in his class. It seemed that every seven years, the quality of writing, verbal, math skills, academic and social skills of Illinois undergraduates got worse.

Dick Hay arrived at the beginning of the semester and threw a party at his home for the geology faculty. However around the department, Dick was subdued. I discovered he gave Hower a list of things he needed done before arriving and Hower did nothing. Dave was sorting it out. I noticed also that when the weather was good, Dick left at 1:00 PM for the Champaign Country Club to play golf. In fact, during the next ten years, I only held one scientific conversation with Dick.

A disturbing incident occurred during mid-September. Cameron Begg, the department's Scottish microprobe technician knocked on my office door and said, "George we've found the body." I asked, "What?" He explained they found John Hower's body. He committed suicide.

The previous night after bedtime, John's wife, Joanne, noticed John was missing and assumed that he took his regular nighttime walk. When he didn't return for breakfast she called his office. No one answered. Around 8:00 am, she called Dave Anderson and he instigated a discrete search. John took his life in the clay mineral prep room. It was September 18, 1983. A Memorial service was scheduled for Saturday at one of the campus 500-seat auditoria.

Immediately, calls were made outside the department to people who were his friends. It turned out I was the only one who knew his Pan Am contacts so I phoned them. I talked to A. F. Frederickson for the first time since leaving Pitt and gave him the news. He thanked me and then said, "George I followed your career since you left Pitt. You really did well."

I also called Frank Stehli (BA. St. Lawrence, PhD. Columbia, Cal Tech, Pan AM Research, Case Western Reserve (Chairman), University of Oklahoma (Dean of Earth Science)) but was told he was in Washington. I left a detailed message with his secretary at OU and she relayed it to Frank's wife, a geologist, who called me. Frank and Hower worked together at Pan Am's Research lab. Frank later hired Hower to join the faculty at Case Western Reserve.

Frank's wife literally talked non-stop for 45 minutes and told me about John's history of depression and other things I didn't know but fit my own observations. She called two hours later after speaking to Frank. They would attend the memorial service and I gave directions. She talked for another 45 minutes. I was disturbed because Frank wrote a letter of recommendation for Hower and disclosed none of these problems.

Dennis Eberl arrived on the Friday before the memorial service. He said, "George none of this surprised me. Back when doing field work in Montana, we used to sit around the camp fire and talk about life. John used to say suicide was a good way to go." Eberl never disclosed this during the search process.

Dennis also disclosed that during his last years at Case Western, John had a woman student, Janet Hoffman. She was somewhat unstable and leaned heavily on John as a father figure. Eventually, they lived together. She finished her PhD and accepted a job in Reston, VA with the U.S.G.S. Shortly after John went to NSF, Janet committed suicide on September 18, 1977 when John refused to divorce Joanne and marry her. John took his life on the sixth anniversary date of Janet Hoffman's suicide.

Jim Kirkpatrick gave a masterful eulogy at the memorial service. About 500 people attended including the Dean of the College of Liberal Arts and Sciences and the Vice Chancellor for Academic Affairs.

Dave called a faculty meeting and asked us to pull together which we did. The administration was concerned because often when events like this unfold in an academic department, it led to chaos. With us it didn't.

Murle Edwards saw me the Monday after the memorial service. She asked if I remembered the auto accident incident two summers before. I told her I did. She asked, "Well, what do you think now?" I told her that I didn't know what to think but suggested she was on to something because I heard many disturbing things from his long-term associates about Hower that go back a long time.

Two new graduate students were admitted to work with me. One was Bruce Phillips (BS, Murray State, MS, Illinois; Martin Marietta Environmental). He came with an outstanding record, except one semester of physics which he took at Illinois. He completed his MS in two years.

The other was David T. Heidlauf (BS, Wheaton College, MS. Illinois, Environ) who applied to work with Carozzi. However, Carozzi, after meeting Dave, refused to let Dave work with him. Dave visited me and although I had reservations about his graduation from a liberal arts college, he had no deficiencies on his transcript and agreed to give him a try. Dave was also an officer in the Army reserve and dressed and groomed accordingly.

When he turned in his draft term paper for the fall semester, he started with a paragraph that was a religious statement. I told him that I respected has rights to his devout religious views, but it was in his own interest not to mix them with either his scientific work, or any duties he assumed as a TA, or a job in the future. I advised he keep his views on religion private. He worked with Albert Hsui and me on Illinois basin subsidence history for his

Master's thesis and it was published in the *Journal of Geology*.

Both wanted jobs in the oil industry but when they finished, none were available, so they made their careers in the environmental field.

At the GSA in 1984, I was approached by Bob Raymond (PhD, UC-Santa Cruz; Sedimentology; Los Alamos) to sign a petition to form a GSA Division of Sedimentary Geology. I agreed and offered to help if needed. In March, 1985, Raymond called again saying they had the requisite signatures but to get the division approved by GSA Council a Chairperson had to be designated. He asked if I would I take it on. I agreed and discovered I would serve through the 1987 meeting, including a year as past-chairperson. GSA Council approved the division. Within a year, we had 1,500 members and it was GSA's fourth largest division in 1986 and 1987.

In 1985, I had my left hip replaced. That fall, I admitted another new student, Dave Watso (BS, San Francisco State, MS. Illinois, Shell, Subsurface Consultants, Unocal, independent consultant, back to Shell). He did a thesis on Cambrian transgressive/regressive cycles and correlated unconformity development with episodes of combined rapid mechanical uplift and subsidence in the Illinois basin. His thesis was published in *Geology*.

While rehabbing from surgery, Albert Carozzi finished a book manuscript I invited him to publish with IHRDC for whom I was series editor of geological reference books. It dealt with Carbonate petrology and microfacies, a field in which he was recognized as a world class leader. I completed the review at home and when finished, I invited him to visit and went over it. I made some heavy duty changes and suggestions and recommended he add a chapter synthesizing the field and tying it to major themes in geology. I completed this three weeks after rehab started.

Apparently my review impressed him. Dave Anderson told me Albert told him that he couldn't believe I was so alert. I said to Dave, "Look, they operated on my hip, not my brains."

Yong Il Lee finished in 1985. He was provisionally promised a faculty appointment at SNU if he could get a post-doctoral fellowship in the USA or Europe. I went to AAPG in 1984 in San Antonio and joined a conversation between Paul Enos (now at KU) and Gerry Friedman. One of Paul's PhD's had just turned down a post-doc with Gerry. I asked if it was still available. Gerry said, "Yes." I recommended Yong Il Lee who was offered one and spent a year with Gerry at RPI. From there he made an outstanding career at SNU participating in various IGCP projects in Asia and became recognized as Korea's most outstanding academic sedimentologist by the International community.

Geology enrollments dropped with the decline of the oil industry in the mid-1980. By 1985, we hired Craig Bethke (BS, Dartmouth, PhD, Illinois; basin geohydrology; Illinois) to replace Domenico. Craig's father was Phil Bethke at the USGS and during high school, Craig worked with Phil as an assistant learning geology. When reviewing his CV in a faculty meeting, my comment was that we were hiring "an experienced, inexperienced geologist."

We also hired Randy Cygan (PhD. Penn State; geochemistry; Illinois, Sandia) who stayed 18 months and left for Sandia. Stephen Altaner (BS, Colgate, MS, PhD, Illinois, clay mineralogy; Illinois) was hired to replace Hower. I suggested that as long as Ralph Grim was alive, Steve invite him to give one lecture every year to his graduate class. However, Steve's wife, Myrna, died tragically in 1990, and Steve never recovered.

During 1985, Heike Ignatius was scheduled as a colloquium speaker. His host was Don Henderson. I visited Don, explained Heike and I knew each other via John Sanders and asked if I could visit with him. Henderson was gruff and told me there was no room in Heike's schedule. I suggested breakfast and Henderson agreed. I was to meet Heike at 7:00 am at his hotel which had a dining room.

I arrived on time but Heike was nowhere to be found. I finally called his room and he was unaware of our meeting. In fact, he was still asleep. He dressed, packed and we had breakfast. To say he and I both were embarrassed was an understatement.

That afternoon, after Heike left, I ran into Henderson in the Hall. By this time, 15 years of frustration had accumulated. I verbally dressed him down for his pettiness and unprofessionalism.

We were outside Dave Anderson's office. Dave came out to stop the argument and I followed Dave into his office. I apologized for the outburst but told him I just had enough of Henderson's deviousness. I also explained that Henderson's failure to notify Ignatius about breakfast likely left a poor perception of the department with the director of an overseas geological survey. Dave told me he would talk to Henderson. I wished him luck.

During December, 1985, Dave Anderson requested my help. Geology

108 (Historical Geology for Science Majors) was taught by John Mann. Dave explained that student evaluations were near rock bottom and the TA's confirmed it. He wanted to relieve Mann and have me teach the course.

I said that I could do it, but wanted to teach it in a different way. I explained that in the past, after a month of background material on evolution, fossil succession, depositional environments, and stratigraphic principals, the course would begin with Precambrian geology and end with Cenozoic geology. I wanted to reverse it starting with the Pleistocene where we knew more, and end with the Precambrian, where we knew less. In short, the approach was working from the known to the less known. Dave approved the plan. On the advice of a senior paleontologist elsewhere, I concluded the course with a three-day traditional overview from the Precambrian to the present to stress the role of precursor and inherited events. It worked.

It also meant that I was to offer Geology 437 (now changed to: Basin Analysis and Sedimentary Geology) and Geology 477 (Depositional Systems) in alternate years. Given graduate enrollment trends, this was a wise move.

During January, 1986, I admitted a Taiwanese student, Wang, and a student from the PRC, Yujin Liu. They took courses in the department and English as a Second Language (ESL) but could not take a course from me until the fall term. I suggested they complete their ESL language requirement by taking two required courses concurrently during the summer session.

The lady ESL coordinator called and explained they couldn't take those courses concurrently; they had to take them sequentially because one was a prerequisite for the other course. I recommended they be given a chance to take the courses concurrently, that in science departments, it was common to take courses currently, even if one was a prerequisite for the other. She wouldn't budge.

I then said something and she went ballistic claiming I was sexist, rude, and shouting at her, none of which was true. She obviously wanted to use the feminist/liberal code words to back me off. She reported the conversation to her department head who filed a letter of complaint with Dean Prokasy, including questioning my ability to supervise graduate students. I wrote back explaining the problem, practices in scientific programs, and then listed all my PhD's and external PhD's, where they taught, books they published, Journals they edited and university departments they were heading.

The ESL department head wanted a letter of apology. Dave Anderson suggested I write one and I said, I wanted to talk to the Dean first and Dave could join us.

I met alone with Prokasy. I said I did not appreciate it that ESL took advantage of foreign students by prolonging their programs. I suggested that the students' academic freedom may have been violated. I reminded Prokasy that two weeks before our meeting, Illinois awarded an honorary degree to an Egyptian Civil Engineering PhD graduate from Illinois who was now Egypt's foreign minister. "You may remember," I said, "that during the Iranian hostage crisis, the Iranian foreign minister was anti-American because he was poorly treated by Georgetown University which flunked him. I don't want to watch TV news in 20 years and see either Wang of Liu spewing anti-American slogans because the ESL program horsed them around."

Prokasy listened and said in general he agreed, he thought my supervisory record with PhD's was exemplary and in fact said, "Both John Hower and Dave Anderson told me you are the best geologist in the department." He offered two choices: either forward the complaint to the Chancellor, or write a letter of apology. I said, "Look, I don't have time to prolong this. I give a colloquium at Wisconsin (Prokasy's alma mater) in three days and from there attend a conference dealing with continental drilling. I'll send you a letter of apology to forward to the ESL program."

I wrote the letter basically apologizing for the misunderstanding and regretted it occurred. But I added that I also thought the students' academic freedom was violated. On my return, Prokasy told me that the ESL people were unhappy with my letter but he informed them I met their request and the issue was closed. In the end, neither Wang nor Liu met my standards. Wang earned a Masters with Marshak, and Liu went to Auburn University.

After hip replacement surgery, I reassessed my career. I knew I was sufficiently disabled that field work would be limited to road cuts and quarries. I decided to move into sedimentary basin analysis and calibrate subsidence models from stratigraphic and sedimentological data. I reviewed this with Hsui and he concurred this would be new, significant and different. We agreed to work together. I was invited by the editorial board of *Sedimentary Geology* to write an article on sedimentary basin analysis and did, including a new basin classification. By 1987, that paper and Heidlauf's thesis was published.

The University of Minnesota held a basin conference in 1986 and I gave a paper. In retrospect, it could have gone better. The organizers were more into code words than science. I met Sierd Cloetingh (BS, PhD Utrecht; basin modeling; Utrecht, Vrije University of Amsterdam) who proposed some joint work. He visited me in 1987 at Illinois.

I also attended the 1987 AAPG in Los Angeles and presented two papers, one on basin classification, the other on the origin of cratonic basins. Dave Anderson and Ralph Grim attended both talks. Dave clearly was wondering if my career change was up to standard. Kevin Burke (PhD, London, tectonics; Univ. of Ibadan, SUNY-Albany, Univ. of Houston, NASA-Lunar and Planetary Institute Director; NAS, Univ. of Houston) stood up after the cratonic basin talk and said he liked the model and mentioned places in Africa where it fit. It was just what I needed Dave Anderson to hear.

One of my former classroom students, Nahum Schneidermann (BS, MS, Hebrew University, PhD, Illinois, micropaleontology; University of Puerto Rico, Gulf Oil; Chevron), organized an AAPG conference on basin modeling in Vienna, VA at the IBM conference center. He invited me to present both my basin classification and cratonic basin talks. Both went well and I got great feedback, so I knew I was on the right track.

One of the featured speakers was Victor Khaim (BS, PhD, Univ. of Azerbaijan, tectonics; Moscow State University), an eminent Russian geoscientist who was the first Soviet scientist to accept plate tectonics. He arrived late and one of the organizers, Greg Ulmishek (PhD, Moscow; regional geology; career in Russia, USGS) asked if I could repeat both presentations privately for Khaim. I did. Greg had given a great paper too and I asked if he could send me a duplicate set of his slides. He gave me the set he used for his talk and I used them in my courses at Illinois and a new short course I introduced.

During 1987, I also gave a one-day basin course at the annual GSA meeting. One person told me it was the most intensive short course he had ever taken. People attending that course included Dag Nummedal and Bill Berry.

I decided to market the short course to universities and presented it at Rutgers University and Case Western University.

It was around this time that we hired Steve Grand (BS, McGill, PhD, Cal

Tech; seismology; Illinois, Univ. of Texas at Austin) to replace Palciauskas. Steve's expertise was seismic tomography. He demonstrated the lateral extent of the Pacific Plate under North America and how the wide extent of that plate influenced regional tectonic events in western North America. His findings fit a regional model of 'far-field tectonics' by Peter Ziegler which was developed in the North Sea. Steve, however, stayed only for 18 months to accept an offer from the University of Texas at Austin. The curse of Vic Palciauskas struck again.

During March, 1987, one of Ralph Langenheim's PhD students, Pius Weibel, presented his public thesis defense lecture and I attended, although I was not on his committee. Pius completed detailed paleontology on Pennsylvanian-age rocks in East-Central Illinois. I heard enough paleontology thesis defenses so expected that either Pius would give an arcane talk about brachiopods that few would understand, or tie his work to regional geology or a general topic. He chose the second approach and talked about the origin of Pennsylvanian cyclothems.

I listened hard to his talk because 30 years before, I did a similar presentation on cyclothems in Dunbar and Sander's stratigraphy course at Yale (Chapter 7). While listening, everything I wrote then came back to me quickly. Except for citing Phil Heckel's 1986 paper on the control of Milankovic orbital parameters on cyclothem deposition, Weibel's talk was no different from mine of 30 years before and the conclusion was exactly the same: The problem remained open.

When Pius finished, I asked many questions in what Phil Sandberg described as my "tough love" style. Pius had difficulty answering. Langenheim was shocked I knew that much about the topic. I closed by saying, "Pius, your summary about the origin of cyclothems is troubling and I'll tell you why. During the last 30 years, we've seen advances in depositional systems, process sedimentology, sequence and seismic stratigraphy, plate tectonics, basin subsidence modeling, and carbonate geochemistry for openers. From what you told us, none of these major advances were used to analyze the cyclothem problem." He acknowledged that was true.

Ralph Langenheim then closed the meeting. I spoke with Pius and asked, "Are you free tomorrow morning around 10:00 am?" He replied, "Yes." I said, "Bring your maps, cross-sections and thesis reference list to the sedimentology lab and we'll go over it." He agreed.

Pius and I met and it was a long discussion. I showed him my Yale term paper. We reviewed everything he had and clearly he covered all bases. But I was puzzled that the topic of Pennsylvanian cyclothems was never evaluated using the newer findings of depositional systems, process sedimentology, sequence and seismic stratigraphy, plate tectonics, basin modeling and carbonate petrology. I realized that this could prove to be a fruitful research topic.

The previous year, I taught a basin analysis course requiring every student to complete a project on a sedimentary basin. One student in the class, a Paleobotany PhD candidate, Debra Willard (BS, Stevens College, MS, PhD, Illinois; paleobotany; USGS) completed a paper on the Central Appalachian basin and I encouraged her to expand it for publication. In her draft, she included a section on cyclothems and although I can't remember what it was, it caught my eye. I suggested she and I re-evaluate the problem and publish.

We reviewed the well-known facies variation of Pennsylvanian cyclothems and observed that the regional trend changed with distance west of the Appalachia-Hercynian orogenic belt. Because we had her tectonic subsidence data from the Central Appalachian basin, Heidlauf's from the Illinois basin, and one by another student from the Forest City Basin, we observed the subsidence rate and magnitude diminished to the west. Cyclothem facies style also changed in the same direction.

We also checked Pennsylvanian paleogeographic maps and preliminary paleoclimate maps, and they provided an interesting pattern. Our findings could by tied to far-field tectonic effects in response to the Hercynian / Alleghanian tectonic belt.

Past arguments about cyclothems revolved around two hypotheses when Debra and I did our research. They were either tectonic or climatic in origin. Our analysis showed it was both, with simultaneous tectonic dominance and climate subordinate in the Appalachian cyclothems and simultaneous climatic dominance and tectonic subordinate influences in the Kansas cyclothems. The Illinois cyclothems appeared to be controlled by nearly equal magnitudes of tectonic and climatic processes. We published the paper in 1989 in *Geology* and it received an almost instantaneous positive response from the geological community. It was also misinterpreted by many.

I wondered about the misinterpretation and realized it arose because of a fundamental flaw in how geologists approach problems. As a student, I was brought up to test **multiple working hypotheses**. The one which fitted facts and observations best was selected as the unique solution.

The Klein-Willard approach to cyclothems used a different paradigm, namely that **geology involves multiple processes which occur simultaneously but function at variable rates and magnitudes. Thus all of the multiple processes (hypotheses) influence a given geological event to a variable degree.** This analysis was misunderstood and the Klein-Willard paper was often misinterpreted to favor tectonic controls only. It didn't.

The next step was quantification. I discussed this with Sierd Cloetingh and we agreed to develop a joint research effort. The plan was that the University of Utrecht would pay me a partial salary to do a pilot study there during the summer of 1988, and I would return for a sabbatical leave in 1989 to complete the project.

In September, I received a phone call from Sierd. He explained he was a candidate for a chaired-professorship in tectonics at the Vrije University of Amsterdam and requested I write a letter of recommendation. I asked for his CV and publication list. My letter, I found out later, helped him get the job. He would start in September, 1988.

In December, 1997, Dave Anderson informed the Dean that he wanted to be renewed for a second five-year term as department head. The faculty elected a committee to evaluate Dave's performance. Steve Marshak, Jim Kirkpatrick and I were selected to undertake the review.

In general, I was satisfied with Dave's performance and concluded he was fair and evenhanded. He served more sanely than any department head or chairman I ever worked for. However, the younger faculty hired since 1981 (Marshak, Bass, Wang-Ping Chen, Chu-Yung Chen, Altaner, Grand, and Cygan) saw it differently. They stuck together and interacted rarely with the older faculty. They did interact more with Jim Kirkpatrick who was closer in age.

The committee was chaired by a Math professor whose name I've forgotten. First, we sent letters to the faculty asking them to write us UNSIGNED letters with their evaluation. We then interviewed the faculty individually.

The results of both the letters and interviews showed a generational split. The older faculty favored Anderson strongly, whereas the younger faculty opposed reappointment.

Kirkpatrick explained he was aware of the younger faculty's concerns. These were (1) Dave appeared to have reneged on commitments made to them for space, startup money and other funds and needs, (2) Grant applications and invoices for ordering new equipment and supplies were delayed by Murle Edwards, (3) the clerical staff was too slow to type their proposals and in several instances, led to a submission of NSF funding requests past deadlines, resulting in cascading delays on notification and approval (usually six months).

We evaluated these allegations. On the reneging, we had no proof. On funding, several of us also experienced problems with Murle. On typing delays, the younger faculty never allowed adequate lead time to get proposals typed and approved. In my case, I allowed at least two weeks and never experienced problems.

We were faced with a tough choice. We decided to recommend that Dave be renewed for five more years, but added conditions and changes he would be obligated to make. Dave decided he couldn't function with those conditions and withdrew, opening up the headship for a new search.

Dean Prokasy, with a Wisconsin and a Utah background, preferred to fill headships internally. With Langenheim, Carozzi and Henderson slated for retirement in 1989, and with the disappearance of their budget lines as per an agreement Hower made with the administration, the choices were limited. Ultimately, we had only one candidate, R. James Kirkpatrick.

Having seen Jim evolve first as a bright graduate student to a junior and senior faculty member, I knew he was up to the job. My only reservation was that he had changed his research from petrology to NMR mineral physics and appeared to be a man of narrow scientific vision. During the search process, I held several conversations with him and concluded this could be a problem, but he fully understood the importance of field work and field instruction, and his geological roots. He was appointed to start on July 21, 1988. I also suggested he ask the administration for a major increase to the departmental salary budget. He succeeded to get such funds effective beginning in the 1989-90 academic year. I was the recipient of one of those special incremental raises.

That summer I went to Utrecht to work with Cloetingh. Returning to the Netherlands was a bit of culture shock for me, having left at age six. When I returned in 1963 for the IAS (Chapter 11), it was unchanged from my prewar life. In 1980, when I taught at the Shell Lab at Rijswijk, I saw marked cultural changes but I was there only for three days. In 1988, I could see how strong they were and clearly, societal standards were disappearing. I recall while waiting for a train an incident at the Utrecht train station where two disreputable-looking people smoked marijuana while talking with a policeman.

Cloetingh welcomed me to Utrecht and I stayed at a small hotel. He tried to find an apartment for my two-month stay. The first was an apartment of horrors. Pieces of oriental carpeting were used as table clothes, and the plumbing was backed up as human excrement flowed into the shower stall from the drain. I turned it down.

The next place he showed was a second-floor apartment in a prestigious location overlooking the town square and the old Utrecht Cathedral. However, the bedroom/living/dining area room was isolated from the kitchen and the bathroom. To access both, one stepped onto the second floor landing to a physically-separate kitchen and bathroom. When the Cathedral bell rang, and despite the beauty of the sound, it was loud, and I knew I'd never get any sleep.

I told Cloetingh it was unacceptable. He was quite shocked and said, "People would die for this apartment." I replied, "I'll die if I live in the apartment. You know, I think I'll just stay in the hotel and eat out." Cloetingh agreed that maybe it was the best solution and the hotel was happy to see me stay.

We tried to do a pilot study but it was difficult. Cloetingh had a lot off-campus obligations, including moving to the Vrije University of Amsterdam. While working on the project, we discovered that critical data was unavailable in the Netherlands and I would have to spend the following year getting it and bringing it with me during a proposed sabbatical in 1989. We were trying to model the crust underneath Pennsylvanian cyclothems in North America from Pennsylvania and West Virginia to Kansas and Nebraska.

Cloetingh and I also visited the Vrije University of Amsterdam because that was where I would spend my sabbatical. We discovered the university had an apartment building in Amstelveen for foreign visitors. I immediately applied for one.

After a weekend in Paris, I returned for two days to Utrecht, packed, and returned to Urbana. Once again, I came to the end of an era at Illinois with a new one about to begin.

LESSONS LEARNED:

1) When things go well for oneself, they may not go all that well for colleagues.

2) When dealing with academic colleagues in other departments or fields, the assumptions with which one operates may not be the same assumptions with which they work. Allowances have to be made.

3) When 'politically correct' complaints are made that are totally bogus, rebut factually and vigorously. If one has to give in to ameliorate the situation, use the opportunity to highlight one or more shortcomings of the perpetrator of the complaint.

4) As one's health profile changes, make plans for career changes and one's research focus (if in academe). Try to select areas where one can be productive.

CHAPTER 24
Illinois: The Kirkpatrick Years (1988-1993)

The Kirkpatrick years were difficult for the entire geology department. The USA entered a recession in 1989 and state tax revenues were down. Grants funds were harder to earn because federal tax revenues were down too. It was a discouraging time in academe.

Challenges were numerous. First, the turnover in faculty continued. Second, in 1991, NHB had a fire in the roof space and the university almost lost the building. After 1990, salary raises were very poor.

During 1989, the College appointed a new dean, Larry Faulkner (BS, SMU, PhD, Texas, Chemistry; Illinois, Univ. of Texas; Illinois: head of chemistry, dean, vice-chancellor of academic Affairs; Texas @ Austin-University; President, Houston Endowment President). Of the Deans I met as an academic, Larry was the easiest one with whom to communicate and clearly the most straightforward. When working with him, you know he would go very far and he did.

To replace Cygan, we hired a geohydrologist, Joel Massman (PhD Univ. of Washington; geohydrology; Michigan Tech, Illinois, Univ. of Washington). He barely unpacked because within 18 months, he left to return to the University of Washington. The pull of alma mater and the Seattle area were difficult for the University of Illinois and Champaign-Urbana to match. The line was frozen until sometime after I left.

I met with Jim Kirkpatrick before the fall, 1988 semester started, gave him my CV and publication list, and reviewed with him the state of the department. Carozzi and Henderson were retiring in 1989 and Langenheim in 1992. He explained a need to restructure the undergraduate curriculum because of these retirements. Jim and I reviewed my sabbatical leave plans for the 1989 fall term and he forwarded my request through channels. I explained I intended to apply for a Fulbright Senior Research Fellowship and Cloetingh would lobby for it with the Dutch Fulbright Commission. Jim knew these were difficult to win, particularly for a scientist.

The semester started with no major problems. I taught Geology 309 and my Depositional Systems (Geology 477) course in the fall and Geology 108 in the spring. Jim was dealing with an acting dean, a classicist, because Prokasy left to become Vice President of Academic Affairs at the University of Georgia. Jim succeeded in getting salaries improved for many of us including me, starting with the fall term, 1989.

In March, 1989, I was awarded a Senior Fulbright Research Fellowship to the Netherlands. The Administration at Illinois was extremely pleased but the financial terms were terrible. It covered my round-trip airfare, apartment rent, a modest per diem, and little else. My salary at Illinois continued at full pay, however.

I made preliminary reservations with KLM which offered direct flights from Chicago to Amsterdam. When I confirmed later, the KLM agent explained that the air fare had increased and asked, "Is that fine, Dr. Klein?" I responded, "Well it isn't fine, but I'll take it because our plans are locked in." I discovered two months later that she put an entry into my reservation record that I was rude to her about the increased air fare.

During July, 1989, the International Geological Congress was held in Washington, DC. I ran two technical sessions, and presented six papers. I also offered a short course jointly with Gerard Bond (PhD. Wisconsin, sedimentology, stratigraphy, tectonic subsidence analysis; UC-Davis, Lamont) and Andrew Miall. The meeting was inefficiently run by a group of USGS scientists and those they deemed politically correct. The chair of the short course committee, a lady geologist at Bryn Mawr College, lacked knowledge as to what materials were needed for course participants and I arranged them through the back-door with AAPG. We planned to start the course at 8:00 am, but participants were told to come at 9:00 am.

Cloetingh and I had dinner at a nice restaurant in the restored Union Station with a fantastic night view of the Capital Dome bathed in floodlights. During dinner Cloetingh disclosed that the Vrije University expected me to pay $10,000 for a facilities fee (overhead) for my stay. He asked if instead I could offer two short courses, one on basin analysis and the other on clastic depositional systems to offset the fee. I realized that in some ways I was trapped and agreed to the offering. I arranged separately to offer my basin course in late November and December at the University of Utrecht, but they offered to pay me to do so. This late request by the Vrije University of Amsterdam and Cloetingh turned out to be a harbinger of what turned out to be the most disappointing sabbatical of my entire academic career.

The Vrije University of Amsterdam (VU) was founded immediately after the Protestant Reformation. Its name was intended to convey freedom from dogmas of the Catholic Church which founded the older University of Amsterdam. VU was associated with the Dutch Reform Church which even in 1989 held a majority of seats on its board of trustees. However, 99 percent of its budget comes from the Dutch government.

The status of geology in the Netherlands also changed. When I visited in 1963, geology departments existed at the universities of Leiden, Groningen, Amsterdam, Utrecht, VU, Wageningen, and one more, now forgotten. The Technical University at Delft also had a strong geophysics program which exists today. However, during the 1960's, the Dutch government discovered that most geology graduates accepted jobs overseas and thus their return on investment via income tax receipts did not balance or yield a gain. They decided to consolidate all seven geology departments into two. Geology programs at the University of Utrecht and VU survived the political machinations that followed. It's hard to believe that the department of geology at Leiden which produced leading structural geologists like deSitter and where Ph. H. Kuenen did his original turbidity current experiments no longer offered geology. That's the problem with a socialistic and bureaucratic agenda: No understanding of the successes of the past and what got an institution where it is.

On arrival in Amsterdam, I rented a car, went to the Vrije University geology office (VU) and obtained my keys to the apartment in the university guest house in Amstelveen. It was spartan. The refrigerator was slightly larger than an office mini-refrigerator.

The apartment, however, was directly below the flight path to Schiphol Airport and approaching Boeing 747's were no more than 1,000 feet above the building. Fortunately, landings and take-offs only occurred between 7:00 am and 10:00 PM, so I got my sleep. Earplugs helped. The apartment was barely usable on weekends.

After unpacking, I took a nap to recover from jet lag. I was awakened 30 minutes later by a phone call from Cloetingh who requested I come to his office immediately because he needed to introduce me to the Dean of the Earth Science School and the Business Manager of the Earth Science School. He explained his family was leaving for a month's vacation that afternoon. I dressed, went to his office and met everyone, received my office and building keys, and wondered why his vacation wasn't disclosed earlier because I could have arrived later. I therefore sat at VU for a month and had to come up with my own project work. Cloetingh gave me some instructions about organizing the data I brought with me for inputting into his computer modeling program.

During the IGC, I met Hans Lippolt from the University of Heidelberg. Lippolt proposed a shorter time scale for the Pennsylvanian based on some geochronology he completed in Germany. I therefore wrote a paper on "Pennsylvanian time scales and cycle periods" while Cloetingh was on vacation, and submitted it before he returned. It was published in *Geology*. When I sent the manuscript, I wondered if it would be the only paper I published from the sabbatical.

Next day, I went to the ABN Amro Bank to open an account. The Netherlands Fulbright Commission sent a letter of instructions. Opening that account was step 1. All American Fulbright scholars, teachers and researchers were required to bank at the Commission's bank for easy transfer of monthly stipends. I then visited the commission office, registered and settled in.

Within two weeks, I saw enough of the Netherlands, their lifestyle, and their deteriorating societal standards and commented, "The smartest thing my parents did was to leave and the second smartest thing they did was not to return when they had the chance. America truly is a great country, but we have to work very hard to keep it that way."

Turn-offs were many. Customer service was generally poor. The best customer service was at Avis. The country was tremendously inconvenient. Public transportation system was great if it was point to point. If it involved transfers, such travel took forever. Environmental standards were below those of the USA and tap water was undrinkable. Adrian Richards (See Chapter 13), who now ran a management consulting firm, told me that in the quaint 14th century village where he lived, a sewer wasn't installed until 1986 because everything went into the canals. Pasteurization standards were weak and I develop permanent lactose intolerance as a result.

For three weeks, I did not receive mail. The office had a mail slot for visitors. I noticed that mail for Cloetingh's predecessor was left there too because, as one secretary put it, "He's no longer on payroll." That predecessor gave up his office and other campus privileges on retirement. This practice contrasted to the USA where usually emeritus professors keep an office and access to facilities. Obviously, socialism as practiced in

Western Europe lacked gentility and decency.

Jan Van Hinte (PhD, Utrecht, geohistory and micropaleontology; Exxon Research, VU) visited me three weeks after my arrival. He just returned from vacation and received about three weeks worth of my mail. The secretary assumed because I was an American I worked with him. He straightened mail deliveries out for me.

And basically, that was the most irritating thing about the Netherlands. The operating paradigm was what a psychologist friend once described as "the myth of assumption" about things that had no bearing on reality. It made getting things done difficult.

A month after arrival, I drove a visiting American to the airport but beforehand, we both visited the KLM ticket office. I reconfirmed my reservation and because of the office layout, saw my reservation on the screen. I asked the agent to print a copy because at the bottom there was a derogatory message that I was rude to the reservation agent in July when finalizing my reservation.

When my guest checked in at the airport, a New York—based doctor ahead of him checked four bags. He was not charged excess baggage whereas when I checked in at the Chicago airport, I was charged for one extra bag. On my return to VU, I drove past the worldwide headquarters of KLM. I called the president's office that afternoon asking who was responsible for this inconsistent customer service. They networked me to the corporate secretary who I called and he scheduled an appointment at 12 noon three days later.

When we met, I showed him the comment on my reservation, complained about the rude service in the states, and the inconsistent baggage fee. I requested that during a round trip back to the US in late October, an upgrade to business class be provided on the return trip, compensation for the excess baggage charge, waiver of baggage charges in future flights, and a waiver of any fees to make changes to my reservation to return to the USA when my Fulbright ended. KLM offered to investigate my complaints. Eventually, they gave me the upgrade and fee waiver on changing flights and waived the excess baggage fee when I flew back in late December.

I returned to VU after that meeting and saw a colleague. Normally, I had lunch with everyone (like UCB; See Chapter 14). My absence was noticed by a colleague. When asked about my absence, I explained I visited KLM corporate headquarters. He asked why and I explained the issues. My Dutch colleague said, "Well George, that's not Dutch." I replied, "You forget, I'm an American and that's how we deal with legitimate complaints."

There were other issues too. The library was extremely limited. They did not subscribe to monograph series like the SEPM Special Publications or AAPG Memoirs. They were bought individually if needed. The librarian explained, "Dr. Klein, the Netherlands is a small country and we have a good interlibrary loan system. We can get anything you need in a week." I filled out cards for items I needed and they were never provided during my stay. However, during a brief return to the USA in October, I found them in the University of Illinois geology branch library, copied them and brought them back.

The work schedule at VU was European. To get a parking place, I arrived around 7:00 am like I always did in Urbana, and had the place to myself except for the janitors from Ghana who spoke better English than the professors and graduate students. Faculty and graduate students arrived between 9:00 and 9:30 am. Lunch was from noon to around 1:30 PM and people left around 4:30 PM. In the evenings, except for students finishing theses, I had the building to myself. On weekends, only Van Hinte, Wolfgang Schlager (PhD, Vienna; Carbonate sedimentology and stratigraphy; Shell, Univ. of Miami, VU) and I were there. I assumed Von Hinte and Schlager acquired this practice while working in the USA.

I usually left at 4:30 PM to buy food for the evening and replenish my breakfast items. I did so because in 1989, stores closed in the Netherlands at 6:00 PM, and on Saturday at 2:30 PM. My refrigerator was too small to stock food for more than two days. It was a time-waster.

I ate out a certain amount. The Amsterdam Hilton featured an outstanding Saturday night Indonesian Rijstafel buffet. The hardest thing to find was broiled steak which was available only at Argentinean and Brazilian restaurants.

Meanwhile Cloetingh and I started our project. However, Cloetingh was far too busy serving on variety of Netherlands' government and European Union panels. He was around only 1/3 of the time. He assigned a graduate student, Beekman, to do plate stress modeling for me and did not supervise him. It became clear that Beekman was not up to the job and I asked him to repeat the computer modeling five times. I was not convinced the last computer output given to me a week before Christmas was accurate and later events proved this to be true.

I enjoyed the Rijksmuseum and the famous Dutch master paintings again, as well as the other art museums in Amsterdam. I also went to a concert at the famous Concertgebouw, but was appalled at the state of disrepair, thread-bare carpeting and seat covers of this marquee concert hall.

A briefing for the entire American Fulbright community was held at the American Embassy in Mid October. The morning was devoted to reports by the political office and the economics attaché. The political officer gave an excellent summary of the current status of Eastern Europe with the changes in Hungary and Czechoslovakia and the turmoil in East Germany.

It was, however, the economics attaché who caught my attention when he said, "You've probably noticed that the Dutch work ethic has evaporated. This has become a major problem for Dutch business houses. Where does one recruit motivated **indigenous** workers? The answer will surprise you. It is the military. There people learned teamwork, meeting deadlines, discipline, and goal setting."

During the coffee break, I talked with the economics attaché. The conversation as I recall was approximately as follows:

Klein: Let me ask you something about your talk.

Attaché: Go ahead.

Klein: I supposed after years of working for corporate Holland, some of the ex-military officers yearn to be of service, run for parliament and get elected."

Attaché: Yes, that's true.

Klein: I estimate that after 40 years of welfare statism, the ex-military officers in parliament must total about 45% of the entire group.

Attaché: That's very close.

Klein: If that's the case, what happens during an economic turndown? Will these ex-military types coalesce and say "Yah, right, let's fix za problems by organizing za country along military lines" and will this lead to the re-emergence of militaristic fascism in the so-called liberal democracies of Western Europe?

Attaché (turning red in the face): We worry about this around here and we worry about it a lot.

The meeting adjourned for lunch. I sat with the director of the local 'Voice of America' office and an American lady who headed the Dutch Fulbright commission. I forgot her name but she grew up in the Chicago suburbs, went to Illinois @ Urbana-Champaign for two years and transferred to Northwestern because she was, as she put it, "sick and tired of classes taught by TA's and I couldn't stand living in that southern town." At Northwestern, she met and married a Dutch international law student. He came from a prominent family that held the contract to print the Dutch currency since 1650, and clearly from her clothes and jewelry, she lived well. Her husband was now that company's president.

The following week I travelled to Germany as a guest lecturer of the German Fulbright Commission at the University of Bonn and the University of Heidelberg. I went to Amsterdam station to take a train and entered a darkened car and compartment. The train was delayed 45 minutes.

Just before leaving, an American came into my compartment and said, "They caught the guy." Ten minutes before I arrived, he entered the compartment and while sorting his luggage, someone tapped on the window. Thus distracted, another person followed him and took his brief case. The 45 minute delay enabled him to debrief the police who recovered the brief case. During the embassy briefing, the cultural attaché warned that Central Station in Amsterdam was crime ridden.

I arrived at Bonn at 11:45 PM and took a taxi to a guest house arriving at midnight. The owner was not pleased at my late arrival.

I gave my lecture at the University of Bonn. While there, I met Paul Wurster (PhD, Munich, sedimentology) again. I met him in 1963 and 1967 at IAS meetings. He was now Dean of Earth Sciences. His door had thick padding on the inside, so I asked, "Paul, what does the dean have to do here to get a padded door?" He laughed and explained that his predecessor was the famous German stratigrapher, Roland Brinkman. I remembered reading Brinkman's books and papers while at Kansas and Yale.

Brinkman started his career just before World War II at the University of Mecklenburg in eastern Germany. A colleague denounced him to the Nazis as a traitor, but the Nazi's were suspicious of the charges. Instead, they posted Brinkman to Poland to head the Polish Geological Survey where he was much loved by the Polish geologists whom he protected and supported humanely. After the war, he returned to Mecklenburg, now in East Germany, but the same colleague denounced him to the Communist government.

Brinkman left and accepted a job at the University of Bonn eventually becoming Dean. During student riots of the sixties, he had the door padded to get peace and quiet.

From Bonn I took the train to Heidelberg via the famous route along the Rhine River. It was beautiful. Lippolt met me and we visited and toured the department. He then took me to the lecture hall where I was to give a talk. On the walls of the lecture hall were numerous portraits of many of the famous professors of geology at Heidelberg. I checked the name plates and recognized most of the names including Werner and Tschermak.

I stopped at one portrait and the name plate read "Saldi". He looked younger, so I asked about him. Lippolt replied with a twinkle in his eye, "Oh, Saldi was here during the 1930's when we started the 1,000 year Reich that didn't last. He left and went to Spain." I realized from Lippolt's tone of voice and nonverbal behavior that the re-education of West Germans by the allied occupation government had a permanent effect.

I returned to Urbana briefly the next week, attended the GSA in St. Louis where I gave a short course, and went to a memorable retirement dinner party hosted by Albert Carozzi's former graduate students at the Adams Mark Hotel. Kathie Marsaglia arranged everything. The wine was superb as was the food. Then there were speeches. I kept mine short but everyone laughed when I talked about my 'cylothomic' relationship with Albert and someone asked whether it was tectonic or eustatic. I replied, "Depending on the weather in Urbana, both."

I returned to VU to give my clastic depositional systems short course. However, I discovered I needed to change it. A faculty member in Wolf Schlager's group was a clastic sedimentologist and offered an identical course. Cloetingh failed to notify him that I would offer such a short course also. We were both upset to discover this duplication so I proposed to focus only on examples from North America and elsewhere other than Europe. Once we agreed, I explained the problem to Cloetingh who said, "Well he isn't nearly as good as you." I replied, "Sierd, I don't know that. He seemed knowledgeable to me. It might have helped if you had discussed this with him beforehand."

I then spent two days a week teaching my basin analysis short course at Utrecht. I shared an office with a Dr. Van Der Linden, a geophysicist who was 'reduced in force' when the Dutch government cut the size of all government agencies, including universities, by 10 percent. Van Der Lingen sued, so while the case was pending, he was given a small office to do research.

I found Van Der Lingen most unpleasant. He asked about my experiences and I was, regrettably, too frank. The Netherlands Fulbright Commission required me to write a letter about my experiences. Mine was extremely frank and detailed my dissatisfactions. Van Der Lingen told Cloetingh. Cloetingh asked to see the letter and then requested I delete half of it. I agreed. However, as he discovered later, I took the part he wanted deleted, cut-and-pasted it, and submitted it as a side cover letter with the official letter. It disclosed the $10,000 charge and my required teaching in lieu of that fee. I discovered later VU policy exempted the fee for Fulbright scholars.

With the holiday season approaching, The U.S. Embassy in The Hague held a Christmas party for the American Fulbright community at the Ambassador's house. I met the Ambassador who was uncomfortable around academics. He was appointed by President George H.W. Bush (41) and was well connected. Numerous autographed pictures of himself and Presidents Reagan and Bush (41) were displayed in the living room and his study.

I also met a psychology professor from VU who completed a sabbatical at Illinois, and later served as one of the rotating academic presidents of VU. He asked about my experience at VU and I was extremely frank. He told me that the business managers ran the university and had more authority than the faculty. He assured me that my perceptions about VU "were not wrong." He also explained that the $10,000 overhead fee was exempted for Fulbright scholars and fellows.

Before leaving the party, I asked the political officer what the ambassador did before being appointed. He was the founder of Pizza Hut. Cloetingh told me later that the Dutch population was upset that President Bush (41) appointed a man who ran a restaurant as the American ambassador to the Court of the Dutch Queen.

I was scheduled to return to the USA on January 8, 1990, but three weeks before Christmas, a notice was circulated at VU. I discovered that between December 22 and January 6, the university would be completely closed. No lights, no heat, no electrical services, no mail, no water. On inquiry, I discovered all museums and government tourist attractions in Europe were also closed then.

I therefore flew back on December 25. On December 26, I was back in my University of Illinois office at my usual time and as I often remarked, "At the University of Illinois, on December 26, 1989, the secretaries are a typing, the radiators are a burping, and yes, there is a partridge in my pear tree." I completed my sabbatical in my office at Illinois because it was the only place I could go to get work done.

In short, my sabbatical in The Netherlands, the country of my birth, was the worst I ever experienced. I discovered later that problems associated with the research weren't over yet.

I settled in and taught Geology 108 and Geology 309. Jim Kirkpatrick brought up the need to modernize the undergraduate curriculum at the first faculty meeting of the semester. He proposed that semester courses in geochemistry and geophysics be added. To balance the number of hours permitted for a major by the LAS College, Mineralogy, Optical Mineralogy, and Petrology were to be merged into a one-year offering, freeing up three hours.

Dick Hay explained that as an undergraduate at Northwestern during the 1950's Krumbein and Sloss taught a combined one semester course in stratigraphy and sedimentology and suggested that Illinois do likewise. I added that I taught such a course at Penn and could easily up-date it for the department if the department chose to go that route. I mentioned that I already added seismic and sequence stratigraphy to Geology 309. That brought out a vitriolic reaction from Langenheim and Mann who were threatened by such a proposal.

Soon after my return, Bob Ginsburg (PhD, Chicago; Carbonate sedimentology; Shell Research; Johns Hopkins, University of Miami (FL)) contacted me. He established an international research initiative with the IUGS (International Union of Geological Sciences) known as the Global Sedimentary Geology Program (GSGP). Their first effort dealt with Cretaceous cycles. He asked me to head a second one on 'Project Pangea." My co-chair was Benoit Beauchamp (BS, Montreal, PhD, Montreal; Geological Survey of Canada). I met Bob and we kicked the project off at the GSA in San Diego that fall. We were to hold a workshop in 1992.

The first decision was a place to hold the workshop. Site requirements were as follows:

1) Accessible field trip stops in Pennsylvanian and Permian rocks,

2) Proximity to a major or medium market airport to reduce

transportation costs,

3) Cost-effective housing and meals, and

4) Access to both a large auditorium and small rooms for breakout sessions.

I narrowed my choices to the University of Kansas, Lawrence, KS, University of Illinois, Urbana, Illinois, and possibly, the University of Colorado, Boulder, CO. Boulder became an expensive place and Urbana was too isolated. Lawrence, KS was most attractive. KU operated a free shuttle bus from the Kansas City Airport to campus and met all other requirements. Moreover, the Department of Geology at KU and the Kansas Geological Survey (KGS) provided enthusiastic support and field trip leadership. KU was my final choice.

I applied for workshop funding from NSF and NATO. In addition I obtained student support from GSA and SEPM. NSF approved funding and between their support and gifts from Mobil Research, Exxon Research, GSA and SEPM, we received sufficient funds to hold the workshop.

The response from NATO was most telling. First, I received a request to change the venue to a European location and was offered a choice of scenic, enchanting, and chi-chi sounding places. However, the availability of nearby Pennsylvanian and Permian rocks was non-existent. I wrote the NATO committee explaining the need for critical geology and their locations did not meet that critical requirement. NATO turned down the proposal probably because Lawrence, KS, was not on their radar screen.

As the semester continued in 1990, I noticed that the Geology 108 students had, as seven years before, incrementally dropped in their academic skills. Not only did they have trouble writing and talking to an adult, but worked only as hard as needed to earn an "A" (if they could) and never did the "something extra" that spelled long-term success. One student, Jennifer Distlehorst, was particularly problematic constantly demanding outlandish things and more than normal personal attention.

She left early before Spring Break and returned late because she participated in Langeheim's Grand Canyon fieldtrip. She demanded to borrow my lecture notes and I told her that was out of the question, suggesting she borrow them from a classmate. She argued and I showed her a page of lecture notes and said, "Does this make sense to you?" She looked at it. It consisted of a column off roman numerals on the left for a coded slide number, and one or two key words to the right. She said, "No" and then asked, "How do you put your lecture together?" I pointed to the right cortex area of my head and said, "It's all here."

Jennifer turned out to be difficult in the stratigraphy course a year later and when the students in the Wanless Room planned a Halloween party, I casually nominated her for Departmental Witch. That generated the usual howls of laughter.

I spoke to Jay Bass about my 108 class because the same students completed his Geology 107 class the previous semester. He talked volubly for 45 minutes, so I knew that student-wise, we were in for a bad time.

The undergraduate curriculum committee started to revise the undergraduate curriculum. I heard that Langenheim was constantly after them to separate stratigraphy and sedimentology into two courses, but neither Ralph nor the committee told me his objections.

During the Geology 108 class field trip, students camped out and I left the two TA's in charge overnight. Because of my hip replacement, the department paid for a motel room. After breakfast, we went to the first outcrop and while the students worked, I asked the TA's how things went in camp. They gave me an earful about the students' cry-baby complaints, their inability to adapt, and their string of verbal abuse. I listened and then said, "Look, you guys can't be more than three or four years older than the students but you sound like the old fogies at the faculty club." The TA's were close in age but far more self-reliant and responsible compared to the generation "X" students.

In May, I received an NSF grant to work on Pennsylvanian cyclothems, attempting to quantitatively separate tectonic from eustatic changes in sea level. I spent the summer in Nebraska, Kansas and Oklahoma getting data.

I had lunch with Louis Dellwig while in Lawrence. We talked a long time. It was, in fact, the longest conversation I held with him since earning my Masters degree 33 years before. He told me what transpired at KU since I left, particularly during the headship of Bill Merrill (PhD, Ohio State, Stratigraphy; Illinois, Syracuse (Dept. Head), Kansas (Department Head; Merrill succeeded Foley). Louis was acting head while Merrill was on sabbatical. Several issues arose that unfortunately involved the department's clerical staff and mail forwarding. It led to Merrill's wife suing for divorce. Dellwig was subpoenaed to testify.

On my return to Urbana, I gave a departmental colloquium about my sabbatical research. I also submitted a manuscript to the *GSA Bulletin* about that research and it came back rejected but the editors advised we try again. Cloetingh and I made revisions via long distance and submitted it again for review.

In October, 1990, I presented a paper at the Dallas GSA annual meeting and John Grotzinger (BS, Hobart and Smith, PhD VPI; Post-Doc, Lamont, M.I.T, Cal Tech) recommended I talk to Gary Karner at Lamont-Doherty. Karner gave me an earful about Cloetingh. I realized I had to find an independent way to check Cloetingh's work.

I discussed it with Albert Hsui and based on his advice, did a simple check and discovered that essentially, Cloetingh's methods were giving me numbers larger than they should be, or 'the sum of the parts exceeded the whole.' I completely changed the methodology and got acceptable numbers that balanced. Our joint manuscript was in review again with the *GSA Bulletin*. I called the editor, explained the problem, and requested he return the manuscript so I could submit a revised one later. His comment was, "George, we are running into this problem a lot with papers co-authored by US and foreign authors."

A Master's student, Jennifer Kupperman (BS, Kansas, MS. Illinois, basin tectonics; U.S.G.S) worked with me. Together, we revised the methodology, rewrote the paper and sent it to the *GSA Bulletin*. I acknowledged Cloetingh and Beekman but removed them as authors. I wrote Cloetingh, explained the problem, and told him he and Beekman were off the project. The paper was published eight months later. I concluded that Cloetingh was a brilliant fraud.

That fall, we reviewed changes in the undergraduate curriculum at a faculty meeting. After Jay Bass, the committee chairman, presented it to the faculty for discussion, Langenheim stood up opposing the combining of stratigraphy and sedimentology into one course. He recommended two separate courses of 2 hours each. He argued that he and Mann could not work with me because of my difficult personality and differences in geological philosophy. He was vitriolic in his comments.

Kirkpatrick reminded everyone that course staffing was his responsibility as department head and if the faculty approved the revised course, he would decide who teaches it.

I raised my hand and was recognized. I said, "First, Jim, I was going to say the same thing about course staffing. Second (turning to Langenheim), Ralph, if you really believe what you just told my colleagues about me, I invite you to file formal charges. Just do so in accordance with university policy and state and federal laws governing tenured faculty."

The faculty sat in stunned silence for a minute and then Phil Sandberg brought up a different issue. The revised curriculum passed.

Jim Kirkpatrick decided the department needed a full-time business manager and the screening committee he appointed recommended Peter Michalove (PhD, Illinois, music composition; IRS tax auditor; Illinois Alumni Association Business staff, Geology Business manager) who was hired. Michalove was competent with handling accounts, grant overhead calculations and so forth. He was utterly colorless and boring and never displayed motivation to get things done in a timely manner. He was far too much the bureaucrat.

I submitted an invited abstract for the Northeast GSA meeting in Baltimore, MD, for a special symposium on cyclic sedimentation in February, 1991. I redid the paper after submitting the abstract because of revisions Jennifer Kupperman and I made and changed the title. When presenting it, I started with a slide saying "Title Change" and gave the new title. The second slide stated, "Reasons for title change" and proceeded to list the problems with Cloetingh's plate stress calculations. Then I gave the revised paper. Bill Dickinson (BS, PhD, Stanford, Basin analysis and sedimentology; Stanford, Univ. of Arizona, President of GSA) told me afterwards, "George that took a lot of courage to admit you were on the wrong track, make the changes, and tell us about it."

Langenheim gave a paper also. In it, he described the Pennsylvanian Period in terms of "Billions of Years." During the question period, I said nothing, but Dan Textoris (PhD, Illinois, carbonate petrology; Univ. of North Carolina) asked, "Ralph, what's this about billions of years? Don't you mean millions of years?" Ralph looked disoriented and answered as if he didn't understand what was asked. Ralph retired in 1992.

Kirkpatrick finally decided that I should teach the new Stratigraphy and Sedimentology course (Geology 341). He discussed this decision beforehand with well known alumni who strongly advised him to have me teach it.

My aunt Olga passed away at age 96 during October, 1991. My father told me that when his time came, I had to keep an eye on her and advise whenever she needed help. I fulfilled that request. She required diplomatic care because Olga was fickle in many ways. When she died, she completely alienated my sister and a large number of other relatives. I was appointed executor of her estate.

I called my sister (we were totally estranged by this time) and asked her to bring a record to be played during Olga's memorial service. Marianne did. We held the service and it was attended by six surviving friends and contemporaries, Olga's sister-in-law, her banker, and three nurses who had provided home care. Ellen Wall, another aunt, who walked with a cane and had braces on her legs, was there too.

When the service was over Marianne said, "I want my record back now." I returned it and she left abruptly. The next day, I visited Ellen Wall who asked about Marianne. I asked why. Apparently, Ellen left early and hailed a taxi. Just as the taxi pulled up, Marianne came and brushed Ellen aside and took the cab meant for Ellen.

During the fall, the tenure and promotion committee reviewed Chu Yun Chen (BS, Taiwan, PhD. MIT, Volcanology; Illinois), the wife of Wang-Ping Chen. Chu-Yun was appointed unilaterally by John Hower in 1983 and I told him I supported it even if the faculty was not given a chance to review her credentials. She already published widely-cited and definitive papers and I thought made a great addition to the department. Jay Bass told me his Cal Tech colleagues were very high in their praise about her work. She took two pregnancy leave-of-absences delaying her tenure decision (new policy).

Between her arrival and 1991, Jim Kirkpatrick soured on her and I never knew why. Before the committee meeting, Jim visited nearly all the tenured people suggesting termination. I found this troubling, particularly because Wang Ping (her husband) had tenure, was doing very well, and was a strong asset to the department.

When reviewing her dossier, the letters of outside experts were negative, the worst group of such letters I ever read. We had no choice but to turn her down. She appealed, and the Dean advised we obtain a second set of letters on her behalf. This second set of letters was extremely positive, so we voted for promotion and she got it.

I discovered later that Kirkpatrick was unhappy with Wang-Ping's departmental citizenship and decided this was a way to get him out. The episode backfired and again raised fundamental ethical questions. Moreover, substantive questions could be raised about why the two sets of letters were so strikingly different. It did not pass the smell test.

The fall ground on and periodically, I made trips to New York because of my duties as executor of Olga's estate. In the process, I rediscovered New York and because it was the first time in my life I visited the city without seeing relatives, I enjoyed it more. I stayed in Olga's apartment during the early trips and enjoyed living like a New Yorker. I wondered if I could find a job opportunity there and leave the corn fields of East-Central Illinois

The Project PANGEA workshop was held in Lawrence, KS, and generated considerable interest with participants from Russia, The Netherlands, UK, France, PRC, Poland, Switzerland and Germany. The first day-and-half, we heard presentations. All were outstanding except for Emil Baud who was at a museum in Neuchatel, Switzerland. Beauchamp recommended him highly, so I scheduled him as the last speaker for an evening session on the first day. His abstract was most intriguing and I thought it would be a great ending for the first day.

Baud used overheads and for the first twenty minutes, talked in a condescending tone about the organization of the IUGS and its committees. He then began his scientific talk which actually was quite good, but interest flagged. When it came time for the five minute warning I said, as I did with every speaker, "You have five minutes."

Baud glared at me and the non-verbal statement appeared to suggest words like 'How dare you clot from a US Midwestern university tell me I have five minutes.' I again said, "You have five minutes." Baud's facial expression froze, realizing I was serious, and he then flipped half his overheads to his conclusions and finished on time.

All participants went to break-out sessions the afternoon of the second day and the morning of the third. Session reports were presented during the afternoon of the third day and we held a banquet in the Adams Alumni center that evening. The KGS organized a field trip on the fourth day and completed the trip at the Kansas City airport so everyone could fly home. It was clearly a success and we published the talks and breakout session reports as *GSA Special Paper 288.*

The atmosphere on the Illinois campus kept deteriorating. Raises averaged about 1.5%, and the incentives to stay were few. The department's size shrank as faculty left. Graduate student enrollments dropped drastically.

I had little incentive to stay. Soft-rock students stopped applying because we had so little critical mass for an attractive program. Consequently, we were unable to compete for them. I sensed not only malaise, but also felt the department was adrift. When faculty left, their lines were frozen. Long term, it was not a pretty picture.

Moreover, most troubling to me was the diminished ability of the Illinois administration to foster research because of financial constraints. During my international travels between 1989 and 1992, the aura of Illinois (See Chapter 18) seemed to disappear because globally, universities were expanding and rivaling American universities in research, facilities, and faculty amenities. Illinois seemed in decline locally and globally.

During the summer of 1992, I saw Kirkpatrick and he said, "Salary raises next year are set at 1 percent." Because I published close to ten papers during the 1991-1992 academic year I said, "That's a disgrace." Jim said, "Well, you could retire." He meant it in jest because he too was frustrated and I knew it wasn't his fault, but the comment stayed with me. I drove east the next day to handle some issues involving Olga's estate and attend a conference at Gerry Friedman's Northeast Science Foundation in Troy, NY. I had a long talk with Gerry who suggested I consider retiring and come east. By now he had a faculty appointment at CUNY. His home campus was Brooklyn College and he offered to try to get me a faculty appointment there.

I served on the University Senate again and learned the boundary conditions for early retirement buyouts at Illinois. Normally it was a two-year phased buyout with annual raises of 15 percent each year to increase the pension payout based on the four highest earnings years. I did some calculations and concluded that I should ask for an 18% raise the first year and 12% raise the second year to front load the proposal. I could meet my retirement goals in one year if another job came along. It also meant at age 60, I would receive a pension identical to the one I would receive if I stayed until age 70. In short, there was neither an academic nor an economic incentive to stay.

On returning, I received a memo to the faculty asking to meet with Kirkpatrick about our annual report and schedule an appointment with Pat Lane. When I saw him, and started to sit down, he said, "George, that won't be necessary. You did great with your publication, your research, and teaching. It would help if you could attract some students but keep up the good work. We're done."

I sat down and said, "No Jim, we aren't done. I've thought about this during my trip east and decided to apply for a retirement buy-out. I'll need your help and instructions what to do."

Kirkpatrick sat there stunned and clearly I had blind-sided him, although it was not intentional. Finally he said, "Well what are you going to do?" I said, "I'm looking for a job in the New York area, and if that doesn't work out, I'll move to Houston, TX, and become a consultant. I've talked with people in New York already." He thought a while and then said, "Have you prepared a draft proposal?" I said, "Not yet, but I can get one to you by tomorrow." He asked to meet me next morning.

I prepared such a proposal. The two year buyout would include an 18% raise the first year, and 12% the second. If I took another job the first year, I would not claim the second year. I wanted the starting date to be backdated to July 21, 1992 (start of the annual pay period). I asked for the indefinite loan of all my computer equipment obtained on grants (It was returned in 1996), a paid leave of absence during the second semester of my second buyout year if I was still there, and some other things now forgotten. Jim said he needed to discuss it with the Dean and let me know.

Later, Jim and I met the LAS College Business Manager who told me the front-end loaded approach had never been tried so he was uncertain if the university would approve it. I requested it be forwarded and it was. In fact, every step of the way, I was told I would not get it. I explained if my proposal was declined, I could resubmit under the phase in of 15% raise per year for two years.

My proposal was accepted. Mort Weir, who was now provost, met me at a party later and said he read it in two minutes and approved it immediately seeing it as win-win for everyone. He said, "I knew you'd get another job by the end of the first year."

To transition out of Illinois, I called several friends for advice. I also called Peter Fenner from Penn days who recommended I read a book entitled "What color is your parachute" and spent the Saturday before Labor Day doing the exercises. When I completed it, I went over it with Peter who told me, "You basically have defined yourself as a consultant."

It took more than two months for the administration to approve my buyout proposal. I started applying for jobs and contacted many of my off-campus contacts. I wondered if word would filter back to departmental colleagues. It didn't. As the 1992 annual GSA meeting in Cincinnati approached, I knew must tell them.

I therefore visited every faculty colleague to disclose my buyout application. I explained that they needed to hear about it from me before hearing about it at GSA from someone else. I recall Tom Anderson saying that he appreciated my being up-front because so often he attended GSA meetings only to hear things about the department at Illinois from someone off-campus and not knowing about a given situation.

The reaction from my colleagues ranged from bored (John Mann), non-verbal happiness (Blake; that changed when he was told he would take over Geology 108 the following year and asked for lab exercises and the course outline), and general understanding (most of them).

The biggest surprise was the reaction of the younger faculty with whom I tended to interact less because their geological interests were somewhat remote from mine. They were shocked and upset. Dick Hay explained that even though they were distant from me, they saw me as a role model because I was actively publishing, applying for grants and working in the building on Saturday's and sometimes even on Sundays, just like they were. They didn't see that in the other senior faculty.

Gerry Friedman arranged two visits to Brooklyn College to meet the administration, the geology faculty, and give a talk. The administration was interested but said that if they could financially swing it, it would not be before 1994. I also thought Brooklyn College was a poor match.

I read an advertisement in *The Chronicle of Higher Education* that the New Jersey Marine Science Consortium (NJMSC) was looking for a new president. I knew nothing about them so called their president Bob Abel (BS, Brown (Chemistry), PhD American University (International Relations); NOAA, National Sea Grant Director, NJMSC, Stevens Institute of Technology). He told me about NJMSC's history. It had 28 member colleges and universities in New Jersey and New York. NJMSC owned five small vessels, the largest being 60 feet. The staff consisted of people working mostly on soft money and support staff, totaling 20 people. They were located on Sandy Hook, a barrier island-spit complex protruding into New York Bay in Gateway National Park within the U.S. Park Service. NJMSC also managed the New Jersey State Sea Grant Program, and the president also served as NJ State Sea Grant Director.

I sent my CV to the head of the search committee, Dr. Joseph Nadeau (BS, Illinois, PhD, Tennessee; geology; Rider College, Associate Dean, LAS College; later Dean). Before GSA, he called because he was attending and we arranged a breakfast meeting. That meeting went very well. Joe wanted to determine how serious I was because most of the Big Ten universities established a system whereby if one wanted a large salary increase, one had to show a better offer from another major institution. I assured him I wanted to make a career change and applied for a buyout from the University of Illinois.

I also applied for a position as a stratigraphic modeler at the Schlumberger-Doll Research Lab in the Greenwich area, CT. I interviewed and we realized I was not the best match. I then suggested that major oil company research labs were cut back or closed and perhaps Schlumberger could hire me to establish a contract research group to meet the research needs of oil companies. Everyone liked the idea but it required approval by the Board of Directors. It eventually happened, but by then, I was unavailable.

Late in December, I received a telephone call to visit NJMSC for an interview. I was to arrive the second Sunday after New Year's, spend Monday at NJMSC, and Tuesday at the new Center for Coastal Marine Studies at Rutgers University to make a second presentation to the New Jersey Marine Science Community at a central location. Before my trip, I received a call from Fred Grassle (BS, Yale, PhD. Duke, marine biology; WHOI, Rutgers), the director of the Rutgers Institute of Marine and Coastal Science. He had an out-of-town conflict and wanted to meet me. Fred suggested that I join him and his wife, Judith Grassle (BS Queensland, PhD Duke; marine biology; WHOI, Rutgers) for dinner on the Sunday evening I arrived. I agreed.

After picking up my rental car at Newark airport, I discovered quickly that driving around New Jersey was not simple. My hotel was in Long Branch, on the New Jersey Shore, and with 55 mph speed limits, I barely had time to check in, unpack, freshen up and drive to New Brunswick to meet the Grassle's for dinner.

We had a productive dinner meeting. Fred Grassle was awarded a Fulbright fellowship after leaving Yale at the Barrier Reef Research Station and met Judith there. We talked about Australia and developments in the field and his hopes for Rutgers and NJMSC. Grassle disclosed that I was one of three candidates for the NJMSC Presidency. One was a lawyer at NOAA who did not interview well, and the third was a marine biologist at Old Dominion University.

Next morning, I drove to NJMSC headquarters on Sandy Hook. As I drove along the shore road, the houses appeared badly damaged by a winter storm. The area suffered during a mild hurricane on Halloween day, 1991. Sandy Hook was cut off from the mainland for three days (a common occurrence after big storms).

My first meeting was with Bob Abel, and his Vice President for Administration, Joan Sheridan (BA, Seton Hall; NJMSC). We spent the morning reviewing the Consortium's current status. In 1989, when Governor Florio took office, their annual state appropriation was reduced from $1 million to $500,000.00. Abel terminated ten people. The professional staff consisting of Jim Nichols, head of Boat operations, John Tiedemann (BS, Ursinus, MS, Florida International; marine biology; NJMSC, environmental consultant, NJMSC, Clean Ocean Action, Monmouth University) head of the education department, a boat safety expert, and Bob Abel raised contract money to keep the organization afloat. Added funding came through federal earmarks, grants and contracts, and sales of boat services.

Clearly, the place had financial problems but was surviving. They survived on a line of credit with PNC Bank whereby they borrowed money as the year continued, and when the new state appropriation was forwarded to NJMSC's bank, the amount owed was paid off. Draws began again, usually five or six months later. Sea Grant overhead and member institution dues also financed operations.

In addition, Bob Abel had a series of grants from EPA, New York Port Authority, US-AID and a few other agencies. Their value was they brought in overhead.

Bob Abel, Mrs. Sheridan, Mrs. Celia Von Oesen, the CFO, and I went to lunch at Bahr's seafood restaurant in Highlands, NJ, a signature restaurant that in time I came to know well. I was surprised Mrs. Von Oesen wasn't on my list of people to meet other than the lunch hour. During lunch, Mrs. Sheridan asked if I was ready to leave the Midwest to move to New Jersey. I assured her it wouldn't be a problem.

Bob Abel and I drove to Bahr's in NJMSC's 'president's car', a 1985 Oldsmobile, and the two ladies came separately. He took me for a drive around Sandy Hook after lunch. It was a cold, cloudy day and as I looked towards Lower Manhattan, I remembered seeing that view before, but couldn't recall when. I was scheduled to meet John Tiedemann but after introducing himself, he told me he was busy finalizing a new Sea Grant omnibus proposal and had to forego meeting me and hearing my talk.

I therefore visited Mrs. Von Oesen and reviewed more about the consortium's finances. I had trouble following much of what she said and her charts and tables were almost incomprehensible. I asked if they had a P&L (Profit and Loss) statement and she could not provide one. I saw warning signals but with my desire to leave the University of Illinois, I ignored them.

Bob Abel disclosed he had a large contract from US-AID to foster scientific research between the US, Egypt and Israel under the Camp David Begin-Sadat peace accords of 1978. That program was ongoing for 14 years, and in fact, Bob had brought the program to NJMSC. Mrs. Von Oesen described a litany of problems associated with the program which involved the Egyptian side whose sense of accountability and documentation was not up to AID accounting standards. Because Fred Grassle said that Abel would move all his grants and contracts with him when he left the consortium, I listened, but said little.

I then gave my talk and suggested a theme of comparative marine sciences, broadening out regionally and globally, using the New Jersey shore as a baseline for analysis and guidance. I stressed the importance of interdisciplinary research and NJMSC's position to capitalize through its network of member institutions. AT 5:00 PM, I returned to my hotel. Bob Abel and his wife took me to dinner at the Molly Pitcher Inn in Red Bank, NJ, another well-known signature restaurant that I also came to know well.

The next day, I went to Rutgers and gave my talk again, although I made changes. The audience consisted of people from other Rutgers science departments, the marine institute, member institutions in Central New Jersey and a Board member, Dr. Robert Tucker (PhD, Duke, marine biology; NJ Dept. of Environmental Protection (DEPE)). I mentioned Joan Sheridan asked whether I could adjust to New Jersey. I explained Bob Abel took me on a tour of Sandy Hook and I looked at Lower Manhattan. I recalled overnight that the view was similar to the morning of March 3, 1947, when I arrived in New York from Australia.

I explained the ship docked at Hoboken, NJ and the first state in the USA I ever visited was New Jersey. I concluded my introduction by saying, "My career here in the USA which you have read about in my CV, started here in New Jersey. I think I can handle living and working here just fine." I then gave my talk.

During the question period, I was asked about Sea Grant, who I knew in Washington, DC, and how to attract minorities into science. On the last question, I proposed meeting the most prominent activist in Newark and jointly trying to raise money to help minority students become marine scientists.

We had a catered lunch at the Marine Institute and Judith Grassle stayed with me. Then Bob Abel suggested we drive to Vineland, NJ to meet the Board Chairman, Frank H. Wheaton, the president of Wheaton Glass, the largest glass company in the world (in 1993). I explained to Bob that I had a 7:00 PM plane to catch so Fred's administrative assistant, a French-born lady drove me there and back. During my meeting with Wheaton, Bob hogged the conversation. I asked Wheaton if the Board was 'participatory' meaning they contributed personal funds to the Consortium. Abel interjected, "No, and I'm proud that it isn't."

On the way back, the French lady spoke critically of Bob Abel. I said, "Look, his answer to my question about a participatory board explains a lot why NJMSC is in such financial trouble. You might want to share that comment with Fred Grassle."

On the return flight, I assessed the interview and concluded that if it happened, I could manage it but it would be difficult. It wasn't the best match but was doable.

The spring, 1993, semester started immediately on my return and I only taught Geology 108.

I returned to New York in late February to visit Brooklyn College again, and settle some business germane to my aunt's estate. On Monday night, Joe Nadeau called. The NJMSC Board met on Saturday and voted to offer me its presidency. He outlined the terms and I said I'll take it. He asked me to meet him at Rider College the following day to sign a contract, and then visit the Consortium the day after.

I was scheduled to meet Gerry Friedman and called early Tuesday morning to explain the situation. He agreed that I should take the job particularly because it paid more than the CUNY system. I drove to Rider College, signed the offer agreement and Joe and I went to Trenton to meet people in the State Department of Higher Education. That agency funded NJMSC. I then drove to Long Branch arriving just as Bob and Mrs. Abel and Joan Sheridan appeared, checked in, and we had dinner in the hotel.

Next day, I revisited the consortium. I agreed to return in March during spring vacation and again in April for 'transition' visits to learn the operation as well as find a place to live. Celia Von Oesen handed me a folder with a lot of financial information that had I seen it earlier, I might have declined the offer.

On my return, I told Kirkpatrick I accepted NJMSC's offer and would leave in May. My resignation was effective July 21 at the end of the fiscal year. I visited the retirement system to complete the paperwork to receive my pension from the State University Retirement System. Jim asked me not to say anything about my departure, but I explained I had mentioned it to Albert Hsui and Phil Sandberg. Albert was colloquium chairman and we agreed that if I left, I would give the last colloquium of the year summarizing 23.5 years of research so I had to commit the date.

Slowly, many tasks took my time. I placed my home on the market the year before and it still hadn't sold. I retained a new realtor who sold it a year later. I pruned my library, maps and other items and gave much of it away to colleagues and students. My winding down from Illinois and life as an academic had begun.

I returned to New Jersey in Mid-March. I discovered finding a rental apartment was difficult. Joan Sheridan networked me to a realtor who could only find two places, one of which the owner refused in the end to rent to me because he wanted to sell it. Meanwhile, Bob Abel wanted me to meet key people and I chose one of them, Tom Gagliano (BS, Brown, JD, Georgetown) former minority leader of the NJ state senate, and Head of the NJ Department of Transportation during the administration of Governor Keane. He formed a public-private partnership, the Jersey Shore Partnership to develop the Jersey Shore. NJMSC had a seat on its Board and I would become a member of their board in June. I chose to meet him first because I felt he would perhaps be a very key contact who could help me get started. We held a productive meeting.

I called another realtor. She was a very tough, high-strung woman but she had two places to show. The first was in a building called "The Shores" which had taken a hit during the 'Halloween Storm'. The lower four feet of dry wall in the entrance lobby was removed and not replaced. The apartment she showed faced the ocean and was also damaged. The parking lot had caved in.

The second apartment faced the ocean too, but was further back and in good shape. I took it.

When I returned to New Jersey in April, I met with the Consortium bankers. It was more social than anything.

Jim Kirkpatrick was serving in his last year of a five-year term as department head and asked to be renewed to serve a second term. That required a departmental review. The review panel consisted of a chemistry professor whose name I can't recall, Albert Hsui, Dick Hay and Phil Sandberg. I concluded that Jim was acceptable except he needed to do strategic planning and resisted it.

I wrote a letter addressed to Dean Faulkner, signed it, and brought it to my meeting with the panel. I suggested Jim be renewed only if he do strategic planning with the faculty, and if he wouldn't do it, find someone who would. I then handed my signed letter to the panel chairman and asked him to forward it to Dean Faulkner. He was shocked I would send a signed letter though the panel. I explained, "Look, I'm leaving in less than two months, I don't mind the Dean knowing it came from me, and for that matter, if the Dean shows the letter to Kirkpatrick, I don't mind if Jim sees it because we discussed this issue many times."

A week later, a strategic planning committee was formed headed by Tom Anderson and included Sandberg, Altaner, Hsui, and Marshak. The committee drafted a plan in terms of faculty replacement (and by geologic field of interest) according to retirement date. I explained this was a poor assumption because they needed to priority rate departmental needs to move forward if people left or retired. Those needs should be identified in terms of where geology was headed in five and ten years.

I then advised that the department return to the core areas of geology and change the graduate curriculum to require students to obtain more breadth.

I presented my 'farewell colloquium' on April 23, 1993, in the auditorium in 229 NHB where I gave my candidacy talk in 1969. Like my PhD dissertation defense, it was a bitter-sweet moment having taught a class in that auditorium, introduced numerous colloquium speakers, and presented a few colloquia. It was standing room only and Ralph Langenheim tape recorded the talk for the University Archives.[4] People came from all over campus, including some from the Administration. I ended with a slide showing the Consortium Logo, views from my new apartment and a few scenic shots from Sandy Hook.

On May 17, the movers arrived and by next morning, they were on their way. I left next morning.

LESSONS LEARNED:

1) Look before leaping from a tenured position.

2) If one holds a job where the situation is deteriorating (as at Illinois during the early 1990's), be sure to estimate the time when one can expect improvement. In 1992, I estimated it would be at least five or more years. That was too long to wait in my case. Hence I chose to retire early. However, during the previous 23.5 years, similar downturns occurred, but I was still young enough to survive a recovery time of five-years.

3) Strategic planning is essential for a program to be successful in academe.

4) When a sabbatical leave is formalized and a last minute surprise, such as VU's $10,000 fee is mentioned, change plans immediately.

5) When interviewing for an executive position, be sure a complete disclosure of financial records is provided, including a P&L statement. If a P&L statement is not provided, don't take the job.

6) Pay attention to scoping out hidden agendas when interviewing for another job. At NJMSC, the hidden agenda was that the new president must be eligible for a senior faculty appointment at Rutgers University (See Chapter 25). Of the three candidates, I was the only one who passed their screen. That criterion evaporated with the election of Governor Christine Todd Whitman in November, 1993 (See Chapter 25).

7) When accepting a job such as NJMSC which was somewhat outside what I knew best, have a plan B if it doesn't go well. Mine was geological consulting (See Chapter 26).

CHAPTER 25
The New Jersey Marine Science Consortium (1993-1996)

"Beware, lest stern heaven hate you enough to hear your prayers"
(or in the vernacular: *"Be careful what you wish (or pray) for"*)
- Anatole France

"It must be considered that there is nothing more difficult to carry out, nor more doubtful of success, nor dangerous to handle than to initiate a new order of things"
- Machiavelli

"..there is nothing you can't do if you are given the tools to do the job"
- William Jefferson Clinton

The New Jersey Marine Science Consortium (NJMSC) was founded in 1969 by several academic institutions in New Jersey seeking to establish a common marine institute. The original proponents were undertaking individual research projects in coastal marine science and needed a common platform to develop joint programs. They also needed regular access to shared boats and field research equipment. NJMSC's mission was to solve the environmental and related scientific problems of the coastal zone.

NJMSC was run informally until 1976 when they appointed Lionel Walford as its first president and established their headquarters at Sandy Hook near Highlands, NJ. NJMSC also became a Sea Grant Institution in 1976. Walford was succeeded by someone whose name I have forgotten. He in turn was succeeded by Robert Abel who served as president from 1981 until 1993. I succeeded Robert Abel.

Over time, the institution grew and by 1993, it had 29 member institutions. Each paid an annual membership fee of $5,000.00 to partially support the enterprise. Of the member institutions, Princeton, Stevens Institute of Technology, Rutgers University, the New Jersey School of

Medicine and Dentistry, The New Jersey Institute of Technology, and Lehigh were, in my view, the strongest institutions. The rest were regional campuses of the New Jersey Higher Education system, private liberal arts colleges, and community colleges.

Leaving Urbana, I drove east on I-74 and stopped for gas in the town of Royal, IL. After using the restroom, I looked west towards Champaign-Urbana as I walked back to my car. When I did, I suddenly felt that my shoulders were more relaxed as the burden of the cumulative weight of 23.5 years of infighting, survival, some great colleagues and students, administrative intrusions, and dealing with unnamed knotheads, was suddenly lifted. I called my realtor in New Jersey that evening to arrange a time to pick up my apartment key, and then slept.

I arrived in Long Branch at 12:00 noon the next day, and checked in at the Hilton Hotel, ate lunch and took a nap. It was Thursday, May 20, 1993

I called Joan Sheridan and she asked if I could visit NJMSC to meet with her and Bob Abel. On arrival, Bob mentioned a problem I would inherit dealing with his US-AID contract. I had difficulty following what he said. He disclosed he would bring it up at the NJMSC Board meeting the next day, at which I was to be introduced. Bob offered to drive me to the board meeting and he and Joan met me at the hotel next morning.

The US-AID issue involved paying the Egyptians researchers (See Chapter 24) for expenses they incurred, but for which they refused to provide receipts and documentation. The project was funded in three phases and the expenses in question were for the second phase with an expiration date in three months. If the funds weren't reimbursed, they would go back to the US State Department. Bob wanted NJMSC to pay the Egyptians from its own funds if the matter wasn't resolved.

Bob offered to let me drive the NJMSC 'president's car,' a 1985 Oldsmobile Cutlass four-door sedan. As we headed for Trenton to one a regional state college for the Board meeting, I discovered that it was no better than the cars I drove as a graduate student and in some ways, worse. Halfway to Trenton, the air-conditioner stopped and summer came early that year. The turn signal would blink twice and cut out. Some of the gauges on the dashboard were giving misleading signals.

NJMSC Board meetings had a sequential structure. First, the Finance Committee met. Bob explained his AID problem and everyone listened.

Then, the Executive Committee met and again Bob explained his AID problem. I sensed his explanation changed. The entire Board met for lunch and the full board meeting was held during the afternoon.

During the regular board meeting, I was introduced and said a few words about how I looked forward to working with them and moving NJMSC into new directions. The board continued its agenda and again, Bob spoke about his AID problem. One of the board members, a banker from a coastal community, said, "Bob, I've heard your explanation of this problem three times today in the finance and executive committees and now to the entire board and your story keeps changing. I suggest our new president, George Klein, address this issue and report back at the next meeting in three months with an outcome."

I was asked what I thought of the problem. I told the Board that I would immediately review the files during the coming week. On June 1, 1993, when I officially started, I expected to immediately contact the people in Cairo, Egypt, regarding what needed to be done. I reminded the board I had US-AID experience in Korea, and based on what I was told, saw a way to resolve the problem but needed time to investigate.

Then I added, "I don't think NJMSC should pay for expenses that can't be documented, particularly when adequate notice to provide documentation was repeatedly given more than a year ago." That comment raised eyebrows. It signaled that some of past practices were about to come to an end.

I drove back with Bob and Joan and we said little. I received a telephone call that evening from the movers explaining they were arriving the next day. I picked up my apartment keys, moved out of the hotel, met the movers at the NJMSC office first to unload my office materials, files and books and then again at the apartment to offload household items and furniture. I spent Saturday night and Sunday unpacking in the apartment.

On Monday, I went to the Consortium and met with Ceil Von Oesen and Joan Sheridan to review the AID problem, and prepared a draft letter to be faxed to Egypt on June 1. Ceil and I also met with NJMSC's attorney at his office in Surfside, NJ and he reviewed my letter.

My administrative assistant, Judy Barrett, typed a Fax, and an identical letter to be sent on June 1. This letter said that I was now president of NJMSC and Bob retired and moved to the Stevens Institute of Technology in Hoboken, NJ. The NJMSC Board instructed me to contact them. The letter stated that we could not reimburse their expenses from AID funds unless receipts and documentation were provided by July 15, 1993.

On June 1, I signed the letter and Judy immediately faxed it to Cairo. Joe Boyd, the janitor, took my duplicate signed letter to the Highlands, NJ, Post Office and sent it registered airmail.

Memorial Day weekend came as a great relief. That Saturday, I drove to New York City to enjoy some of the museums and buy some additional suits and sports jackets. I now had a job requiring a different suit every day, a new experience for me.

I ate dinner at a Japanese Restaurant and was seated next to a table with two ladies who were speaking Korean. When I ate my appetizer, one of them said, "You use chop sticks very good. Where did you learn that?" I explained that I lived in Korea and Japan. The lady who asked me was single, her friend was married. The single lady was Suyon Cheong, and she and I got married 15 months later (See Chapter 28). Meeting and marrying Suyon was the best thing that happened to me during my entire life.

Returning to the consortium, I began to get the feel of the place. I was required to travel to meet presidents of member institutions. A week later, I met Frank Wheaton to review my three-year contract. I discussed selling the 'presidential car' and using my own claiming mileage reimbursement. I said, "Frank, an organization with the problems NJMSC has doesn't need a car for the president." He understood and agreed immediately.

A troubling pattern emerged as I reviewed finances. The professional staff were doing a good job of raising grants for projects to keep themselves employed but NJMSC was only collecting overhead from Sea Grant and a few of Bob Abel's trailing grants that couldn't be transferred. The majority of grants did not generate overhead. I reviewed this with the professional staff and they seemed interested in remediating the problem by adding overhead to their proposals. I discovered in time they didn't understand it, or didn't know how to bill grants for direct costs in lieu of overhead. I addressed it again a year later.

John Tiedemann met with me because during early August, the National Sea Grant Program was holding its biennial meeting of all directors, educators, communicators, marine agents and the Washington office in Honolulu, HI. He planned to go but because officially I was now NJ State Sea Grant Director, he offered to step aside to let me network to the entire

National Sea Grant community.

Before I arrived, the Sea Grant Director in fact was Bill Gordon who handled it part-time from his home in central Pennsylvania. I met Bill during a transition trip and he retired permanently in May, 1993. John Tiedemann ran the day-to-day operations which the Executive Committee of the Board of Trustees found troubling. Bill Gordon and Bob Abel assured John verbally, but NOT IN WRITING, that his duties continued.

Joe Nadeau telephoned early in June, 1993, telling me that he and Jim Henry, another Board member on the executive committee (representing the NJ Department of Education), wanted me to take over Tiedemann's duties. I explained this to John and asked for an orientation which he gave, but it was a hostile meeting. John never accepted the fact that this change came from the Board of Trustees and not from me.

Before I arrived, Bill Gordon and John submitted the 1993-95 NJ Sea Grant Omnibus Proposal, starting July 1, 1993. I read it and felt the projects were poorly defined. On June 30, the Sea Grant monitor in Washington, Captain Bob Norris, called and told me that the national office wanted me to eliminate two proposals and require NJ Sea Grant to use the funds for development grants awarded annually. Because state Sea Grant directors RECOMMEND projects for funding, I had no choice but to accept Norris's decision, although I could have negotiated the projects back in. The two projects they eliminated were extremely weak so I concurred. Unfortunately, the PI was the same person on both projects and obtained matching funds from a private group, The Hudson River Foundation in New York. The proposal dealt with dredge spoils and its effect on fisheries, a particularly hot local environmental issue.

I notified the PI, a Rutgers professor, who asked the University to sue NJMSC. That request was turned down. The award letter he received failed to mention that the approval of funding was a RECOMMENDATION CONTINGENT ON A FINAL REVIEW by the National Sea Grant office. I discussed this with Fred Grassle, who understood, and the PI, who refused to accept the explanation.

The Science Director at the Hudson River Foundation, Dennis Suszkowski (PhD, Delaware, Coastal sedimentation; EPA, Hudson River Foundation), also failed to understand this critical point and his hostility towards me increased over time. He was also a Board Member of NJMSC. I told the PI I'd try to get him some other funding, and suggested he apply for a development grant in the fall to get better data to formulate a stronger proposal.

The environmental community was not happy with this decision either. Cindy Zipf, the director of a local coastal activist group, Clean Ocean Action, never contacted me but wrote the local congressman, Frank Pallone asking for intercession. Pallone was a former marine sea grant agent handling marine law before going into private legal practice and later was elected to Congress.

Early in July, I received a fax from the Egyptian scientists who worked with Bob Abel on his US-AID grant. They provided an accounting and attached copies of receipts. Ceil and I reviewed the receipts and decided to pay the remaining money owed to them out of the residual funds of the Phase II grant to Abel. The issue was now closed and when I reported this to the Board at their August meeting, they were relieved.

I went to Hawaii for the National Sea Grant meeting and met fellow directors. All were helpful. I spent the time listening and spoke only when I felt I was on strong grounds. I discovered later that the directors were comfortable with my appointment because like them, I had experience as a faculty member in a Land Grant University, whereas Bill Gordon and Bob Abel did not. Two directors were geologists who knew me.

The Sea Grant Directors elected me for two years to the executive committee of a companion group, the Sea Grant Association, a lobbying group. I also met the Director of the Illinois-Indiana Sea Grant College Program for the first time. He was a faculty member in the Department of Leisure Studies at Illinois (I had to become a state Sea Grant Director to learn such a unit existed on the campus where I taught for 23.5 years).

One of the people at the meeting told me that the NJ Sea Grant program was being watched and in danger of losing certification. I had a year to turn it around.

That watch arose because NJMSC was isolated from the daily flow of information about trends and changes in grant accountability and related requirements. Consequently, they were not following procedure as changes occurred. That was easy to solve. I asked Fred Grassle to network me to the Research Grants office at Rutgers who agreed to keep me in their information loop. *The Chronicle of Higher Education* was also a great resource for this information.

The second reason NJ Sea Grant was on the national office's watch list was that they questioned Bill Gordon's and Bob Abel's reviewing procedures and project recommendations. I attended a farewell party for Bob Abel on Frank Wheaton's yacht and met a lady representative to the NJMSC Board from the NJ Institute of Technology. She submitted a proposal which was declined and no reason was given. She asked if I could check the files and let her know the reasons, and I agreed.

Her proposal was initially reviewed by Bill Gordon together with a NJ Sea Grant Advisory Board of scientists. No minutes of their meetings existed. The only thing I found in the lady's file was a torn strip of paper from a secretarial dictation pad with a one-sentence comment. No external reviewer's forms or letters existed in the file or were found after asking for them.

I discovered that Bill Gordon summarized the Advisory Board's decisions with one-sentence statements on a secretarial dictation pad and then tore sections of his notes and put them into each proposal file. It was my professional and personal opinion that Bob Abel and Bill Gordon had unwittingly and unintentionally distorted the review process.

The solution was easy. I designed a review form similar to one used by NSF. I added and required reviewers to rank their assessments in terms of a percentile distribution of all proposals they read and all scientific projects with which they worked. That provided a semi-quantitative ranking of all proposals received.

In August, I went to Silver Springs, MD where Sea Grant had its national office. The meeting provided an orientation with two other new Sea Grant directors, Anne McElroy (NY Sea Grant) and one other person, meet everyone in the office, and hold private discussion with our monitor. I reconnected with the national director, Dave Duane (PhD, Kansas, coastal sedimentology; ONR, Sea Grant) and Ned Ostenso (PhD Wisconsin; marine geophysics; ONR, NOAA Science Director) both of whom I knew. It was a useful experience and enabled me to do my job better.

I drove to Silver Springs in the 'presidential car' and it was my last trip in it. The air conditioning went out again as I crossed the Delaware River Bridge. When returning to the office, I gave the keys to Joan Sheridan and told her to sell it fast for whatever she could get. I explained I was now driving my own car and claiming mileage. She was quite shocked, but I told her I already received Frank Wheaton's approval to sell it.

NJMSC sold it two months later for about $1,500.00 and once Ceil cancelled the insurance, we saved an additional $2,000.00 per year. This was offset by $3,000.00 in repairs that I requested.

The National Sea Grant office announced a special initiative in marine biotechnology and I circulated notices to the member institutions. We received seven proposals. When the awards were announced, NJ Sea Grant received the fourth highest number of awards and dollar amounts of all 28 Sea Grant programs in the USA. The new review form worked well and led to our success which was noticed throughout the program.

During the fall meeting of all the Sea Grant Directors in Washington, The Illinois Sea Grant Director told me that he read my name on a list of retirees which was published in *Illini Week,* a campus newsletter. He disclosed that 150 people retired at the same time. He sent me that issue and when I read it, I was shocked how many people I knew retired at the same time as I did. They were world-class scientists, engineers, business faculty, and leading scholars in the Humanities and Social Sciences. Clearly, I left Illinois at a good time.

When I accepted the NJMSC offer sheet in February, 1993, I met with two people in the NJ Board of Higher education. They mentioned that Rutgers built a successful Coastal Marine Science Center and raised the possibility of merging NJMSC into it. They explained that one of the reasons I was hired was that I was qualified to be appointed to a tenured professorship at Rutgers. In fact, later I received a courtesy graduate faculty appointment there. If the merger occurred, the NJMSC president would be tenured and the salary would come from the Consortium appropriation.

Other member institutions were hostile to the proposal and suspicious of Rutgers' motives. Consortium staff concluded they would all be fired if such a merger occurred.

Rutgers University had concerns too. I met their Vice President for Academic Affairs a month after arrival and developed a friendship with him. He too was a visiting fellow at Wolfson College at Oxford. I later met with him, the Rutgers Vice President for Business Affairs, and the Rutgers University attorney. Their concern was the liability associated with NJMSC's southern NJ facility at Seaville, NJ. When the merger with Rutgers was reviewed by the NJMSC Board in February, 1994, it was rejected by one vote. If it had passed, my life might have been different.

During the fall of 1993, we sent an RFP for Sea Grant development grants and received one from the Rutgers PI who the National Sea Grant Office turned down at the last minute in June. I arranged reviews from the advisory committee before the meeting and on a five point scale it rated around 2.2, at the bottom of all proposals. When the advisory board met, Dennis Suszkowski went to bat for that Rutgers PI because implicitly we committed funds by encouraging him to apply for development funds. I brought the discussion back to science and the two marine biologists confirmed the proposal was weak. One referred to it as "bioweird".

Bob Abel then suggested we award the funds as a compromise to appease Suszkowski, whose organization was willing to provide some matching funds. The advisory board voted to do so.

Over the weekend, I reviewed it again and concluded this was a poor way to run a science program and unilaterally declined it on the grounds that the final decision is the director's and not the advisory board. The advisory board recommends to the State Sea Grant Director only. As State Sea Grant Director, I still had the final say on which development projects to approve. That detail was emphasized during my orientation in Silver Springs, MD, in August, and sustained later by Anne Studholme, a marine biologist who was director of the NOAA fisheries research lab on Sandy Hook. She also was a member of the NJ Sea Grant advisory board.

When word of my decision reached the National Office and after our success with the biotechnology competition, NJ Sea Grant was removed from its watch list and considered to be in good standing. This was achieved within eight months.

A major issue arose that required immediate attention. It dealt with NJMSC's Southern New Jersey facility at Seaville, NJ. The member institutions in southern New Jersey resented NJMSC's location at Sandy Hook after NJMSC was established and expressed interest in a satellite location in their area. During 1978, the US Air Force Palermo Radar Station in Seaville, NJ was closed. It was the main surveillance site protecting the US East Coast from impending Soviet attack which never came. The Air Force removed everything and placed the site for sale.

NJMSC purchased the land, the site, including an abandoned, stripped radar tower and all the buildings for $30,000 under a special program relinquishing such facilities to non-profit organizations. However, the site came with a heavy price, namely environmental clean-up, a major financial drain on NJMSC. The year before I was hired, Bob Abel proposed closing the facility and the local county commissioner was strongly opposed. A demonstration was held and Bob's effigy was hung and set on fire.

Part of the property was used by local sports teams. The rest was used for summer courses for southern New Jersey member institutions and to offer a local education and outreach program. Moreover, early after acquisition, a group of dedicated ladies were given free use of a building as a hospital for injured birds. For unknown reasons, no rent or utilities were charged to the bird hospital.

I drove to Seaville in early September, 1993, to inspect the site and meet the full-time custodian and part-time educational coordinator. The disrepair was appalling and I wondered why the county health department hadn't closed it. The radar tower had no immediate use. I finally visited the bird hospital and although the well-intentioned ladies were nice, the place was filthy and hadn't been cleaned in years. NJMSC had no option but to close the Seaville facility during the winter months, open it only for three months for summer courses, and terminate the educational coordinator. The bird hospital would have to move. Eventually full closure and sale would be required.

On leaving to return home, I decided to call on the local county commissioner, the owner of the largest liquor store in New Jersey, 100 yards from the southern city limits of Oceanside, a dry community. Because it was a cold call, I decided to introduce myself and let him know I wanted to call him for an appointment in two weeks to discuss NJMSC's Seaville operations. I waited five minutes after asking to speak with him and he came from his office. I gave him my business card, introduced myself and told him Bob Abel retired and I replaced him. His non-verbal reaction was one of complete surprise that I took the initiative of going to his store to meet him, knowing about his past history with Abel. We chatted amiably and I told him I would call in two weeks which I did.

A week later, NJMSC held a board meeting and I proposed partial closure and limiting activity at Seaville to the summer months as a cost cutting measure. The board approved my recommendation.

Joan Sheridan and I returned to meet with the local county commission to inform him of the Board's decision. I explained that the custodian would stay and access to sports fields could be arranged through him. Joan explained the extent of the financial burden and he understood. He asked about the bird hospital and I said I would write them asking them to move their operations at the end of December because on January 1, access would be barred. He said nothing further and we left. He left me with the impression that he appreciated my taking the extra effort to keep him informed about our decision.

Around the first week of December, I visited southern New Jersey again to meet with the Congressman Hughes, a senior congressman with considerable influence. The bird hospital asked him to write me about the closure and requested his intercession. When we met, he told me he advised the bird hospital years before to find another location, anticipating closure at Seaville. He explained a congressman is asked to intercede by writing letters and usually they do so expecting nothing much to happen. Hughes assured me he wanted to work with me. I gave him a copy of NJMSC's mission statement and he was most interested because it coincided with his concerns too. We had a good meeting but I was sorry to see him retire from Congress the next year. Access to the kind of influence he had is not easy to come by.

During my service as NJMSC president, I was responsible for state and federal legislative relations. At the state level, I had easy access to legislators. They were part-time and were easily reached at their regular local offices. All were within a day's drive from Sandy Hook.

At the Federal level, Bob Abel advised I go to Washington and meet the critical legislative aide in a Congressman's or a U.S. Senator's office and deal directly with them. Chances of reaching a Congressman or Senator were remote and if one was fortunate, a chance meeting could occur locally, or the aide would make arrangements for a meeting. Mail was to be routed via an aide to insure it was read and answered.

I visited Washington in late June on other business and spent an extra day making the rounds of New Jersey Congressional and Senate offices. I called ahead and made appointments. Those I could not reach, I cold called. If they were in, we met and talked. If not, I left a business card with a message on the back to call me. I would arrange to return to meet them. Some aides were absolutely tops. I recall the aide to Congressman Gallo, Ed Krenek, was outstanding and helpful. The aide to Congressman Zimmer also proved to be helpful. Senator Lautenberg's marine affairs aide was knowledgeable, but somewhat distant. The most disorganized congressional office amongst the NJ delegation was Senator Bill Bradley's.

Two examples illustrate how these relationships are beneficial. NJMSC rented its boats to an environmental firm in Flemington, NJ. That firm owed NJMSC $50,000 and Ceil Von Oesen called to my attention it was unpaid for ten months. I called the company president who explained his company had rented NJMSC boats to complete a contract with the US Navy. The Navy owed $250,000. Attempts to collect their money proved fruitless. I asked who was his Congressman and he told me it was Congressman Zimmer. He explained he wrote Zimmer a letter and heard nothing. I suggested he call Zimmer's aide, gave him the phone number, and to let the aide know I suggested he call. Within two weeks, they received their money from the Navy and the company president sent his secretary to my office to personally deliver a check for $50,000.

Not all dealings with Congressional offices are that smooth. I recall contacting all the aides in the New Jersey Congressional delegation to obtain their support to reauthorize the Sea Grant program and its funding. Sea Grant and NOAA are housed in the Department of Commerce. All appropriation bills for Commerce are combined into one package with the Department of State and the Department of Justice. During 1995, I received a call from the aide to Congressman Andrews who represented Camden, NJ. The aide explained that the Congressman was a strong supporter of Sea Grant but had to vote against the bill because the 'cop in the street program' was reduced. This was unacceptable to the congressman and his constituents. He apologized and I said, "Look, I understand the congressman's situation. Tell him I'll look forward to working with him again in the future." The aide was surprised and told me normally when making such calls he gets yelled at, sworn at and threatened. I responded, "There is always another day."

I personally met some of these elected officials. I briefly met Senator Lautenberg at two functions and found him distant and cold. I met Senator Bradley at a fund-raiser for the mayor of Highlands who was running for the state senate. Bradley was a terrible speaker at this event and after playing basketball at all levels, his knees were gone.

The local Congressman was Frank Pallone and in late December, 1993, he called and I spent two hours at his office reviewing Sea Grant issues. I also tried to get some earmarks for NJMSC but got nowhere. Congressman Zimmer came to NJMSC on a site visit.

It was Congressman James Saxton with whom I dealt the most. After 1994, he chaired the House subcommittee responsible for Sea Grant. The National Sea Grant Association asked me to meet with him regularly to assure Sea Grant funding. I worked with his aide and arranged a meeting in Washington. After thanking Saxton for his support of the Sea Grant Program and mentioning a marine agent project in his district, we discussed Sea Grant's needs. He said he would discuss it with Congressman Don Young. Suddenly, the vote bell rang; a bell no different than a high school classroom bell. We quickly concluded our meeting, and his aide asked me to wait until she returned. She accompanied Saxton explaining what the vote was about, returned and told me, "You did well today. When a congressman says he'll talk to another congressman on your behalf, you've done well." I met with Saxton later in his district office during a congressional break and it was more casual.

In the fall of 1993, New Jersey held an election and Christine Todd Whitman was elected the new governor. One immediate outcome was that NJMSC was removed from the state's budget line and the Department of Higher Education, from which NJMSC's budget came, was abolished. I arranged a grass-roots letter writing campaign and called all the coastal state legislators and senators to whom I had become networked. We succeeded in restoring the budget line. In subsequent years, our requests went through the state Treasurer's office.

That spring, I testified before the NJ State legislature on behalf of NJMSC's appropriation. That hearing and response went well and we received the annual appropriation of $500,000.

I also made preparations for the next Sea Grant omnibus funding cycle with our proposal due in February, 1995. I held an orientation meeting at the Rutgers Marine and Coastal Sciences Institute and asked Tiedemann, Kim Kosko, Sea Grant communicator, and all four marine agents to attend. I outlined the procedures that we would follow and emphasized the importance of incorporating research with extension and to include an extension component. When the meeting ended, I left the building and saw Tiedemann and the four marine agents head for lunch. I regret in hindsight not joining them and footing the bill. It may or may not have mattered long term. However, the first steps were taken to initiate a new set of proposals for the next omnibus proposal and to increase research funding as well.

As a Sea Grant Director, I dealt with my share of what in the South we call "Knothead" PI's. Problems ranged from unwillingness to follow guidelines and timelines, failure to keep NJ Sea Grant in the loop concerning key issues such as incoming earmarks, demanding special treatment based on a misunderstanding of a variety of labor and affirmative action laws, expecting NJMSC's accounting to do what their own campus research accounting office provide, and failing to pre-plan scientific missions. In short, some PI's failed to meet their own grant research and management responsibilities.

Four people stand out as the worst offenders. An oyster researcher in a Rutgers marine biology field lab in the Delaware Estuary could never meet deadlines and questioned our decision to submit the omnibus proposal without his because the submittal deadline had to be met. I later discussed this individual with Fred Grassle who admitted the man "was difficult."

Dr. Kenneth W. Able (PhD Georgia, fisheries biology; Rutgers), the director of the Egg Harbor Rutgers Fisheries field lab, is unquestionably an outstanding, highly-regarded fisheries researcher, but in dealing with him on an earmark he arranged, a poor manager of project accountability. In Mid-December, 1994, the National Sea Grant office notified me that Able received an earmark which was being passed through Sea Grant and NJMSC to Rutgers. It required Able to write a formal proposal and submit it through Rutgers back to NJMSC for transmittal to Washington by January 6, 1995.

I called Able's office and his secretary explained he was out-of-town. I asked for phone contacts and she explained he left no itinerary. Rutgers, unlike the University of Illinois (and most state universities), did not require faculty to leave itineraries with their offices when on travel. I left a message that he call immediately about his earmark appropriation.

Not hearing from Able in a timely manner, I double checked. Able's secretary explained she still hadn't heard from him. On December 30, he walked into my office and asked casually what the fuss was all about. Ken had just visited Anne Studholme at the NOAA lab next door and called his office. I explained that he was to receive an earmark and we had a week to turn it around and get his proposal to Washington. I mentioned he needed to budget NJMSC's direct costs to handle the accounting over and above the $250 pass-through fee. He agreed we had costs and got them included. I explained we needed the proposal no later than January 5 to meet the national office's deadline. He then returned to Egg Harbor, wrote the proposal, and presumably was sending it through channels.

January 5 arrived and no proposal. I called again and Able was out-of-town. The secretary explained they gave the proposal to a graduate student to deliver to the Rutgers grants office. After calling that office, Able's secretary discovered the proposal was delivered to the wrong office and the grants office just got it. Because we would pass the deadline, I called Able's secretary again to ask Able to call the National Sea Grant Office in Washington to request an extension. She was not pleased with my request but eventually he received his funds.

Dr. Bonnie McCay (PhD Columbia, Anthropology; Rutgers University) received an omnibus award during 1991-1993. I have forgotten project details but funding was included for a contract co-investigator to jointly prepare an informative publication to be published by NJ Sea Grant. The co-investigator bailed out and she did nothing to complete it.

The protocol at Sea Grant is that the last 20% of a grant funds are withheld until deliverables are received. No deliverables, no final 20% payment. She kept asking NJMSC to release the funds and Ceil and I reminded her we needed deliverables before we could do so. She came up with all the lame excuses an undergraduate would give, including that as a female faculty member, she deserved special consideration, but we stood firm. Joan and Ceil were particularly incensed with her gender-based request for special treatment as were all the women on staff at NJMSC.

Three months before funds would revert to the national office, I received a call from the Rutgers University grants office inquiring why we had not released the remaining 20% of Dr. McCay's funds. I replied, "I'd like to release the funds to you and wish we could, but Dr. McCay has failed to comply with Sea Grant's deliverable requirement." The person at the other end told me they would get back with me.

Six weeks later, we received her deliverables and released the funds. Kim Kosko arranged publication as stipulated in the grant. The irony was that Dr. McCay's booklet became NJMSC's best selling publication and even won an award. It would have been much easier if she had done her job instead of pushing the bureaucratic system to the limits and having her own university administration force her to complete the deliverables.

The last difficult person was, regrettably, Dr. Gail M. Ashley (BS, MS, Univ. of Massachusetts, PhD. Univ. of British Columbia, sedimentology; Rutgers University; second woman president of GSA and SEPM) who I had known since she was a graduate student. Gail completed some definitive research in Olduvai Gorge for which she received considerable acclaim. Gail ran afoul of NJMSC in two ways.

The New Jersey Geological Survey contracted NJMSC to vibracore several sites on the New Jersey Continental Shelf. Gail asked to join the team to examine core and recover samples for her own research. She joined the cruise and once cores were recovered, opened and slabbed, she was upset with their quality and filed a formal complaint. I investigated and discovered she failed to specify coring requirements for her research and instead assumed that the people writing the contract specifications would do things the way she wanted them done. When I explained this outcome to her, she was quite hostile and refused to assume her share of the responsibility.

Gail also ran into difficulty with Ceil demanding copies of her Sea Grant project financial statements from NJMSC. Normally, these are distributed to PI's by their home institutions (See Chapter 12) and according to Ceil, Gail became argumentative. I finally called Gail and explained that just like NSF grants, all this came from her own grants office and more likely their data was a month more current than ours. Again, she became argumentative.

During the spring of 1994, I dealt with a complaint made against Kim Kosko, the Sea Grant Communicator (Communications Officer) by one of the accounting clerks who also was Ceil Von Oesen's daughter. It alleged Kim misused travel funds by failing to attend a meeting but claimed mileage expenses. The complaint was countersigned by Teidemann's assistant director of education, a California-born Mexican American graduate of USC married to a JAG lawyer at Ft. Monmouth. That assistant director encouraged Ceil's daughter to file the complaint.

I examined the expense report, called the marine agent who mentioned Kim was missing and confirmed her absence. I also discovered that the complaint was the talk among the ladies working at NJMSC who kept Kim in the dark.

I talked to Kim. On the way to the meeting, she left early to run several errands and encountered serious traffic problems. She realized she couldn't attend the meeting in a timely fashion so returned to NJMSC. Having encountered similar driving problems in New Jersey, I understood. Moreover, the claim was only $25.00. I decided the claim should be paid, sent a memo to the consortium staff that I completed the investigation and found it was a minor infraction. I then wrote a separate memo to the education assistant from USC stating that she had exceeded her authority when filing the complaint. She responded with a quote from the Catholic Bible. I discovered that all the ladies at NJMSC were disappointed with my ruling and I could not understand why they were so unforgiving of Kim.

While at Illinois, I had my share of frustrations with the non-academic staff, most of whom were woman, and so did other colleagues. I witnessed some of the verbal abuse heaped on these women by faculty colleagues when things went wrong. However, the ladies were very forgiving of these transgressions. The women at NJMSC were not. The difference, I regret to say and even generalize, was that in Illinois, the non-academic ladies were mostly Protestant, whereas at NJMSC, they were mostly Catholic. As I worked with NJMSC staff, I discovered they were very rigid about some human issues.

Judy Barrett went on disability leave in January, 1994, and passed away in April. She was a heavy smoker and cancer caught up with her. I hired Michelle (forgot last name), the wife of a Coast Guard Officer, who was college educated and she was one of the best people I ever had working with me. But, she found the pace hectic and resigned. She was replaced by a younger lady who was totally hopeless and quit on her own.

I then did a search again and finally found someone who was excellent, Priscilla Gettis. When we were organizing the Sea Grant Omnibus proposal, she was outstanding. We rehired Michelle temporarily to help Priscilla and met the deadline with three days to spare.

When interviewing people, I discovered a major flaw in NJMSC's hiring procedures. Joan Sheridan gave each applicant a generic form to complete on which candidates listed references. During my interview with Michelle, I discovered she listed family and friends, but no references from prior work experience. I discussed this with Joan who told me it worked well in the past. I insisted on work references as a matter of policy and I telephoned past employers of each candidate.

When openings existed, past practice was to let the managers hire whoever they liked. I established a policy whereby I interviewed every candidate because if hired they were given the key to the building. My concern was building security. Managers still could make a final selection after I reviewed candidates with them including the rare times I had reservations.

When Ceil's daughter quit, I decided NJMSC needed someone with an accounting degree. None of the people in Ceil's office, including Ceil, earned one. I hired Frank Ngo, a Vietnamese immigrant who earned all his degrees and worked in the USA.

I met the managers that winter, and told them we needed to improve overhead recovery. I instituted a policy of personally reviewing and countersigning all proposals before they were sent and made it clear if there was no provision for overhead, or added direct costs in lieu of overhead, I would not release the proposal. Prior to this change, no proposals were internally screened or reviewed. This created problems with Tiedemann who told me that the state appropriation should cover his overhead charges. I explained that his math didn't add up.

Tiedemann turned out to be a problem employee. I often described him as a New York Indian, with hair boxing the compass, and a voice louder than the Sandy Hook lighthouse in a winter fog. When he brought his first new proposal for my review, I turned it down because he did not include overhead. Moreover, he refused to add to his direct costs items such as clerical support, accounting support, fees for space and so on, in lieu of overhead.

I offered a constructive solution. I explained with his track record as a Science Educator, he qualified for NFS science education support and networked him to the relevant program director in Arlington, VA. I offered to have NJMSC pay expenses to visit NSF. He never made the trip. He preferred to rely on knitting together small grants from various organizations which did not permit overhead charges, but permitted direct billing costs for supporting clerical and accounting services, and facilities fees. Unfortunately, his projects cost NJMSC more than $100,000 during the previous three years. In retrospect, I should have fired him.

At the November, 1994, NJMSC Board meeting, Cindy Zipf (BS, Univ. of Rhode Island, marine biology; Clean Ocean Action), executive director of Clean Ocean Action, was elected to the NJMSC Board. Because she was an environmental activist, the appointment was viewed as a surprise by member institutions, Joan Sheridan and Ceil Von Oesen. All considered it a conflict of interest. Substantive questions arose about her appointment to a consortium board managing a federal program viewed as an "honest broker."

Her hidden agenda was the NJ Sea Grant College Program because it became too much of a 'science program.' When we discussed this, I reminded her that the strong peer review system I instituted saved NJMSC Sea Grant from decertification. I assured her that the proposals relevant to her interests that survived peer review would lead to better outcomes for her and her organization. Although that slowed her down, it did not totally satisfy her.

Her membership on the board, however, gave me a close view of the goals and methods of her organization and the environmental activist movement. First was the politics of destruction used against those whose position was not theirs. Second, it became obvious after attending an event her organization sponsored that it was a cover for class warfare by elitist, mostly white affluent individuals.

During the spring of 1995, Clean Ocean Action invited me and my wife to their annual fund-raising spring fashion show which featured a presentation of an award to Steve Corodemus, a local legislator who I personally knew. The event was held at a period piece Jersey shore hotel.

When Suyon and I arrived we discovered quickly she (as a Korean) was the only person there who wasn't white. Second, no one talked with us, except Cindy said a warm "hello" when welcoming us to the event. Suyon and I sat at a table and then Steve Corodemus arrived. He went to the head table first to let them know he was there, and scanned the audience like a good politician to start working the room. He spotted us and headed straight for our table and spent five minutes talking with us. He met Suyon before and was comfortable talking with her and any other citizen. Naturally, this was noticed, and after he left our table, we were quickly joined by others because Steve basically gave us his "stamp of approval" to the other the attendees.

But it was obvious from things stated during the event that the environmental movement didn't care about the economic consequences of their agenda, saw it to keep the poor, minorities and working class 'in their place', and supported its own agenda because as affluent people, they could afford it, whereas the rest of society could not. I discussed this with Tom Gagliano who not only confirmed my assessment, but also told me the federal government was watching these movements closely and it influenced federal funding decisions.

A TV report by a New York City TV station placed this squarely into context for me. The report involved Dr. Angela Cristini (PhD, CUNY, marine biology; Ramapo College), a recipient of NJ Sea Grant funds. Dr. Cristini studied the fate of pollutants in fin-fish and discovered that pollutants preferentially accumulated in the fish's glands, a delicacy for many minorities. She recommended filleting fish and disposing of the glands as a health guide. The response of the NJ Department of Environmental Protection (DEPE) to her findings was posting advisories in heavily polluted coastal areas.

Ramapo College issued a press release about Dr. Cristini's work. A New York TV station reported the story after sending a TV crew to a DEPE posted pier where three men were fishing and interviewed them. The first person interviewed was an 80-something unshaven white man who responded to a question about the advisory, "I ain't dead yet" while casting his fishing rod into the polluted water.

The second person interviewed was an African American who told the TV reporter that he read the advisory but had a family to feed and fishing saved money. The third person interviewed was a Hispanic man who said something about not understanding or reading "gringo rules." Clearly, these DEPE rules were well-intended but adversely affected racial minorities and the poor.

We submitted the Sea Grant Omnibus proposal in February, 1995, and our total grant was increased by the National office. Of the approved projects, we had more good ones than we could support. I could not renegotiate budgets, so ended up funding science projects and Tiedemann received no funds. It was the only way I could run the program. I offered him development funds in the fall if he applied, and we approved such funding later.

During March, 1995, I held a late afternoon meeting with Tiedemann. He yelled and screamed and called me a variety of unprintable names (He was less civilized than Dennis Wood). At 4:45 PM, I stood up and said, "John, I suggest you go home and discuss this with your wife. Let's meet tomorrow morning when you come in." I started to put on my overcoat and he left. I left 5 minutes later.

When I started my car to drive home, I said, "I gave up geology for this!! How do I get back in?"

I met John next morning and concluded the meeting at 9:00 am. I told Priscilla to hold all my calls, let people know I would see them the next day, opened my Rolodex file, and proceeded to telephone everyone I knew around the country to get advice about returning to geology. Their advice: Go into consulting. Everyone said that with my publication record, my short course teaching, and all my former students working in the petroleum industry, I had a network that could help me. I took it under advisement because my contract at NJMSC did not expire until July 31, 1996. I knew that NJMSC turned out to be a very bad match for me. Likely, it would have been a bad match for any of the three candidates interviewed in 1993.

In Late May, the biennial National Sea Grant Meeting met in Puerto Rico at a coastal resort and condominium complex. I found Puerto Rico not to my liking and regretted voting for this venue. By this time, Dave Duane retired and a temporary national director had been appointed. Sensing problems at the Consortium, I applied for that position when it was advertised.

Because of dissatisfaction with the overhead issue, Tiedemann went around me to several Board Members including Dennis Suszkowski and Cindy Zipf. During the June 1995, Board meeting, the Board went into executive session to discuss John's complaints. They appointed three board members, Noel Blackman, the president of a seafood importing company, Dennis Suszkowski, and Bob Tucker to review them and report outcomes.

In preparation for my meeting with this special panel, I decided I would be willing to forego renewal of my contract in exchange for certain considerations. I prepared such a list.

The panel met in late July. They interviewed Ceil Von Oesen first and then met with Joan Sheridan. Ceil, to her credit, explained that Tiedemann's project grants cost the Consortium additional funds, estimating the loss to range from $100,000 to $150,000. Joan apparently spent time on the "soft" issues about personal interactions.

I met with the panel next. Noel mentioned that Tiedemann received close to $600,000 in grants in three years. He considered that impressive. I explained that it was impressive but it led to a drain on the Consortium's finances because he never added overhead either as a surcharge or as additional direct costs. Consequently, NJMSC lost at least $100,000 which on an annual basis was 20 percent of the total state appropriation.

I then explained that in 1991 when the state appropriation was cut 50 percent, Bob Abel did not screen proposals. Any proposal that might generate income was justified to keep the consortium going whether overhead was collected or not. I said, "As NJMSC's fiscal steward, I had to bring financial stability to the daily operations. I therefore imposed a system to screen proposals to assure some level of overhead recovery or that billable direct costs were recovered in lieu of overhead." Blackman understood and even Suszkowski said, "George, it's good you are doing that. We use a similar system at the Hudson River Foundation."

Tucker, however, argued that because the projects did so much good for education in New Jersey the consortium should exempt overhead charges. I pointed out that if the Board could not sustain my screening for overhead recovery, NJMSC risked financial failure in the future.

They raised several personnel issues which I rebutted. However, Dennis Suszkowski picked up on some of the staff problems that I reviewed and said, "You know, we may have to fire everyone at NJMSC to set things straight." I replied, "That may have to be a Board decision down the road." At least Dennis realized that the problems were two-way streets and involved more than me.

They also brought up Tiedemann's complaint that I had "stolen" his Sea Grant duties when I arrived. I said, "First, I'm sorry John feels that way. In fact, I can understand his disappointment. However, within a month after arrival, Joe Nadeau called and explained that on behalf of the Executive Committee, I was instructed to take over these duties immediately."

Noel said, "Why wasn't the board informed?" as did Suszkowski. I replied, "I recall an announcement was made. Perhaps Joe Nadeau can tell you more." I explained that I reported the change at the September, 1993, Board meeting, and the restoration of NJ Sea Grant to good standing at the national level at the May, 1994 meeting. No one raised questions about it.

I then said, "You may not have known, but when I came, New Jersey Sea Grant was on a watch list for possible decertification by the national Sea Grant office. Actions I took within my first eight months saved NJMSC from Sea Grant decertification. Perhaps Joe and Jim knew a problem existed and wanted me to pay attention immediately."

Blackman wanted to know why the National office was considering decertifying NJ Sea Grant, I explained it was a combination of improper accountability, poor reviewing procedures, and recommending substandard research projects for funding. Tucker responded, "Oh, you're just attacking Bob Abel and Bill Gordon personally." I replied, "I don't think I mentioned their names. If you don't believe my assessment, why not call Captain Bob Norris at the National office? I can provide his phone number later."

The discussion continued and it was clear I successfully answered all their questions and rebutted all arguments. I saw frustration on their faces because they realized I was doing my job, perhaps my style was not their style, but I was doing what was required to keep NJMSC afloat and meeting all responsibilities. In short, they had no reason to discipline or sanction me. However, they were under pressure to find a way to get me out because NJMSC was no longer a happy ship. I only succeeded in surviving until my contract expired.

Finally, Blackman said, "You know, George, you must realize that if things don't improve, we may not renew your contract when it expires in July, 1996." I replied, "Sir, I recognize that's your right and in fact, I've thought about it. I'm prepared to forego a review for a renewal in exchange for certain considerations. Here's a list of what it would require" and handed them my list. The three of them were completely surprised and sat without saying a word. Then they read my list.

Blackman said, "George, these are reasonable requests and I think we might be able to meet them. I'll check on the item about health insurance. However, I think the rest can be approved. I have to check with the Executive Committee and poll other Board members." I replied, "Noel, I understand that. Technically, you have until July 31, 1996. However, some items require an early decision to be implemented, including the item about paying an executive placement firm to find me another opportunity. Also, the Board should settle this before conducting a search for my replacement." He agreed.

I informed the panel that I was leaving for Korea in early October and returning at the end of the month. I said, "It looks like we may not be able to settle this before early November"." Noel responded that was acceptable.

The outcome was that I had held my ground, and at the same time provided everyone with a reasonable solution. I could tell Blackman, Suszkowski, and Tucker understood that certain professional obligations were required, but we were all uncertain if other board members and NJMSC staff, once informed, understood that, and would function in a professional manner.

I went back to my usual duties trying to act as if nothing had changed. I was aware the rumor mill was active, but heard no further from Blackman or the other panel members.

My situation reminded me of a story Pat Domenico shared at Illinois about the ants in a cow pasture in West Texas. One day, two ants were busy enjoying the sunshine when a cow came by, raised its tail and buried them in a cow flop. The older ant sat still. The younger ant moved around and more 'doo-doo' covered his body. He asked the older ant why he sat still. The older ant explained that the cow flop would dry quickly, crack open, and they could escape. He concluded, "When you're in deep shit, don't move."

In mid-September, I completed annual personnel evaluations of the managers and Priscilla. By this time, I transferred the jurisdiction of Tiedemann's education department to Joan Sheridan to diffuse tensions. However, I had to complete an evaluation for John and concluded I couldn't do one. When I met him, Joan Sheridan was present and I told him that I found his actions unethical, reprehensible and unprofessional and therefore I would not evaluate his performance. I gave him a letter to that effect which he read, handed back to me and left. However, he came back and asked to have it back and I gave it to him. That may have been a mistake.

He called Suszkowski and Zipf who called Blackman. Noel called and suggested I suspend any further evaluations and I agreed. Tucker called and said I misused personnel evaluations to make personal attacks and that I should use them as a tool to bring about improved performance of the staff. I told him that was my goal, but when people were undermining the financial integrity of the organization and acted unethically, I needed to take stronger action.

I left for Korea, via Phoenix, AZ, to attend the fall meeting of the Sea Grant Association. The Acting National Sea Grant Director told me I failed to make the short list for that post. I assured him that was not a problem.

I also met Anne McElroy, the NY Sea Grant Director. She too was under fire with the New York Sea Grant community for reasons similar to mine. Anne grew up in suburban Chicago, and earned all her degrees in marine biology from M.I.T though their co-op program with Woods Hole. She accepted a job at SUNY-Stony Brook and rose through the academic ranks quickly because of her definitive research. She was appointed NY Sea Grant Director in 1993.

We sat down to analyze our situations and after reviewing problems we encountered, I said, "You know Anne, what surprised me about coming to New Jersey from Illinois was how weak the state higher education system is. I had a hard time understanding the consequences of that weakness." Her eyes widened and she said, "That's it! It's the same in New York State. It's unbelievable how weak it is." She went back to her full-time academic duties at the end 1995.

From Phoenix, I flew to Los Angeles and flew to Korea the next day. The conference at Cheju, where I gave a keynote talk, went very well and it was a positive experience to reconnect with the scientific community there. I also spent time meeting Suyon's family because she (See Chapter 28) flew to Korea ahead of me.

I had lunch with Blackman and his wife on my return. He brought an agreement I had previously screened and approved. I reviewed the agreement again and signed. He counter-signed and his wife signed as witness. Frank Wheaton resigned as board chairman and Noel was slated to take his place at the next board meeting.

While talking, Noel asked about my trip and then turned to his wife and said, "You know, George has a very young wife." I said, "Noel, how much younger than me do you think she is? Do you really know? It happens to be 13 years. She's blessed with excellent health." The Consortium staff clearly misread her age and fed Blackman gossip.

Next day, I met with the Consortium staff and announced my decision not to have my contract renewed in exchange for certain considerations that the board agreed to provide. I said I would stay on until July 31, 1996.

The November Board meeting was held ten days later and the agenda listed a search for a new president. When people arrived, many of the representatives from member institutions were angry at the change and asked what happened. I explained that to turn NJMSC around, I made tough decisions that did not go down well with staff and some former Sea Grant PI's.

During the meeting, many of the representatives from member institutions, including successful Sea Grant PI's, spoke in my favor. They chastised the board for not supporting me, crediting me with instituting increased accountability at NJMSC, increasing Sea Grant funding by 37%, and putting NJMSC on a more professional footing. One of them said that although he didn't always agree with my approach, he felt I did what was necessary to save the Consortium from a downward spiral that I inherited from the previous administration and made the difficult decisions to improve it. He advised the board to reconsider its decision.

Another one made it clear that I was the first NJMSC president who understood their concerns as academicians and this led, in his view, to improved relations between NJMSC and some of the member institutions. A third person stated that NJMSC went through a major and costly effort to recruit me and now the Consortium Board will have to do this again.

Blackman, Suszkowski, Tucker, Cindy Zipf and a few others were surprised at the strong opposition to decisions made since the June board meeting. I finally said, "Folks, I appreciate the support and want to thank you for it. However, given what transpired and the acceptance of my proposed considerations by the board, it's time to move forward and look to a different future. I tried to institute needed changes and when a new president arrives, there are bound to be more." That ended the discussion and I could tell by Blackman's non-verbal reaction that he appreciated my helping him out of a sticky situation. But clearly, I knew how to talk with the member institution representatives and Blackman and most board members did not.

Blackman asked me to run all decisions I made past him before taking any actions in the future. This, however, proved to be unworkable because he was either travelling or unavailable and sometimes not even reachable in the evenings at home. I unilaterally decided to only take major decisions to him running the risk that disagreements might arise. With half the issues, I decided that the next president can make the decision after July 31, 1996. When people objected, I advised patience.

After partial closure, the future of the Seaville facility was never resolved. At one Board meeting in 1994, it was suggested NJMSC give it away. Noel Blackman missed that meeting. I made inquiries with Ocean County but they had no interest. At a subsequent Board meeting, Blackman asked, "Why would you want to give that property away? This is an asset and it has value." I explained it was a Board decision at the previous meeting.

He proposed that NJMSC find an environmental clean-up firm and establish a joint venture partnership whereby the firm cleaned up the site during periods when their staff was inactive, and then sell the property and split the proceeds 50-50. I made inquiries and received expressions of interest, but Joan Sheridan, inferring that I could use a successful outcome to reverse my agreement with the board, undermined my efforts. Eventually such an agreement was executed and several years later, Seaville was sold to a developer.

I became suspicious about Joan for some time. One of the more senior member representatives told me that before Bob Abel became president,

Bob's predecessor was ousted by a combination of Bob and Joan. I therefore called Bob's predecessor and asked what transpired. He refused to tell me much.

Moreover, starting in November, Ceil stopped sending me financial statements on instructions from the NJMSC Treasurer. I was unaware that spending started to exceed income by a large margin.

In late January, I received notification from the State of New Jersey that our budget line was eliminated in the proposed budget. I immediately organized another grass-roots effort to reverse this decision. That effort was successful.

However, I made a major error in late January. Joan Sheridan brought a folder about a cooperative agreement between NJMSC and the Puerto Rico Sea Grant College Program. It was set to expire. I first thought that the next president should handle it. However, the expiration date was in June, 1996, so I couldn't avoid it. I examined the file to see what joint projects arose from the agreement. There was only one involving the lead marine agent, Alex Wypizinski. I decided because Alex could continue his work on a bilateral basis, I would not renew the agreement and wrote Puerto Rica Sea Grant accordingly.

My actions set off a fire storm. First, the Puerto Ricans charged I terminated the agreement because of an alleged bias towards Hispanics. Second, Dennis Suszkowski called and explained that Puerto Rico Sea Grant complained to Governor Whitman. I said I found it strange the Governor's office didn't call me first before calling him. Noel Blackman chewed me out for not running the decision past him, and ordered me to write Puerto Rico Sea Grant retracting my letter of termination. I explained I could but NJMSC never benefited from the previous arrangement. He suggested I write and I did.

That episode brought out more demands for my removal and several wanted me fired at the next Board meeting. Joe Nadeau explained later that Blackman told the Board in executive session that NJMSC had a contract with me and it was best to buy me out and appoint one of the board members, Dick Dewling, as Acting President. Blackman reminded the board that he did not want to run the risk of seeing NJMSC exposed in open court and the press about its operations and therefore NJMSC must avoid a potential breach of contract suit.

Blackman, Suszkowski, Dewling and one other board member met with me at the end of the Board meeting. The Board decided to buy out my contract, I was not to return to the office and they would mail me my paychecks. I explained I had a project contract to complete and was told to do it at home and file expense reports and time sheets by mail. I then explained my professional library had to be moved. We set a day to arrange for movers and I could come to NJMSC and move things out. I then requested the area be sealed off until I was ready to move. Blackman agreed.

The following Wednesday, I arranged for a local moving firm to pack and take my books home. Blackman and Dewling were there. I found out later, Blackman was trying to get an additional loan to tide the Consortium over until the end of the year. When the last load was ready to leave, I gave my keys to Blackman and left.

However, that was not the end. Two weeks later, Dewling and Blackman fired the accounting office except Ceil Von Oesen, one of the boat crew, a secretary, and one of Tiedemann's education staff to cut costs. Moreover, they cut everyone's salary 10%. I was notified by certified letter and called Frank Ngo who explained what had happened.

I reviewed the letter with my attorney, a specialist in labor law. He immediately saw a breach of contract and gave me two alternatives: sue for a large sum of money and spend three years in a court room, or because the difference in lost salary for five months was close to $5,000, sue for that amount in a small claims court and resolve it faster. I chose the second route but told the attorney, "Let's not file until August 1. I don't want to give them an excuse to fire me." The attorney agreed. We later sued and I collected the funds. My lawyer said, "A judge in a court of law in the State of New Jersey has ruled in your favor and the people at NJMSC now know that."

I completed my report on the small project I subcontracted from Rutgers on the Jersey Shore Master Plan and the $20,000 costs offset their financial losses considerably. Between my early departure at the end of February and July 31, 1996, I laid the ground work to start the next phase of my life.

Looking back, I realized I was as much to blame for what happened as anybody. It was a bad match from the outset although I enjoyed the Sea Grant role because it was closer to what I knew. The finances were a bit beyond me. Moreover, because my style was more focused and closer to "tough love" as Phil Sandberg said years before, it did not sit well within an organization viewed as a place that left everyone with a "warm, fuzzy feeling."

However, I learned some things too. I learned the legislative relations system and to develop political grass-roots efforts, both at the state and federal level. I learned much about business finances and applied them later. Above all, I developed a screen for toxic people to a level I never needed before and in time, this proved helpful also. I learned that when things are not rosy where one is, one should still be more discerning about where one wishes to go next. But above all, I survived the experience knowing I kept my integrity intact even if it cost me my job.

Ultimately, it boiled down to NJMSC and me being a very bad match. They should never have offered me the job, and I should never have accepted. Hidden agendas on both sides resulted in my appointment and changing circumstances and paradigms undermined the effort. Moreover, they did not provide all the tools for me to be successful. Both of us were at fault. It was that simple.

LESSONS LEARNED:

1) If during an interview or follow-up visits a bad match is indicated, decline the opportunity. The signs are usually there. In my case, I discounted the signs.

2) When interviewing for an administrative position, demand complete financial statements, including a P-and-L statement. If one is unavailable, request one be prepared. If such statements aren't produced, do not accept an offer.

3) If an employee is insubordinate over a key issue such as overhead recovery and fails to follow instructions, fire that individual. The financial integrity of an organization may depend on it. I didn't in Tiedemann's case, and should have done so.

4) In an organization where close to 75 percent of the employees were women, be more sensitive to their style. In my case, I dealt with them as professionals failing to realize they wanted something else.

5) Always plan an exit strategy. It limits surprises which will come.

6) If tough decisions about budget, personnel, priorities, and focus need to be made, implement them early (within six months). It becomes more difficult if one delays beyond that time.

7) Always hold to one's core values of integrity and accountability and be uncompromising about it. Doing so in this case steered NJMSC away from decertification of its Sea Grant program to an increase in funding of 37 percent in two years and the respect of fellow state Sea Grant directors.

CHAPTER 26
Consulting (1996-Present)

My career as a geological consultant began when I physically left NJMSC in late February, 1996, although I was invited to interview for two academic administrative positions. One was potentially attractive but it didn't materialize, and the second was a bad match from the minute I arrived at their campus.

The first position for which I interviewed was the Deanship of the Graduate School at the University of Maryland, Baltimore County. The university was more new age and politically correct than I anticipated and delved into new areas of scholarship with which I lacked familiarity. I met the president, an Illinois Math PhD from the 1970's and we did not communicate well. I also met the Provost, a lady with a PhD from George Washington University. She was one of the best people I met in such a position. We communicated well and I thought I could work with her. They hired a gerontologist from Boston University.

The second position was the Deanship of the College of a branch campus of the Long Island University (LIU) in the Hamptons, NY. That community was richer than my bank balance could accommodate. It was a beautiful place, however.

When arriving on campus, I was first interviewed by the search committee. Gerry Friedman told me that one of his students was Dean at the LIU main campus and assured him I was the administration's top choice. The branch campus had a marine institute which was represented on the search committee.

The faculty was unionized, the chair of the search committee was an English professor, and they immediately asked me to address sensitive internal problems without providing a briefing about their needs and concerns. Asking questions concerning sensitive internal problems of candidates is, in my view, unprofessional and inappropriate. Whether this was an ambush to counter the main campus's choice or just their institutional culture, I'll never know.

I concluded it wasn't going to happened and it didn't. I was totally unsuited for the place. The visit reminded me of Edward Albee's "Who's afraid of Virginia Woolf."

To start my consulting business, I first contacted my entire geological network to let them know I was in business and looked forward to working with them and to consulting leads.

Next, I talked with the Consortium banker. He advised when my contract at the Consortium officially ended, I should apply for social security benefits because I was eligible and to continue to draw my pension from Illinois to provide me with an income floor. He advised to also draw on my TIAA-CREF pension from my days at Penn and NJMSC.

That banker also explained that any income I derive from consulting was subjected to Social Security Payroll tax, but the benefit was that it would increase my annual payout to reflect added earnings. The rule of thumb in consulting is that one needs at least six months to a year of income as a reserve to keep going between consulting jobs. The two pensions and the Social Security stipend provided that financial base.

The local high school offered an evening class on starting a small business. For $20.00 it was a good two-hour investment. At the beginning, the instructor asked the class to show him our business cards. Mine read "George D. Klein & Associates" following a time-honored tradition I observed among AAPG members who became individual consultancies.

The instructor looked at me and said, "This card is uninformative. It doesn't explain what you do. You need to put that on your card." The person at the end seat in my row owned a pest-control business and his card had an ugly red cockroach image. The instructor handed me the card and said emphatically, "This card tells me what the business owner does."

I took his advice and changed my company name to "SED-STRAT Geoscience Consultants, Inc" because people could know quickly I do sedimentology, sequence stratigraphy, and basic petroleum geology relevant to their needs

During an AAPG in 1999, I had breakfast with Dag and Alicia Nummedal. Roger Walker (BS, PhD, Oxford, sedimentology; Post-Doc at Johns Hopkins, McMaster University, Petrocanada, Consultant) visited and gave us his card which said "Roger Walker and Associates". After Roger left, Dag asked why I didn't use my name the same way because I was 'widely known.' I told him about the above experience.

The 1996 AAPG Meeting was held in San Diego, CA in late May. I attended to reconnect with people and expand my network. I succeeded on both counts. I met with Dag Nummedal who was at Unocal and gave me lots of advice. Ed Clifton (BS, Ohio State, PhD. Johns Hopkins, sedimentology; USGS, Conoco) also gave me excellent guidance.

In addition, I met Jory Pacht (BS, Ohio Univ., MS, Wyoming, PhD, Ohio State; stratigraphy; ARCO Research and Exploration; Consultant, Seis-Strat Services, Energy Quest) again. He not only consulted, but developed a staffing division in his company and agreed to market my services. I talked with others especially to learn current trends and needs to see where I fit in best to market my services. The entire experience energized me to move forward as a geological consultant. In addition, I developed some leads for contract work.

During July, I realized that I needed start-up money. I did something that all pension fund managers automatically tell everyone they should never do. I obtained approval from the University of Pennsylvania Administration to withdraw $16,000 of principal from my TIAA-CREF account which was vested through them when I worked there. Doing so reduced my TIAA-CREF monthly pension by $50.00. I concluded it was worth the gamble. Having earned approximately $3/4 million in gross consulting income since 1996, the return on investment paid for itself many times over.

It took five months to receive approval to work, and then another three months to actually start work on my first consultancy. Gerry Friedman networked me to Longleaf Energy in Brewton AL. They requested I complete core description, a facies analysis of the Frisco City Sand, and recommend new locations for future drilling. Longleaf was a subsidiary of a family-owned lumber company in southern Alabama. During the 1940's, it formed an oil exploration subsidiary to find oil on their land.

I learned a valuable lesson during that consultancy. Sometimes one earns one's fee by the money one saves a client as well as by the money the client can make from a consultant's recommendations that lead to new oil or natural gas production. I noticed they obtained cores from red beds which are very poor reservoirs. I asked why and was told that they wanted more information. I explained that most redbeds are formed by dissolution of iron-bearing silicates that form clays which occlude porosity and permeability. Thus they likely would not contain oil or natural gas. I recommended they stop coring redbeds. The owner told me that recommendation alone saved his company approximately $35,000 per drill hole.

A lot of the client work I did was short term and included screening deals, advising a client whether to move in a new direction or a new play, core description, briefs for expert witnesses, and short term mapping.

Early in 1998, Dag Nummedal called. He was promoted to geological manager of the technical services group at Unocal and needed to hire a structural geologist. He was totally frustrated with the candidates referred by 'head hunters" and thought that through my academic network, I could do a better job. The Unocal higher management agreed. Unocal would pay a finder's fee.

The plan was to contact all the academic structural geologists I knew and find out who amongst their recent PhD's were unhappy where they were and would consider a change. I also screened any graduating PhD's that might qualify. I collected resumes, completed a phone interview with the candidates, and then forwarded all credentials and my assessment to Dag. He hired a structural geologist that way. In fact, Unocal subsequently hired both a sequence stratigrapher and a regional tectonics petroleum geologist through my efforts.

It was obvious that to be successful as a petroleum geology consultant I had to move to Houston, TX, and did so in May 1998. Moving to Texas was a good fit. Within two days I said, "It's great to be back in America." I found New Jersey had many amenities, but it is one of the most over-regulated states in the USA where one can't pump one's own gasoline, garbage disposal units are not allowed in a home (too much overload for antiquated sewer systems), and garbage was segregated into a variety of coded recyclables.

Within a month I found a consulting opportunity with Halliburton in Maracaibo, Venezuela. It involved core description for an EOR (Enhanced Oil Recovery) project but as the project moved forward, they found more work for me to do. I extended the visit. The project suffered from management turnover.

I made two more trips to Venezuela on behalf of Omni Laboratories, describing cores and made a presentation to their client, PDVSA. However, I have not returned because since 1999 Venezuela became an even more dangerous place to visit.

Many people were laid off by major oil companies during the economic downturn of 1999. I met many when attending regular meetings of the Houston Geological Society. One, Jeff Aldrich (BS, Vanderbilt, MS, Texas A&M, petroleum geology; Maxus, Forest Oil international, Consultant, South African National Oil Company, Evergreen Resources) and I developed a good relationship and he got a job within a month as chief geologist, Forest Oil International. That led to four consultancies involving work based in Houston on projects in Romania and South Africa.

I also agreed to offer a presentation at the 1999 AAPG Annual meeting sponsored by SIPES (Society of Independent Earth Scientists) and AAPG's Division of Professional Affairs (DPA) on starting a consultancy. I recall meeting a young geologist with Mobil who was preparing for an uncertain future. Only about six others out of a group of 25 from the Houston area stayed in Texas as consultants. The rest disappeared. I discovered later that after lay-offs occur from major oil companies in the Houston area half the people return to their childhood communities and do something else.

In 2000, Jory Pacht received a subcontract from Schlumberger to staff a project in Veracruz, Mexico, on Schlumberger's contract with PEMEX, the Mexican national oil company. He retained me to go there to undertake sedimentology, basin hydrological modeling, and seismic interpretation. It was an experience from which I benefited.

I returned to Villahermosa, Mexico, in June, 2002, to work on a PEMEX project that taught me much, but should never have been undertaken. It was to determine whether a producing gas field had remaining potential. I asked how much the initial estimated reserves were when the field was discovered in 1970 and how much was produced. 99 percent was produced by 2001! Eventually, the company terminated the project after I left. PEMEX initially rejected my recommendation to do so.

That trip turned out to be my last doing consulting work on-site overseas. I found Mexico became more dangerous where literally one did not go out on the streets from one's hotel at night or on weekends. Because Houston is the oil capital of the world, I continued to work on overseas projects in Houston. With improved digital data acquisition, one can undertake projects anywhere in the world without leaving the Houston area.

In 2001, Phillips Petroleum retained me to complete a three-month seismic and sequence stratigraphy study of the tight gas sand play in the Bossier Shale (Jurassic) of the East Texas basin. The play was discovered by

Anadarko Petroleum in 1998. Phillips had pipelines in the area and their goal was to find Bossier production to keep the pipelines supplied with gas.

The project moved forward and I developed lead areas. I proposed extending my contract to develop these leads into drillable prospects. There was a lot of interest in having me do so. However, during my last two weeks, I noticed the atmosphere changed, and the project was terminated. When it was, I asked if the science part could be released for publication and presentation, and my request was approved.

I gave my final presentation followed by a memorable lunch by my co-workers and the project management. Three days later the ConocoPhillips merger was announced explaining their sudden declining interest in retaining me.

During 2002, I completed projects for Samson, Forest Oil, N. S. Neidell and Associates, and Omni Labs. Norman Neidell (BS, mathematics, NYU, MS, Imperial College, PhD, Cambridge; geophysics; Gulf Research, several of his own companies, Neidell and Associates) is an eminent geophysicist who received a contract from Vanco to evaluate a concession in Senegal. My task was seismic stratigraphy and comparing my results with a holographic method Norm designed to highlight high amplitude and velocity signals ("bright spots") indicating the probable presence of hydrocarbons.

We recommended several locations around a dome of unexplained origin but Vanco relinquished the concession.

2003 was one of my better years. Late in March, I received an email from the geology manager at Angola LNG, a multi-partner organization including Chevron, BP, ExxonMobil, Total and Sonangol (National oil Company of Angola). He needed a sequence and seismic stratigrapher to work on the Congo basin and asked about my availability. I responded with an email and resume and followed up with a phone call. They interviewed me a week later. I started working until the end of the year.

The data set consisted of well logs, biostratigraphic reports, and three combined 3-D seismic surveys as well as several 2-D lines. Using biostratigraphy, I established a seismic and sequence stratigraphic framework and mapped it on their seismic template. The focus was on an unassociated gas discovered in 1970 by Texaco for an LNG project. The group was interdisciplinary with reservoir, drilling and chemical engineers, geologists, organic geochemists and geophysicists.

The seismic data enabled the team to map the changing history of the Congo River. It was a great example of the global application of the fluvial axes concept developed in the Gulf Coastal Plain of the USA.

That LNG project also provided a glimpse of how less developed, resource-rich countries are manipulated into projects by European countries, particularly in Africa. After the 25-year Angolan civil war ended in 1995, Angola was cash-strapped. The Angolan government applied for a development loan from the IMF (International Monetary Fund), and a team flew from Europe to Luanda for a site visit. About an hour before landing at night, they flew over offshore oil platforms which were flaring gas. In fact, the flames from gas flares were so high that the night sky was as bright as Las Vegas at 2:00 am.

The IMF group immediately scolded the Angolan government in a patronizing manner about their poor global citizenship and the negative impact of gas flaring on global warming. They recommended that an LNG project be undertaken to reduce gas flaring and generate income for the country. The Angolan government agreed and that's how Angola LNG was established. Angola also received IMF funds for agreeing to do so.

Since then, the Angolans found another source of development funding that is not agenda driven but focuses instead on Angola's immediate needs, China. China, in fact, made significant inroads commercially throughout Africa because they focus on trade and meeting needs of African countries rather than seeking an agenda-based approached to global problems. During 2003, the *Houston Chronicle* reported that a 'China Town' was being built in Luanda, Angola for at least 5,000 Chinese.

I entered a slower period after my stint with Angola LNG doing very short term projects for various clients. However, with the increase in oil prices beginning during the spring and summer of 2005, my consulting fortunes improved. Between September 15, 2005 and October 31, 2008, I had my best consulting years since starting consulting activities.

In September, 2005, after recovering from a laminectomy, I received a call from Pat Donais (BS, Tulsa, Geophysics; Phillips, consultant) at Dune Energy. We met at a SIPES luncheon several years before. He worked half to three-quarter time at Dune Energy as a consulting Exploration Manager. He explained their contract half-time geologist passed away. I interviewed and met the president, the CFO, and the operations manager, a petroleum engineer, and was placed on a retainer for a half-time commitment on site.

They asked me to develop a prospect in the Gulf Coast in the Yegua and Upper Wilcox plays. They also were networked to an affiliated company which had Illinois basin production and asked me to be on call to them whenever they needed my expertise.

The arrangement worked out reasonably well and I spotted one location for them. It had a barely commercial show and then discovered later, the lease was not in good order. We mutually agreed to terminate the relationship at the end of January 2006.

The day after concluding my assignment with Dune Energy, I received a call from Mike Stearns (BS, Indiana, Geophysics, Tenneco, consultant, Mohave Oil and Gas) at Mohave Oil and Gas which is active in Portugal. One of their partner companies in Calgary, Canada, Dual Energy needed an independent estimate of Mohave's Portugal reserves for their IPO on the Toronto Stock Exchange. Mike inquired if I was available to calculate estimated reserves. I agreed, the Dual Energy people contracted me and I worked on site at Mohave evaluating all their reports. I spent six weeks completing calculations and probabilities of success, and my report. Dual ultimately got listed and their IPO was a success.

During 2002, I attended a SIPES meeting and met a geophysicist, Wulf Massell (BS, Minnesota, PhD Texas; geophysics; various oil companies, consultant, Epic Geophysical, Fusion Petroleum Technologies, consultant). Wulf joined Fusion Petroleum Technologies (referred to herein as 'Fusion') in 2004. Fusion is a geophysical processing firm which planned to undertake integrative projects. Wulf networked me to the president, Alan H. Huffman (BS, F&M, PhD, Texas A&M, rock mechanics and pressure regimes; Exxon Production Research, Conoco, Fusion) who I met.

During 2004, Fusion was reorganizing and redirecting its efforts to grow. The company was originally founded by John Castagna (BS, Brooklyn College, PhD, Texas; geophysics, ARCO Research, Univ. of Oklahoma, Univ. of Houston). John was a pioneering geophysicist who improved the analysis of AVO (Amplitude VS Offset) as a hydrocarbon indicator. He and his wife owned Fusion and when things did not go well during 2002, they brought in Alan Huffman who invested and owned a minority (40%) share of the company. The company was headquartered in Norman, OK, with a Houston office in the Woodlands, TX.

Alan called during 2005 to include my name on projects on which Fusion bid but nothing developed. Early in 2006, he called again and brought me in to consult for one month on a project involving a data set from Russia. My task was interpreting seismic stratigraphy and depositional systems. We recommended that the client not invest in the opportunity.

Alan also explained that Fusion successfully bid on a six month's project to work on two offshore blocks off the Louisiana Continental Shelf in the Gulf of Mexico and retained me for that project too. Because I lived 55 miles from their office, Fusion rented a two-bedroom furnished apartment, I paid half the rent and shared it with short or medium-term consultants or affiliates from Norman, OK working on projects in The Woodlands. Alan also raised my daily rate to offset billing for a per diem. It was a great arrangement.

My project manager was Allen Bertagne (BS, Imperial College, Univ. of London, PhD, Texas; geophysics, Exxon Production Research, PGS, Fusion, Shell Oil). He was one of the best people to work with, always generating ideas, keeping the project moving forward, and maintaining focus on a common goal. I was surprised he left to go to work for Shell Oil on December 1, but remotely completed the project and attended the final client presentation.

When Bertagne left, so did a skilled petrophysicist and geophysicist in the Norman office. I later learned that the Castagnas decided to sell Fusion. Alan Huffman bought an additional 20% share and brought in an investor/partner as a minority shareholder. Mrs. Castagna stayed as Vice President for Administration.

Mike Stearns at Mohave Oil and Gas called again early in 2007. Mohave decided to get listed on the Toronto Stock Exchange also and do an IPO. I spent six weeks during the beginning of the year recalculating reserves, adding areas that did not involve Dual, but as rules tightened in Canada, periodically made changes later. Their IPO is pending.

I returned to Fusion from April through November, 2007, working on three projects, one in the Permian basin, another in the San Joaquin basin, and a longer project in Galveston Bay. Work, however, was not continuous; a few weeks on, a week off, two months on, a month off, and back for another month or two. Again, Fusion rented an apartment which I shared with one of their employees.

The company also changed. During 2006, it had about 20 employees. During 2007 it grew to at least 30, and an office was opened in Libya, which would employ 30 people. Moreover, a comptroller with extensive international experience was added as was a new group that licensed, sold and supported Fusion's proprietary software. Since 2006, Fusion acquired two additional companies and is now renamed FusionGeo, Inc. During 2008, Wulf Massell left. I did not return because Fusion hired a young geologist from Algeria as a cost-saving measure.

Reviewing my experience with Fusion, I enjoyed the work and the people. With their reorganization in 2007 came some staff turnover and the goals of the company appeared to change as well.

During 2008, I worked on and off with Norman Neidell, completed more work for Mohave's IPO, and also worked briefly with Mike McCardle (BS, MS, Stanford, geophysics; Tenneco, consultant). McCardle was a captain in the US Army Special Forces in Vietnam before working as a geophysicist. We developed good rapport. He asked me to help with consultancies in the Maverick basin and the East Texas basin on the Bossier play and its supposed age equivalents for Blue Star Petroleum.

Norman Neidell also networked me to Hermann Homann, the founder and owner of Fronterra Geoscience, a company he formed in 2004. It now employees at least 40 people with offices in Buenos Aires, Vienna, Oklahoma City, Dallas, Denver, Midland, and Oman. They specialize in processing and interpretation of borehole image logs. Hermann, a German immigrant, worked for Dresser and Baker Hughes before going on his own.

Hermann now is expanding the scope of Fronterra to be more integrative. He already contracted me to develop a seismic and sequence stratigraphy short course and to present it. The first course offering went well.

2009 started as a year of great uncertainty with changes in Washington DC, and a global economic turndown. I noticed my colleagues in Houston who consulted exclusively on Gulf Coast and Gulf of Mexico projects were experiencing difficulty finding work. I expected a difficult year for me as well. However, all the activity in the oil industry during 2009 was international and because of past experience on international projects, I completed a three month project on East African data for the US division of an Australian mining company, and am presently working on a project with BPZ Energy, LLC, in Peru.

Reviewing my consulting history since 1996, I wish I had done this sooner. I met a high class of ethical people in the Houston petroleum geology community, and made some new friends. The consulting community in

Houston is willing to help newcomers, help each other over the long term, network each other to opportunities, and displays a higher level of collegiality than I found in academe.

A good support structure exists in Houston enabling a consultant to be successful. First, there is the Houston Geological Society, the largest local geological society in the world, with a membership larger than the international SEPM. It sponsors general dinner and luncheon, International dinner, and North American dinner meetings with many excellent technical talks that allow one to stay up-to-date and learn about geology all over the world. Moreover, it provides an excellent venue for networking and meeting people. It sponsors short courses enabling one to stay current.

SIPES has a chapter in Houston and meets monthly in the Petroleum Club. It is a business networking organization devoted to the advancement of the independent consultant and independent petroleum geological operator. Their monthly meetings focus on networking and include a technical talk. I've met many people through SIPES who I would not have met any other way.

Living in the Gulf Coast, I attend annual meeting of the GCAGS (Gulf Coast Association of Geological Societies), the Gulf Coast Section of the AAPG. The meetings are great for networking and the technical talks are of high quality and focused on the region's geology. Through it, I met many of my counterparts in other parts of the Gulf Coast

Last, I should mention the GCSSEPM (Gulf Coast Section, Society of Sedimentary Geology). They sponsor an annual three-day research symposium that is one of the best I ever attended. The networking opportunities are excellent. Technical talks are state-of-the art. They publish their annual symposium volume and it surprises me how few geologists worldwide are aware of its existence. Some of the best papers on deep-water exploration geology, sedimentology and sequence stratigraphy are published here. It's the best thing SEPM provides its membership.

With this support structure of kindred people, a person can do well as a consultant in the greater Houston area. Add to it the ability to network and to develop a few elementary marketing skills make it possible. My dad taught me the art of networking (Chapter 2). I learned elementary marketing 'on the job.'

Last, a word about the City of Houston. It has matured into an international city with a world class medical center, outstanding museums, outstanding performing arts centers, and world class restaurants as good as any I've enjoyed in the New York area, and more cost effective.

It was not always that way. During one of my earliest trips to Houston in 1963 to attend an AAPG meeting, the venue was the 'Cow Palace.' When the AAPG members arrived on Monday morning, a cattle show obviously was held during the weekend and Pinesol could not mask the odor of cow flops. I returned frequently during the 1970's and 1980's to attend AAPG conventions and teach short courses and saw the city develop and mature with a convention center, new buildings, a park system, improved museums, and improved restaurants and hotels.

Affordable good quality housing is available. Air Conditioning modulates the summer humidity of its tropical location. The traffic is not nearly as bad as in the New York City or New Jersey. Houston is no worse than any major metropolitan American city I ever visited, and in many ways, has better amenities.

When I moved to Houston in 1998, I sent emails about my relocation to everyone I knew. Bill Fisher emailed me back saying, "Eventually all geologists come to Texas." I'm happy I did.

LESSONS LEARNED:

1) When marketing services to clients, always make sure to spell out what one can and cannot offer or undertake. It may mean the contract is awarded to someone else, but they generally comeback later to ask you to do different work.

2) When meeting a client, often they don't know what they want. It is best to help them define their goals, and help them move in a direction where mutual interests match. Again, it may mean someone else gets the contract, but they come back with a second opportunity later.

3) Meeting clients is about establishing long-term working relationships.

4) Be very aware of the scope of work for which one contracts. If a change in scope arises, be prepared to negotiate in a friendly way for its inclusion and extending work deadlines and commitment of funds for the extra work.

5) Sometimes, one earns one's fee by saving the client money.

6) Be prepared to do 'good deeds' for other people and to serve professional societies. However, limit service to no more than two professional society commitments at one time.

7) When meeting new clients or the public, or one's compatriots at professional meetings, always wear a suit and tie or coat and tie. If a client's office is "dress casual", one can adjust when working on site. It's hard to adjust in the other direction.

8) When undertaking reserve calculations, screening deals, or work involving offerings for an IPO, increase one's fee because the exposure and risk is greater.

9) Keep abreast of the industry by subscribing to Harts' publications and reading the business pages of the Houston newspaper, even online.

10) Maintain the highest standards of ethics when dealing with clients, compatriots, team members and office staff.

PART II:

Additional Events and Experiences

CHAPTER 27
First Marriage

The reader who has known me a long time already is aware that my marital history was less than exemplary. Regrettably that's true. My first marriage was to Joan E. Larsen (BA, Michigan, PhD, Radcliffe College of Harvard University, Renaissance English literature; taught at Wisconsin, Pitt, Duquesne, Bryn Mawr, Illinois) who I met in Pittsburgh in 1961. I had warning signs early that this may not go well and completely ignored them.

I prepared two courses from scratch my first semester at Pitt. Often I returned to my campus office after dinner to continue preparation and arrived home around 9 or 10 PM. I lived in an apartment building on a hillside with a parking lot in back of the building. That lot was a paved terrace cut into the hillside, oriented parallel to the apartment building, and sloped about 2 degrees towards the building. Parking spaces were marked and each apartment received one. Stripes delineating parking spaces were marked in the downhill direction.

During September, 1961, a Volkswagen with Wisconsin plates was parked across two parking spaces at right angles to the marked stripes (in geological terms, parked across slope instead of downslope or upslope). By 10:00 PM all parking spaces were filled, and I parked at the end of the lot in the weeds.

After two nights, I asked a neighbor if she knew who owned the VW. She explained it belonged to a lady who was at Pitt as a Mellon Post-Doctoral Fellow and told me her apartment number. I knocked on the door and the VW owner was Joan Larsen. I explained the problem and she expressed concern her VW might roll downhill having recently learned how to drive. She had neither learned to use the parking brake or set the gears in reverse or both to prevent such a calamity. I later demonstrated it for her.

She asked what I did and when I told her she asked where I earned my PhD. I replied, "Yale." Her response was, "I'll recognize Yale." She then disclosed that she earned her PhD at Radcliffe College which was the woman's division of Harvard. Harvard awarded graduate degrees to women only through Radcliffe College until 1962.

After three years of a very limited social life at Yale, a disappointing social life in Tulsa, OK, and not yet meeting any interesting ladies in Pittsburgh, a lady PhD in English Literature seemed like someone I ought to know. I offered to take her to dinner and she accepted.

We dated regularly until December and by that time, I was blindly smitten, and proposed marriage. She was driving back for Christmas to Menominee, MI, to visit her parents and wanted to think it over. On her return, she accepted. Joan said her mother told her, "You don't want to keep that nice boy waiting."

We got married in April, 1962. She received a temporary faculty appointment at Pitt for the 1962-63 academic year, and then accepted a permanent position at Duquesne University. My career at Pittsburgh was on a downward spiral (Chapter 10) by then. When I accepted Penn's offer, we commuted for a year on weekends.

She moved to Philadelphia in 1964 and taught half-time at Bryn Mawr College until we moved to Illinois. She was successfully appointed to a tenure-track position in the Department of English at the University of Illinois in 1970 and taught there until retirement around 2002.

We had several disagreements about a range of things early in our marriage. We could never resolve them because she always deflected discussions so closure was impossible. Life for her was an endless seminar on any topic that could never be resolved. Occasionally she responded directly, but still, nothing was resolved. She clearly was a combination of passive-passive to passive-mildly aggressive. It became increasingly difficult to live with her.

We adopted a child in 1967 after failing to conceive any. The best option in the Philadelphia area was through an Episcopal family agency which provided excellent prenatal care to expectant unmarried mothers at their own facility. That assured some advance knowledge of an adoptive child's prenatal health history, an important consideration we discovered later.

That agency developed a unique way of screening candidates. At a public meeting attended by approximately 200 people, the major screening paradigm was the following question: "When do you tell your adopted child that they are adopted?" Joan and I sat and listened as others argued but the only answer the agency would accept was: "From the very beginning and all the time." That was a paradigm Joan and I could live with and we filed a formal application.

Screening consisted of several evening interviews, and a site visit to our home. They placed a 6-week old boy with us and we named him Richard Larsen Klein. Richard has grown up to be a first class citizen probably due to the excellent prenatal care he had and his own ability to scope through problems. In some ways he showed he learned things from me and in other ways he didn't.

Richard earned a BA in political science at Michigan State in January, 1990, worked at Quantum Research in Washington DC, for three years, and accepted a job with NDI, the National Democratic Institute for International Affairs where he still works today. He started as a field officer in Namibia for three years. Richard then worked in other southern African countries (South Africa, Zambia, Malawi, Zimbabwe) while based in Johannesburg, South Africa, and slowly moved through the ranks at NDI. He completed a Master's degree from the London School of Economics in 2007. Richard is now NDI's Senior Adviser for southern Africa and lives in Washington, DC. I'm very pleased with and proud of how he has grown.

After moving to Urbana, we adopted another child, Roger Nelson Klein. He was placed in August, 1970, through the Illinois Department of Children and Family Services the only agency in the region. Being new to Illinois, we were unaware that this agency had a very spotty record with adoptions. Roger's birth mother appeared to have received minimal prenatal care and in time, we paid the consequences.

The marriage continued to run into difficulties and finally I sued for divorce. We were divorced in April, 1973. Richard was nearly five years old and Roger was almost two. The decree stipulated my visitation rights and when in town, I tried as much as I could to see both boys on Saturdays. I also paid monthly child support until they were 18, and was responsible for half of their undergraduate college education. Joan's parents moved from Menominee and lived with her to provide free child care and housekeeping.

Joan's mother was polite whenever I picked up my sons for visitation. Her father, however, refused to speak with me. His attitude carried into the boys' life and he was undermining everything I tried to teach them. With Roger, this stuck because as Joan once told me, "He never really knew you." With Richard, however, some things I taught him lasted, perhaps because I was around him longer.

Roger, after listening to his grandfather negate things I suggested, often refused to join Richard and me for visitation. Regardless of circumstances, I did what I could to meet my responsibilities to the best of my very limited child-rearing abilities. I made sure the monthly child-support checks arrived on time, even when travelling or on research missions far from Champaign-Urbana.

When my mother passed away in 1977, Joan called to inquire if she left anything for Richard and Roger in her will. I told her to contact my father. The will stipulated that $50,000 was to be taken out of my share of the estate when my father died and put into two trust funds of $25,000 for a college education fund for each of my sons. My father established the trusts immediately after my mother passed away to maximize gains from high interest rates.

When my father passed away, Joan called again to inquire if anything additional was left for Richard and Roger. I told her that was all they would get. She was most unhappy.

As my financial circumstances improved, I took both boys for a week to Colorado in August, 1979, and a week to San Diego in June, 1980. I used both trips to show great scenery and different types of restaurants, museums, and opportunities for children's activities that were unknown or unavailable in Champaign-Urbana, IL

The excrement hit the fan, to put it politely, in February, 1984. Joan called to explain that Roger was having academic and behavioral difficulties at Urbana Junior High. The school recommended he be tested at Northwestern University's medical center, which she arranged without my involvement or advance notification. Test results showed Roger had a learning disability which allegedly explained both problems.

I later met the people at Northwestern's medical center for confirmation and asked questions which they could not answer. They explained how lack of structure at home, self-indulgence of Roger's whims by Joan, and the role of his deceased grandfather may have contributed. They recommended Roger enroll at a special private school for children with learning disabilities. I asked whether Roger, who experienced a rapid growth spurt, had a related physiological disorder that may partly explain his problems, having experienced something similar myself during my early teens. Again, they had no answers.

When the bill arrived from Northwestern, Joan called and I agreed to pay half. When payment for his private special education arose, I told her, "Joan, I'll pay half the first year only. We should then evaluate to see if it is helping Roger, or if there are other root causes and other alternatives."

Roger enrolled in the fall of 1985 at the Landmark School for Language Learning Disabilities in Prides Crossing, MA. It was costly (average annual tuition around $10,000 in 1985 dollars). Moreover, I became dubious of the school director, a minister. His methodology to raise money and add fees for every one of the students' activities appeared capricious at best.

Joan contacted me in March, 1986, about a second year at Landmark. I asked her to wait until June so we could evaluate the first year. She claimed that the school wanted a commitment in March for August registration, including a deposit, and she needed to know now. I told her to wait until June so we could complete a full evaluation of the effectiveness of Landmark's program and how much improvement Roger showed. During vacation visits to Urbana, I hardly saw any.

Her response was to file a lawsuit in April 1985 for additional support. I retained Robert Isham Auler to represent me. He correctly described the case as classic 'post-accretal divorce' to clarify loose ends from original divorce decrees.

The case lasted for three years. Compounding it were numerous discovery requests by her attorney for my financial records. We requested hers and discovered she falsified her disclosure form for Richard's financial aid package at Michigan State. Moreover, she often refused to produce Roger for visitation.

I may not have acquitted myself that well in the eyes of the court, probably by appearing argumentative. When the trial began, Joan's allies were in court hoping to see me smashed by her attorney. I came prepared and not only answered his questions, but also asked him to be more specific and successfully rebutted him. Although I exercised my legal right to do so, the Judge was not pleased that a defendant out-debated an attorney in his court room.

During the spring of 1985, I received an announcement that the director of the Landmark School was to be interviewed on the morning NBC "Today Show" by Bryant Gumbel. I contacted the Today Show's producer on Bob Auler's advice. She allowed me to send questions for Gumbel to screen for possible use. I prepared ten questions and watched the show.

The show started with scenes from the school and the usual questions such shows carry. As the show ended, Gumbel asked "You know, the school provides a good program, but isn't it rather expensive?" The school director responded, "Yes, but it is deductable from Federal income tax" and the show ended.

Gumbel apparently understood my concerns and narrowed them to the cost issue. Moreover, the school director's response was totally inappropriate and undermined his position. He failed to use the opportunity to extol the cost benefits the school offered one more time, in my view. In short, he exposed himself, and during my visit to the school that fall, it was clear he knew it.

The issue was further compounded in February, 1987, when I arrived for another court hearing and sat with my attorney. Both attorneys then entered the Judge's chamber and left Joan and me in the courtroom. She said nothing but looked extremely agitated. Both attorney's returned, told us the Judge wanted to meet with both of us and our attorneys in his chambers. Bob Auler disclosed that Roger was just expelled from the Landmark School for poor behavior. In short, their program did not improve his behavioral problems and failed our educational expectations.

We met in the Judge's office. He explained that Roger was expelled from Landmark and already on an airplane home. The judge wanted both Joan and me to meet him together at the airport to reassure him he is "OK." I agreed, but I also told the judge that I was uncertain that Roger was properly evaluated regarding his academic, learning disability and behavioral problems. I wanted to know what was really wrong with him so that we could all make an informed decision. I requested the right to have him independently evaluated before any further choices or decisions were made about schooling. The judge agreed and ordered me to arrange such testing.

With Bob Auler's help, I arranged both an educational psychological and a psychiatric evaluation. However, Joan failed to produce Roger for one of the educational psychologist's appointments and when Bob and I returned to court, the judge ordered Joan to produce Roger for such testing, but failed to cite for contempt of court.

During one of these court deliberations, Auler stood up and said, "Your honor, this case is crying out for a solution" and proposed that the court consider placing Roger in a foster home. The judge responded in two ways. First, he said he would consider it and give an opinion at the next court hearing (he declined the recommendation), and second, he admonished both Joan and me NOT to disclose this recommendation to Roger.

Roger was scheduled to meet me at my campus office three days later. When Roger didn't arrive, I called Joan. She said she dropped him off at NHB but admitted that she told Roger about the foster home option.

I called Bob Auler and we were back in court. The judge castigated Joan in no uncertain terms but again did not cite her in contempt of court. It appeared that the Judge was reluctant to order a ruling against a tenured professor in the county's largest employer.

Joan then unilaterally enrolled Roger in the Brehm School in Carbondale, IL. Its mission was identical to the Landmark School. Again, Bob Auler and I filed a complaint in court. The judge did nothing.

The judge finally ruled I pay 55% of Roger's costs and Joan pay 45%. He also ruled Roger did have a learning disability, a finding that pleased Joan but not me. I was uncertain what the future held for Roger to be so defined.

Between these costs and Auler's legal fees, this case hurt financially and it took a long time to recover.

Roger was expelled from the Brehm School in March, 1988, for misbehavior and graduated from Urbana Senior High School. His class ranking, surprisingly, turned out to be higher than Richard's!

Roger tried a series of Art Schools after initially starting at Parkland Community College in Champaign and periodically dropping out. He now works in Chicago as a commercial artist.

In retrospect, I bear the responsibility for having made an extremely poor personal choice. I had my clues from her comment about recognizing Yale. I should have been more astute and wasn't.

The good news is that my oldest son, Richard Klein, turned out well, has learned the need to do the 'something extras' that spell success, and is a productive member of society. For that I'm grateful.

Richard visits me once or twice a year depending on his schedule. I haven't seen Roger since 1991. He told Joan to let me know that because he was of age, he didn't want to see me ever again.

LESSONS LEARNED:

1) When meeting people and observing negative or unusual behavior that makes you uncomfortable, be more discerning, without necessarily ruling out the potential of a good working or personal relationship. Just be more guarded. When I met Joan Larsen, I wasn't.

2) If in a protracted lawsuit, it might be best to seek a settlement sooner rather than prolong the problem out of blindness, consequences for personal relations, and one's financial situation. Neither Joan nor I attempted to do so and in retrospect, that's what we should have tried.

CHAPTER 28
Suyon Cheong

Memorial Day Saturday, 1993, was my first free weekend day since arriving at NJMSC to assume its presidency (Chapter 25). By then I attended an NJMSC Board meeting and unpacked my personal belongings at my apartment and NJMSC office.

Memorial Day weekend is extremely busy on the New Jersey Shore. I decided to visit New York City on Saturday. I needed to buy clothes, and wanted to enjoy the cultural amenities of the city, being deprived of them during most of the previous 23.5 years in Champaign-Urbana. After spending the afternoon at the Metropolitan Museum of Art, I went to a Japanese Restaurant.

The restaurant hostess seated me next to a table where two Asian women were talking in Korean. I was soon lost with my own thoughts and when the appetizer arrived, I started to eat with chopsticks. One of the Korean ladies next to me said, "You know how to use chopsticks really good. Where did you learn that?" I told her politely I lived in Korea and Japan. She asked why I lived in Korea. I explained about my visiting professorship at SNU, and subsequent visits to raise research funding. The lady who asked about my chopstick skills was Suyon Cheong and she was single. Her friend was married.

The three of us continued to talk. Suyon's married friend then left but Suyon chose to stay. I offered to buy Suyon a drink at a nearby cocktail lounge, and she accepted. We continued to talk, exchanged phone numbers and agreed to stay in touch.

Suyon impressed me very quickly. She was well spoken, had a reasonable command of English, was smartly dressed, and had a presence about her that I found attractive. Suyon radiated a combination of serenity, spiritual strength, and self-discipline. I never met a woman before who had such command of all three. We dated on subsequent weekends, and late in June, I invited her to spend a weekend with me. She accepted. I took her on a three-hour brunch cruise in the Navesink River.

As we dated, I found out a lot about her. Suyon was the oldest of four children and was born in 1946. Her father was an architect but died during the Korean War. He also was a Tae Kwan Doe champion. As a young man, he visited festivals in rural communities and competed. Often he won. The first prize was a cow which he would sell.

Suyon's mother had four children so when widowed, she arranged for her oldest brother to look after Suyon and her sister. Her uncle lived on the East Coast of Korea just south of Samchok in Kangnam Province. He owned a distillery and from his earnings invested in the fishing and the food business. He had the means to raise Suyon, her sister, her numerous cousins and his own eight children during very difficult times. Her two brothers remained with her mother.

Her grandmother also lived with Suyon's uncle and helped raise her. It wasn't until 2007, however, that I truly understood the root of Suyon's inner and spiritual strength. We visited her aunt and because it was my first visit to Samchok, Suyon gave me a tour of the important places from her childhood.

We visited her uncle's tomb, her elementary, junior (middle), and senior high schools, and then drove from her aunt's house to a Buddhist Temple on a hillside where the family worshipped and where her grandmother's remains were buried. The temple was 5 km from her aunt's home. Suyon explained (after 13 years of marriage) that her grandmother wanted Suyon to become a strong woman. Every Friday after school, they both walked the five km to that temple and stayed overnight through the weekend. Monday morning at 4:30 am, Suyon woke up and walked to school. Korean winters are known to be brutal. But Suyon and her grandmother walked to and from that temple every weekend regardless of weather conditions. It explained a lot about her.

Suyon returned to Seoul after graduating from high school to live with her mother who owned a grocery store. Suyon wanted to attend nursing school but had no tuition money. Fortunately, the pragmatic Korean medical system arranged a job for Suyon as a nurses' assistant during the day. She went to night school to earn her RN Certificate from the Kyungi Medical Center with qualifications in Oriental and Western medical nursing. She paid her way and obtained valuable experience simultaneously. She worked as a nurse for three years and then opened a retail store selling household linens.

Suyon told me much about Korea I did not know. Because her father was no longer living, it was difficult for her to marry under the Korean tradition of matchmaking whereby not only are a bride and groom united, but so are their complete families. She met and married a senior American executive widower who worked for a large multinational corporation when she was in her late twenties. In 1980, she immigrated to the USA. After eight years of marriage, she got divorced, and opened a retail clothing business in Manhattan. When I met her, she had sold the business.

We spent nearly every weekend together that summer. One weekend she insisted on cooking dinner and after completing the grocery shopping together, she ordered me out of my kitchen and served a great meal. In fact, I noticed over the weekends she stayed with me that she took as good or better care of me than my mother. At the end of July, I proposed marriage and a week later she agreed. She moved in with me in late September.

We shared a lot of common interests particularly in art and music. We also enjoyed the out-of-doors. Our love continued to grow.

My two-bedroom apartment was too small, especially when she moved everything she owned into it. We looked for a larger place to rent but were unsuccessful in finding one. We found a town-house complex in Matawan, NJ that seemed possible. It was a 22 mile drive through eight New Jersey townships from my office. When I sold my house in Champaign, IL in April, 2004, we opted to buy a townhouse there and closed on June 1, 1994.

Suyon and I were married in my attorney's office in Little Silver, NJ, on September 23, 1994. Her best friend Mrs. Song, and my oldest son, Richard, who came from Washington, served as maid-of-honor and best man, respectively.

Despite our compatibilities, we had adjustments to make. We generally communicated well. One thing about Suyon, her meaning was clear and unambiguous. We delayed our honeymoon until the end of October combining it with an Annual GSA annual meeting in Seattle and toured the Oregon and Washington Coast.

That was a great trip for both of us. We visited Siletz Beach, OR, resort, ate at their five star restaurant, and drove to Newport, OR to see some of the dune country and ORST's aquarium. We toured Astoria, OR, the Lower Columbia River and the southwest Washington Coast.

We arrived in Seattle on the Saturday before the GSA meeting and stayed at the Embassy Suites which provided a fully equipped kitchen. Suyon read about the Seattle fish market. There she bought freshly cooked Alaska King Crab, two nutcrackers and lobster picks and we ate dinner that night in our room.

Suyon had two conditions when we got married. First, I had to take her to a Buddhist Temple in Queens, NY, once a month. Other than driving around Queens, NY, this was not a problem. During their service, I found a place to read a book or journals. The service was followed by an excellent pot luck Korean meal that was worth the drive. Because most of Suyon's friends also lived in Queens, we met with them regularly. The Korean and Chinese restaurants in Queens are outstanding and we visited often.

The second condition was that we were not buying any big ticket Japanese products. I agreed but explained to her that Dutch people also suffered badly during World War II. Therefore we should apply the same principle to German products. Suyon agreed before realizing she negotiated away the chance to buy a Mercedes (not that I could afford one).

Suyon's family in Korea was anxious for both of us to visit and to meet me. We arranged to do so in October, 1995. That trip coincided with a quadrennial meeting of the Asian Marine Geological Congress at Cheju Island and they asked me to give a keynote talk and covered expenses.

Suyon flew earlier to reconnect with her family and I arrived ten days later. The first Sunday I attended a round of three parties to meet her relatives, not realizing she was part of a large extended family. I lost count but estimate I met at least 200 that day. I then flew to Cheju Island and Suyon joined me on the last day of technical sessions. We both went on a post-meeting geological field trip let by Kyung Shik Woo (BS, SNU, MS, Texas A&M, PhD Illinois; carbonate sedimentology; Kangnam University).

We spent that Saturday touring Cheju Island. However, Suyon said we had to return to Seoul on Sunday to attend two more parties. One was at the home of one of her cousins who married Kwang Soo Lee. He works for IBM and is now their East Asia regional procurement manager. He spoke great English and we also established great rapport.

The other party was at the home of another cousin who married Chang Shik Lee, the president of a major Korean brokerage house. They lived in a condo converted from the 1988 Olympic Village and he showed me his art and coin collection. He also spoke good English and we also established good rapport.

Chang Shik Lee clearly is a man of vision. He arranged for his wife to live in Canterbury, New Zealand, so their three children, a daughter (Soe

Kyong) and two sons (Hanjuk and Hansuk) were educated in English schools and universities. Chang Shik realized his children needed to obtain an English education to function in the global economy. All three graduated from the University of Canterbury

When the trip ended, it was difficult to leave, especially for Suyon. I had not realized how close she was to her family, but the trip drove it home.

Shortly after I left NJMSC, Suyon opened an Asian Wellness facility in Rutherford, NJ, and incorporated. We sold our townhouse because of the commuting distance. She financed the corporation using the traditional Korean 'curb loan' system through her network in the Korean community in New York and northern New Jersey, and paid the loan off in a year.

When I moved to Houston with her concurrence, we lived a commuting marriage for close to nine years. That meant nightly phone calls, and visiting each other once per month. It also meant that I became involved long distance with some aspects of her business. Often it required me to ease the bureaucratic hassle for her when dealing with the onerous state and county bureaucrats on issues related to annual licensing, health inspections and ever changing regulations, some almost on a whimsical basis.

Our marriage grew stronger, I think, because whenever we visited, it was like a honeymoon all over again, relaxing together, going out to good restaurants, visiting friends, or just sight-seeing.

Sometimes we combined such visits with other trips. The GSA Sedimentary Geology Division awarded me the Laurence L. Sloss Award in 2000, in Reno, NV. I rented a car and showed her Pyramid Lake and Silver City. The meeting had some strange downsides. We stayed at the Pepper Mill Hotel, one of the designated convention hotels. They kept us waiting for three hours before checking us into our room.

The Pepper Mill Hotel had a signature restaurant and all my former students attending the meeting offered to take Suyon and me to dinner there to celebrate after the Sloss Award ceremony. The food was excellent. The staff was Iranian, and did not take too kindly to our international marriage. I recall they asked everyone for their order including Kathie Marsaglia and Alicia Nummedal. When they came to Suyon, they asked me "And what does the missy want?" I told them to ask her directly and they repeated their question.

I filed a discrimination complaint with the Nevada Gaming Commission.

It was the ONLY time during our 14 years of marriage that an incident of this sort arose. Contrary to national thinking, we never encountered such problems in The South.

Since our marriage, some of her relatives visited us. Kwang Soo Lee and his wife and children were posted to IBM's Mt. Kisco headquarters for three years. I saw them often on visits to New Jersey. Their two children did well in high school. Their oldest son graduated from Mt. Kisco High School.

Soe Kyong Lee, Chang Shik Lee's daughter, visited after graduating from the University of Canterbury. She accepted an internship with Lehman Brothers in Tokyo and then was hired full-time in their Hong Kong office. She was in New York for training and stayed over. Soe Kyong is an attractive woman and extremely bright.

While in Hong Kong, Soe Kyong met Oh Kee Kwan, a Korean-born American whose family immigrated to Queens, NY. Oh Kee's parents worked hard and struggled economically but made sure he studied hard. He was awarded a full scholarship to Harvard and graduated Cum Laude in economics in 1997. Oh Kee accepted a job with Morgan Stanley, first in New York, then in Hong Kong, and then in Korea as the head of that office. He was transferred back to New York, returned to Korea and has a senior management position in Morgan Stanley's Hong Kong Office.

Soe Kyong moved to New York to earn an MBA at Columbia which was awarded in 2006. When she and Oh Kee decided to get married, her father wanted an independent opinion about Oh Kee. Therefore, Suyon, Soe Kyong, Oh Kee and I had dinner in New York. It was my task to interview him and report back to Chang Shik through Suyon. I found Oh Kee a great guy, intellectually bright, and knowledgeable, but also detected some New York 'street smarts' which he controlled to the point one barely noticed.

I realized also that if Oh Kee and his family had never immigrated to the USA, he would never have met Soe Kyong.

Soe Kyong and Oh Kee were married in 2006 and Suyon and I attended their first (or US) wedding in Queens, NY. They repeated the marriage ceremony three weeks later in Seoul.

That trip gave me a chance to reconnect with her family and come to know them far better than during my first trip.

In 2007, we returned, this time for the marriage of her oldest niece, Soe

Youn, in late October. We again reconnected with her family and spent two days in the Samchock region where she grew up.

We attended a class reunion and party with her Samchok classmates at a rural Korea restaurant with great food. What impressed me was that all her classmates grew up on small farms and moved into positions of importance in Korean society. One was a university president, one was the owner/director of a regional hospital, one was retired military having served in a high ranking intelligence position, and others owned businesses. Her girl friends seemed to have married well. It demonstrated that Korean society was changing and developed a system of upward mobility.

Suyon closed her business in late 2006 and now lives with me at our home in Sugar Land, TX. She is very active with the local Korean Buddhist temple, the Korean Community Center, does part-time work for my consulting business, and continues to take good care of me. Marrying her was the best decision I ever made in my life.

LESSONS LEARNED:

1) When meeting someone with common interests and with whom one communicates well, don't be shy. Instead forge ahead and see where the relationship takes you.

2) Cultural differences can be bridged and it's not that hard to do. Mutual respect and a willingness to learn new things are the keys. There is a commonality to all cultures making the transition easier. Most important, look beyond the surficial appearance of a person, get to know them, and discover who they really are.

CHAPTER 29
Giving Back

I believe I was blessed as I look back on my career as a geologist. That career provided a rewarding life by meeting many outstanding people, enjoying their company and respect, as well as pursue geological research of interest to me. I also encountered my fair share of irrational disgruntled students and colleagues, some of whom were mentioned herein. I probably met the complete range of people that exist in society and they were as variable as the many rock types I examined.

As a child I could only dream of the paid travel opportunities that came my way and allowed me to see so much of the world. My career track far exceeded my expectations when I started as an undergraduate geology major. Timing helped. I started a PhD program in 1957 when the US government panicked with the launch of Sputnik and poured financial largess for 15 years in the form of research grants on the scientific establishment. Like others, I encountered the highs and lows of life. The previous chapters described the most interesting and challenging.

Some of these experiences were used as a nucleus to write a novel "*DISSENSIONS*" published in 2004. One reader, Raymond C. Murray, was extremely kind when he wrote a review in *The Professional Geologist*:

"George Klein is a most distinguished geologist with a list of contributions to sedimentary geology as long as your arm. A Yale PhD, research at a major oil company laboratory and teaching appointments at three research universities were the making of a great career"

Achieving that career was not easy. It depended on the good will and encouragement of former professors and many friends and colleagues. It benefited also from funding by granting agencies, a network of contacts from classes I taught in universities and petroleum industry short courses, and a network of people who heard me give talks at meetings or university colloquia. I enjoyed the goodwill of consulting clients and personal associations with people I met in the Houston petroleum community after moving there to work as a geological consultant. It required patience and focus to overcome adversity during periods when things did not go well.

I once thanked one of my undergraduate professors for a letter of recommendation to graduate school. He said, "George, you don't need to thank me. Gratitude is second hand. In time, you will extend that gratitude by the letters of recommendation you will write for other people." This was true. Ultimately, all I could offer my former colleagues and students was the integrity of my recommendation. I did everything to keep it that way.

Another way of saying thank you is to leave a permanent record of one's contributions. The books and publication one publishes during an academic and scientific career are part of an evolving continuum of the best progress reports one writes about a nearly endless set of changing scientific hypotheses and interpretations. Writing the definitive benchmark paper is a rare event that very, very few people achieve. Over time, newer data, and interpretations displace older ones. Science is ever changing.

DISSENSIONS was published late in 2003. After publication, I exchanged emails with a former colleague, Ralph L. Langenheim at the University of Illinois at Urbana-Champaign, where I spent nearly two dozen years on the faculty. I mentioned a file[4] that I used for background to write the novel. Ralph suggested donating it to the University of Illinois Archives, which I did.

In 2005 I had major back surgery. I reviewed contingencies with Suyon (my wife). She asked what she should do with my books if I didn't survive. After some thought I donated an entire set of all the books, monographs, reprints of journal papers I had written, and my MS and PhD theses to the University of Illinois Archives. Suyon enthusiastically embraced this donation. It assured that a near complete set of my published work was adequately curated, barring a major catastrophe. It was an appropriate place to house these publications because most of the papers I wrote involved research completed at Illinois. It's a way of saying thank you to the University of Illinois for fostering the opportunity to complete research I chose to undertake and come in contact with some truly outstanding students and colleagues.

The remaining books will be donated to the AAPG Publication Pipeline Committee, a group I helped form through the leadership of Martin Cassidy (BA, Harvard, MS, Oklahoma, PhD Houston; organic geochemistry; Amoco, Univ. of Houston) and Rick Wall (ConocoPhillips, Samson Oil and Gas). These books will go to university libraries in the developing world.

During 2007, I receive a general email from Daniel F. Merriam asking me and all alumni of the University of Kansas Department of Geology to contribute information about their student career. He was writing a book about the first 100 years of the department[2]. His questionnaire included one about "Turning Points." I responded by describing critical colloquia presented by E.D. McKee, and Harold Fisk (Chapter 6).

I realized when responding to Dan's request that there was a way to give back something to both my profession and the University of Kansas which not only accepted me in 1955 as a graduate student when no other graduate program did, but also for providing a strong foundation for my PhD program (Chapter 7). Those two colloquia presented at KU provided me with a road map for my career. I decided to endow a colloquium speakership at the University of Kansas enabling future students to possibly experience a similar turning point that would give them a road map for their future research and possibly their career.

With the help of KU's Geology Chairman, Bob Goldstein and Kathleen Brady, Senior Development Director, College of Liberal Arts and Sciences, at the KU Endowment, I established such a fund. They knew I was not wealthy, but was living a middle class life. Through their efforts and assistance, I established a speakership within the "KU Geology Associates Fund" and started contributing from consulting earnings.

I also proposed that the KU Foundation contact all former students and living faculty colleagues. Out of 80 people contacted, 13 donated a total of $12,000.00 to this fund. I still contribute today and a provision was added to my will to complete fully funding it. The fund is called the 'George Devries Klein Colloquium Lectureship in Clastic Sedimentary Geology' at the KU Geology Department.

Many have asked me why KU? My earned degrees came from Wesleyan (BA), Kansas (MA) and Yale (PhD). I taught at Pittsburgh, Pennsylvania and Illinois (Urbana-Champaign). During my lifetime, geology departments changed and evolved in multiple directions aiming to ride the wave of new research or fashionable societal trends. Wesleyan, Penn and Pittsburgh became focused on environmental geology. Yale became a strong, but somewhat narrow, center of geochemistry and geophysics. Illinois, at one time a powerhouse in sedimentary geology, evolved into a department of specialized people who were considered 'excellent' in mineral physics, seismology, petrology, and geohydrology, and now is slowly developing, through endowment bequests, programs in sedimentary processes and geomorphology. But frankly, during my career at Illinois, it was always a department without adequate strategic planning while searching for an undefined and elusive goal.

So why KU? Over the years, the KU department of geology not only emphasized continuously the core areas of geology but brought in people who were strong both in geochemistry and geophysics and other specialties and blended those specialties with the major core themes of geology. They have done it extremely well.

That's the type of geology department that will prove to be successful long-term and that's the type of university geology department I decided should house my colloquium speakership. It is the legacy I leave from my career as a way of saying "Thank you."

LESSONS LEARNED:

1) Never forget where one came from and who and what got you where you are today.

2) When the opportunity arises, leave some of what you earned to help others or a program in a special way.

Epilogue

Rereading and revising this book made me realize again that to experience the life I described herein, I have numerous people and organizations to thank for all they did for me that made it all possible. Many were mentioned in the book, some in an understated way. Now is the time to recognize them.

First, my parents, Alfred R.H. and Doris Devries Klein. They opened a world of possibilities to me, made the early years a delight, made enormous sacrifices for my well-being, and taught me much including the importance of doing good deeds for others.

Second, Joe Peoples at Wesleyan University who wrote letters of recommendation for jobs and graduate school in addition to mentoring and showing me the possibilities of what lay ahead.

Harry H. Hess at Princeton for networking me to my first paying summer job as a geologist.

Raymond C. Moore at the University of Kansas for teaching me the art of scholarship, the importance of doing one's homework, the importance of high standards, and above all, to continue to believe in oneself.

Louis F. Dellwig, my master's advisor at Kansas, for showing me how to organize my thoughts into decent prose.

Robert R. Shrock of M.I.T. for encouraging me to pursue my interests and made it possible for me to earn a proverbial second chance in a top-level PhD program.

John E. Sanders at Yale who taught me the importance of taking advantage of all research possibilities an area has to offer.

Richard F. Flint at Yale for being willing to take a chance on me as a graduate student in the department of geology there and being supportive of my efforts.

B. Clark Burchfiel, Stephen C. Porter, and Charles A. Ross for a lifetime of good friendship.

The Geological Society of America for awarding me one of the largest

Penrose Research Grants in its history (1959) to complete my PhD thesis field work, and later (2000), granting me the Laurence L. Sloss Award for Excellence in Sedimentary Geology.

Bernie Rolfe at Sinclair for teaching me how to understand and face the real world.

Agnes Creagh of the Geological Society of America for showing me how to put my scientific work into proper format for publication and asking me to do the same for my students.

Aaron C. Waters who taught me the importance of always taking sabbaticals, even going into debt for them if one must.

Howard A. Meyerhoff at Penn for teaching me that I don't know everything and could be wrong, for giving me an opportunity to excel, and for showing me what it means to be a wise man in our profession.

W. Stuart McKerrow at Oxford University for encouraging me to think globally about my research, facilitating my research in the UK to experience global approaches, and arranging a Visiting Fellowship at Wolfson College at Oxford in 1969.

The University of Illinois for facilitating and fostering my research, for teaching me the art, structure, and form in academe, and allowing me to grow intellectually in my field.

Fred A. Donath at the University of Illinois who, despite our disagreements, gave me the opportunity to work in one of the best research environments on earth.

The SEPM (Society for Sedimentary Geology) for awarding me its 1972 Outstanding Paper Award in the *Journal of Sedimentary Petrology*, the first career award I received for my published research.

Gerald M. Friedman, Robert H. Dott, Robert N. Ginsburg, Raymond C. Murray, William R. Dickinson, and Gerard V. Middleton for being outstanding friends, writing countless letters of recommendations, supporting letters for research funds and sabbatical leave support, and being the decent human beings they truly are.

Noriyuki Nasu is thanked for his friendship and guidance as well as arranging a Japan Society for the Promotion of Science Fellowship at the Ocean Research Institute of the University of Tokyo (1983).

The Fulbright Commission of the Netherlands is thanked for awarding me a Senior Research Fellowship to the Netherlands in 1989.

Amongst my many students, I must include Dag Nummedal, Roscoe Jackson, Gordon Fraser, Paul Schluger, Yong Ill Lee, Debra Willard, Jerome P. Walker, Margaret S. Leinen, John L. Shepard, Kathie Marsaglia, Jay Scheevel, Carl Steffanson, Dan Lawson, William C. Dawson, Tricia Santogrossi, J. Stephen Tissue, Pius C. Weibel, Frank R. Ettensohn, amongst many, for asking the key questions that led to better understanding of key geologic issues and opening new geologic vistas to outstanding problems to solve. Some taught me how to run better field trips, one taught me the importance of being more "radial" instead of linear, and all taught me how to view and accept them as colleagues.

In the greater Houston area, I owe special thanks to Norman S. Neidell, Robert Pledger, Carl Marrullier, Jory A. Pacht, Janet M. Combes, Walter W. Wornardt, Wulf Massell, Barbara Radovich, Hermann Homann, Fernando J. Zuniga y Rivero, Kenton Holliday, Hugh Hay-Roe, Mike McCardle, William C. Dawson, Allen Bertagne, Sam LeRoy, Charles and Linda Sternbach, Rosemary P. Mullin, Clint Moore, Craig E. Moore, Jeannie F. Mallick, Patrick T. Gordon, James L. Wilson, Gerald J. Kuecher, Allen F. Mattis, Tricia Santogrossi, Marc Helsinger, Marc. B. Edwards, Robert G. Hickman, and Deborah K. Sacrey for their helpful guidance during the early days of my consulting career, their encouragement, support, leads to consulting opportunities, and their friendship.

The Houston Geological Society is acknowledged and thanked for providing excellent technical talks at its various meetings, together with outstanding short courses which enabled me to understand and stay current with modern petroleum geology, and for providing an excellent venue for networking.

Similarly, the Houston Chapter of SIPES (Society of Independent Earth Scientists) is thanked for providing a strong technical venue to stay current as well as networking opportunities. The national SIPES is thanked for inviting me to be a distinguished lecturer during 2004 and 2005.

My son Richard L. Klein, for developing a model career and pursuing his own dreams and success.

And last, but not least, to my lovely wife, Suyon C. Cheong, for her love, devotion, care, and commitment to this Rocknocker.

End Notes

1) **Kuecher, G.J.,** 2008, Fruitcake Hill: a history and memoir of life on the hill of a family of 15: Vancouver, CCB Publishing, 84p

2) **Merriam, D.F., 2007,** Raymond Cecil Moore: Legendary Scholar and Scientist: Univ. of Kansas Dept. of Geology and Paleontological Institute Spec. Pub. 5, 169 p.

   Merriam, D.F., 2009, Geology at the University of Kansas: The First Century (1866- 1966) and a bit beyond: Univ. of Kansas Dept. of Geology and Paleontological Institute Spec. Pub. 6, 210p.

3) **Fisher, W.L., 2008,** Leaning Forward: A Memoir: Texas Bureau of Economic Geology.

4) Most of what appears in this Memoir is based on memory. However, the reader is advised that a file I submitted to a special panel (Applequist Panel) to review the Donath headship in 1976 is now located at the University of Illinois Archives and is available on site only. I did not keep a copy. For more details, see:

http://www.library.uiuc.edu/archives/archon/index.php?p=collections/controlcard&id=2148&q=George+Devries+Klein

www.ingramcontent.com/pod-product-compliance
Lightning Source LLC
Chambersburg PA
CBHW031610160426
43196CB00006B/84